WATER-SAVING IRRIGATION TECHNIQUES FOR ARID AND SEMI-ARID REGIONS

WATER-SAVING IRRIGATION TECHNIQUES FOR ARID AND SEMI-ARID REGIONS

Dr. Hazem Shawky Fouda

Delve
PUBLISHING

www.delvepublishing.com

Water-Saving Irrigation Techniques for Arid and Semi-Arid Regions

Dr. Hazem Shawky Fouda

Delve Publishing
4164 Lakeshore Road
Burlington, ON L7L 1A4
Canada
www.delvepublishing.com

Email: orders@arclereducation.com

ISBN: 978-1-77956-400-9 (Hardcover)

ABOUT THE AUTHOR

Dr. Hazem Shawky Fouda has a PhD. In Agriculture Sciences, obtained his PhD. from the Faculty of Agriculture, Alexandria University, 2008, MSc. In Agriculture Sciences from the Faculty of Agriculture, Alexandria University in 2004, Post-Grade Diploma in Cotton, 2001, BSc. in Agriculture Sciences, from the Faculty of Agriculture, Alexandria University, 1997, worked in Cotton Arbitration & Testing General Organization (CATGO) from 1999 till 2018 was working in the International Cotton Training Center (ICTC) – Cotton Arbitration & Testing General Organization (CATGO) from 2000 till 2015, as a Lecturer & Classer's Trainer for Egyptian and foreigner classers, technicians, ginners, spinners & traders in all cotton aspects. Besides that he was an editor and active member in the Research & Translation Committee, participating in issuing weekly, monthly and annually issues about the international & local cotton market including price trends and direction, recent developments & researches concerning cotton production, protection, harvesting, ginning, fiber testing, spinning & weaving since its foundation in 2000 till 2014 and from 2015 till 2018 he worked as an inspector, since 2018 till present works as a consultant.

TABLE OF CONTENTS

LIST OF FIGURES

LIST OF TABLES

PREFACE

Quite possibly of the direst issue confronting our planet today is water deficiency, particularly in parched and semi-bone-dry regions where there is a restricted stockpile of water normally. Compelling water the board strategies are progressively required as populaces proceed to rise and climatic patterns shift. To keep up with farming, save livelihoods, and safeguard environments in regions with scant water supplies, this book is given to analyzing the numerous parts of water preservation and compelling water system frameworks.

With a specific spotlight on water system strategies, the work you are holding endeavors to offer a careful handle of the issues and arrangements related with water use in parched and semi-dry locales.

One of the main themes of this book is the role of irrigation in agriculture. We provide an in-depth investigation of different irrigation methods, starting from traditional systems like distribution channels and rainwater harvesting structures to more advanced technologies like drip irrigation and precision irrigation systems. Each method is discussed according to its suitability for arid regions, with a focus on its efficiency, cost-effectiveness, and sustainability.

In addition to discussing irrigation methods, we investigate the soil-water-plant relationships that are important to understanding how crops use water and how efficient irrigation can improve crop productivity. Key practices in water-saving irrigation, such as mulching, water management, and system maintenance, are highlighted as vital components for sustainable agricultural practices in these water-scarce areas.

In the later chapters, we present some of the most advanced irrigation technologies, including drip irrigation and subsurface irrigation, which have a great impact on water use in agriculture. These systems offer unique efficiency, delivering water directly to the root zones of plants while minimizing evaporation, infiltration and runoff. Also, we discuss the role of precision agriculture, where practices such as remote sensing, GIS, automation, and artificial intelligence are used to optimize water use in real-time.

Present the importance of water harvesting techniques, stressing on the positive impact of rainwater harvesting, what it can offer and the potential of treated wastewater in irrigation. These methods are becoming increasingly important as alternative water sources to address water scarcity. The reuse of treated wastewater, in particular, offers a promising solution, but it must be carefully and properly managed to avoid health and environmental setbacks.

In this book, we provide a general view of both traditional and modern advancements in irrigation. The solutions we present are not only dedicated for large-scale farming but also include low-cost, small-scale methods for small farmers in developing countries, where water-saving techniques can make a significant difference in their productivity and livelihoods.

Finally, this book is a reference for researchers, students, and agriculture engineers in the field of water management, as well as anyone concerned with the sustainable use of water in agriculture. By presenting a wide range of practices, technologies, and solutions, we aim to offer practical and doable solutions for addressing the increasing challenge of water scarcity in arid and semi-arid regions.

I hope that this work will inspire more discussions, further research, and the development of innovative solutions for the vital issue of water management in some of the most vulnerable regions of the world.

In the first chapters, we will establish the groundwork by investigating the climatic attributes of these areas and their exceptional ecological difficulties. From addressing the reasons for aridity to showing the elements of soil, temperature, and vegetation, we offer bits of knowledge into the particular requirements and impediments that these biological systems face. As we dig further into the significance of water preservation, we look at both conventional and current answers for tending to water deficiencies, like reservoir conduits, desalination, and water reuse.

–Author

Chapter **1**

Introduction to Arid and Semi-Arid Regions

Contents

INTRODUCTION

By definition, arid regions receive little precipitation, typically less than 10 inches (25 centimeters) annually. 10 to 20 inches (25 to 50 centimeters) of rain fall annually on semi-arid regions. The tectonic features and rocks that underlie arid regions may not be different from those of other regions, but the landscape is unique. Erosion is the primary force that shapes the land surface because there is little vegetation and a lot of loose surface material.

Normal hot summers are a feature of the hot arid and semi-arid regions. It rains less than 400 mm on average per year. The annual growing season is less than 90 days, and there is a significant water deficit as a result. The depletion and pollution of groundwater are increasing. Most of Rajasthan's districts generally experience a drought every third year. A moderate amount of calcareous and alkaline reaction can be found in most sandy soils. Traditional monocropping agriculture using rainfed practices is practiced in the area. Non-saline areas are used to grow crops that withstand the short duration of the rainy season, like pulses, pearl millet, and "Chari" (fodder). Globally significant plants and animals make up the region's traditional biodiversity, which is well-known. The area is home to a wide variety of plants, including cucurbits, sorghum, pearl millet, moth beans, cluster beans, cowpeas, black and green grams, Brassicaceae, sesame, and grasses. With typical management techniques, the yields are low. Tropic thorn forest, which is rare and irregular, makes up the natural vegetation. Various desert shrubs and trees, such as Ker, Ber, Phog, rohida, and khejadi are indigenous. Rapeseed is the most produced crop in the nation, along with mustard, coriander, cumin, fenugreek, cluster beans, and moth beans. The diversity of indigenous crops must be preserved and used in order to increase the sustainability of production systems and livelihoods.

1.1. OVERVIEW OF ARID AND SEMI-ARID CLIMATES

Climates classified as "arid" have low levels of precipitation and are therefore unsuitable for the growth and development of plant and animal life. Drier areas, sometimes referred to as deserts or arid lands, lose more precipitation to transpiration and evaporation than they gain from the weather. Whatever rainfall these areas receive is swiftly evaporated by their arid, sandy, rocky, or bald surfaces, rendering them uninhabitable by organisms. Under Köppen's Climate Classification System, arid climates are divided into semiarid steppes and hot or cold deserts. Desertification, a form of human degradation that reduces vegetation productivity, and drought are two threats that affect semiarid and arid regions. Erroneous farming methods, deforestation, excessive grazing, and other disturbances to the organic vegetation patterns can all lead to desertification.

When there is such a severe lack of water in an area that it prevents or severely hinders the growth and development of plant and animal life, the area is referred to as arid. There is no agreed-upon definition of an arid region other than aridity because arid environments vary widely in terms of soil, climate, water balance, flora, fauna, and human activity. Denoted as p/ETP, aridity is expressed as a function of rainfall and temperature, where:

P = precipitation

ETP = potential evapotranspiration, taking into account, wind, solar radiation, and atmospheric humidity.

Hyper-arid zones have an index of less than 0.03 and cover 4.2% of land on the planet, arid zones have an index range of 0.03–0.20 and cover 14.6% while semi-arid zones have an index range of 0.20–0.50 and cover 12.2%.

Rainfall in arid climates is highly variable and typically amounts to less than 25.04 cm annually. The soils, wildlife, and flora of arid environments are varied, nevertheless. About one-third of the land area is made up of arid landscapes. Irrigation is usually necessary for farming to occur. These landscapes have very little vegetation, mostly in the form of sporadic shrubs, grasses, and small trees. Large daily temperature variations that can cause the physical or mechanical breakdown of rocks or wind erosion of parent material are among the processes that form soil in arid climates. Because there is little vegetation and little water, chemical weathering occurs slowly. "Islands of fertility" arise where vegetation is present, increasing organic matter, root development, and nutrient content beneath the vegetation. Dry soils' ability to retain water is typically restricted by their coarser texture, lack of structure, and low levels of organic matter. In dry soils, impermeable layers can frequently be found, which can affect plant viability by causing problems with water storage and infiltration. A common occurrence in arid areas is salt accumulation, which can hinder plant growth. Aridisols, which are found in cold and hot deserts and frequently accumulate salt, gypsum, and carbonates, are the soil order common to arid climates.

1.1.1. Characteristics of an Arid Climate

One characteristic of an arid climate is heat and dryness. The majority of the arid regions on Earth, which make up about one-third of the planet, are located 30° north or south of the Equator. In order to survive the harsh environment, both plants and animals have evolved adaptive strategies.

- **Low Precipitation:** The majority of dry areas experience yearly rainfall of less than 10 inches. The Polar Regions are regarded as having arid climates despite the fact that they receive little precipitation in the form of snow, even though most arid regions are hot and dry. In a dry climate, rainfall occurs infrequently and primarily during thunderstorms. There are flash floods and excessive evaporation because the soil is compacted and dry, making it slow to absorb water.

- **Extreme Temperatures:** There are seasons in some arid climates and not in others. While other places, like the Gobi Desert, have hot summers and freezing winters, the Sahara, for example, experiences year-round heat and dryness with no discernible seasonal changes. But because bare earth cannot hold heat, days are always warmer than nights in arid regions. Average temperatures in arid regions, with the exception of the poles, range from 78° to 94°F.
- **Wind:** The lack of vegetation in arid climates causes windy conditions, which in turn reduce air movements. Plants and soil release damp air into the wind, which increases evapotranspiration. Additionally, wind erodes the topsoil, which stops plants from sprouting. Sandstorms are sporadic events that pose a significant risk to the adjacent areas because they cover fertile soil with sand, causing the desert to spread.
- **Vegetation:** Due to high winds, sand-filled soil, and little rainfall, flora is scarce in arid regions. Either ephemeral annuals, succulent perennials, or non-succulent perennials can thrive in these circumstances. They grow long roots and spiny leaves and require very little water. But while Antarctica is too cold for plants to grow, some arid regions, like the Atacama Desert, are so dry that they cannot support plant life.

1.1.2. Semi-Arid Climate

One subtype of dry climate is semi-arid climate, also known as semi-desert climate or steppe climate. It is found in areas where precipitation is less than potential evapotranspiration, but not as little as in a desert environment. Depending on factors like temperature, there are various types of semi-arid climates that give rise to various biomes.

1.1.2.1. Attributes of Semi-Arid Climates

The Köppen climate classification provides a more accurate definition by treating steppe climates (BSh and BSk) as intermediate in terms of ecological traits and agricultural potential between humid climates (A, C, and D) and desert climates (BW). Because they typically cannot support forests, semi-arid climates favor short, prickly, or scrubby vegetation that is dominated by either grasses or shrubs. Establishing the precipitation threshold is a prerequisite to identifying whether a place has a semi-arid climate. This is how the precipitation threshold (measured in millimeters) was determined:

- multiply by 20 the average annual temperature in degrees Celsius and then
  - add 280 if at least 70% of the total precipitation falls in the summer half of the year (April–September in the northern hemisphere, October–March in the southern hemisphere)
  - add 140 if 30–70% of the total precipitation falls in the summer half of the year
  - add nothing if less than 30% of the total precipitation falls in the summer half of the year

Steppe, semi-desert, or semi-arid climates are categorized as BSs if the region receives less than the threshold but more than half or 50% of the threshold in millimeters of annual precipitation. Furthermore, a mean annual temperature of 18°C (64.4°F) is used as an isotherm to distinguish between hot and cold semi-arid climates. If the mean temperature of a place with a BS-type climate is higher than this isotherm, it is categorized as hot semi-arid (BSh); if not, it is categorized as cold semi-arid (BSk).

1.1.2.2. Hot Semi-Arid Climates

Hot semi-arid climates (type BSh) are found in the high teens to mid-30s latitude range of the tropics and subtropics, usually near areas that experience humid subtropical or tropical savanna climates. Summers in these climates are typically hot, and occasionally very hot, and winters are mild to cool, with little to no precipitation. The edges of subtropical deserts are typically home to hot, semi-arid climates. Most frequently found regions with hot, semi-arid climates are South Asia, Australia, and Africa. In Australia, the hot semi-arid climate zone encompasses a sizable portion of the Outback, which surrounds the central desert regions. South Asia's monsoon season is felt in parts of Pakistan and India, where there are brief but distinct wet seasons. However, the region's overall precipitation is insufficient to be classified as a humid subtropical climate or a tropical savanna.

In addition, there are regions of hot, semi-arid climates in the United States Southwest, including the Central Valley of California, the sertão, the Gran Chaco, and the poleward side of the arid deserts. These regions receive a Mediterranean precipitation pattern, with summers usually being dry and winters being wetter. These regions include most of northern Mexico, the ABC Islands, the rain shadows of the mountain ranges in the Dominican Republic and Haiti, and parts of North America. In some parts of Europe that surround the Mediterranean Basin, they can also be found. SE Spain is the region in Europe with the highest concentration of BSh climates. Also, it is mostly found in areas of south Greece, but it is also found in the majority of Formentera, the marginal areas of Ibiza, the marginal areas of Thessaloniki and Chalkidiki in north Greece, and the marginal areas of Sicily, Sardinia, and Lampedusa in Italy.

1.1.3. Cold Semi-Arid Climates

Generally found in elevated regions of temperate zones, cold semi-arid climates (type BSk) are closest to a humid continental climate or a Mediterranean climate. These latitudes range from the mid-30s to low-50s. Additionally, the interior of continents far from major bodies of water are usually home to them. While not quite as hot as those in hot semi-arid climates, summers in cold semi-arid climates are typically warm to hot and dry. regions with cold semi-arid climates typically experience chilly, possibly freezing, winters in contrast to hot, semi-arid regions. During the winter, these regions typically receive some snowfall, but far less than in places with more humid climates and comparable latitudes.

Regions with cold semi-arid climates are typically higher altitudes than those with hot semi-arid climates. These regions also typically experience significant daily and nighttime temperature variations, sometimes by as much as 20°C (36°F). In hot, semi-arid climates, such significant daily temperature variations are uncommon. Higher latitudes with cold semi-arid climates typically have dry winters and wetter summers, whereas lower latitudes with cold semi-arid climates typically have precipitation patterns more similar to those of Mediterranean climates, with dry summers, relatively wet winters, and even wetter springs and autumns. The Middle East, various regions of Asia, and central Asia are the most common locations for cold, semi-arid climates. Still, other regions where they may be found include parts of South America, Northern Africa, South Africa, and interior southern Australia (e.g., the regions of Europe, and southern New Zealand (Alexandra and Kalgoorlie).

1.2. THE ARID ENVIRONMENTS

When it comes to their land forms, soils, wildlife, plants, water balances, and human activities, arid environments are incredibly diverse. It is impossible to derive a useful definition of arid environments due to this diversity. But aridity is the one thing that unites all arid regions. The typical way to express aridity is as a function of temperature and rainfall. The climatic aridity index provided below is a helpful illustration of aridity;

p/ETP

where; P = precipitation; ETP = potential evapotranspiration, calculated by method of Penman, taking into account atmospheric humidity, solar radiation, and wind.

This index can be used to distinguish between three types of arid zones: hyper-arid, arid, and semi-arid. The world's arid zone accounts for 14.6%, the semiarid zone for 12.2%, and the hyper-arid zone for 4.2% of its total land area. As a result, dry land makes up nearly one-third of the world's total area. The dryland regions that make up the hyper-arid zone (arid index 0.03) are devoid of vegetation, save for a few sporadic shrubs. There are many instances of true nomadic pastoralism. Rainfall is little and seldom exceeds 100 millimeters per year. The rains are erratic and infrequent; occasionally, years may pass without any rain at all.

Pastoralism is the main feature of the arid zone (arid index 0.03–0.20), and farming is not permitted beyond irrigation. The native vegetation is mostly made up of shrubs and small trees, as well as annual and perennial grasses and other herbaceous plants. There is a lot of variation in the amount of rainfall that falls each year, varying from 100 to 300 millimeters. Rain-fed agriculture can be supported in the semi-arid zone (arid index 0.20–0.50) with somewhat sustained levels of output. Livestock production is also sedentary. Many species, including fortes and half-shrubs, shrubs, trees, and grasses and grass-like plants, are representative of the native vegetation. With summer rains, annual precipitation ranges from 300–600 to 700–800 millimeters, and with winter rains, it varies from 200–250 to 450–500 millimeters. In the sub-humid zone (arid index 0.50–0.75), there are also arid conditions. The hyper-arid, arid, semi-arid, and sub-humid zones are all referred to together as the "arid zone" in this context.

1.2.1. Causes of Aridity

The presence of dry, descending air causes aridity. Consequently, aridity is primarily found in areas with persistent anticyclonic conditions, such as those under the subtropical anticyclones. When cool surfaces are present, subtropical anticyclones have a greater effect on rainfall. In the lee of large mountain ranges, arid conditions also exist, which cause cyclones passing over them to lose structure and produce rain shadow effects. Significantly heated land surfaces also prevent rainfall, which is why there are significant stretches of dry climate found far from the coast.

1.2.2. Arid Zone Climate

Extreme heat and insufficient, erratic precipitation are the hallmarks of the arid zone; nevertheless, there are climatic variations. Generally speaking, variations in temperature, the season of precipitation, and the level of aridity are the causes of these climate contrasts. When characterizing the arid zone, three primary climate types are distinguished: the Mediterranean climate, the tropical climate, and the continental climate. The rainy season falls in the fall and winter in a Mediterranean climate. Winters are mild, and summers are hot and dry with no rain. The five-month dry season follows a wet season that begins in October and ends in April or May, as depicted in Figure 1.1 of the Mediterranean climate.

Summertime brings summertime rainfall in the tropical climate. The length of the rainy season decreases with increasing distance from the Equator. Long and arid winters prevail. The wet season in Sennar, Sudan, which is typical of a tropical climate, lasts from mid-June to late September, and is followed by a nearly nine-month-long dry season (Figure 1.2). Rainfall in a continental climate is spread equally throughout the year, with a tendency toward more precipitation in the summer. The dry season lasts the entire year in Alice Springs, Australia, because the monthly precipitation is less than twice the mean monthly temperature (Figure 1.3).

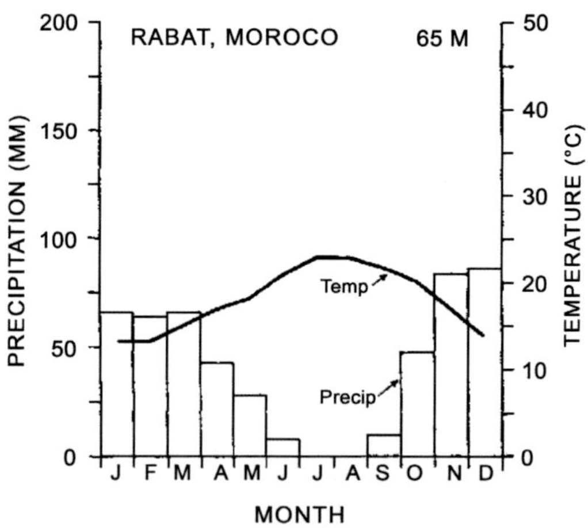

Figure 1.1. *Annual precipitation and temperature in rabat, Morocco.*

Source: https://www.fao.org/4/t0122e/t0122e00.gif.

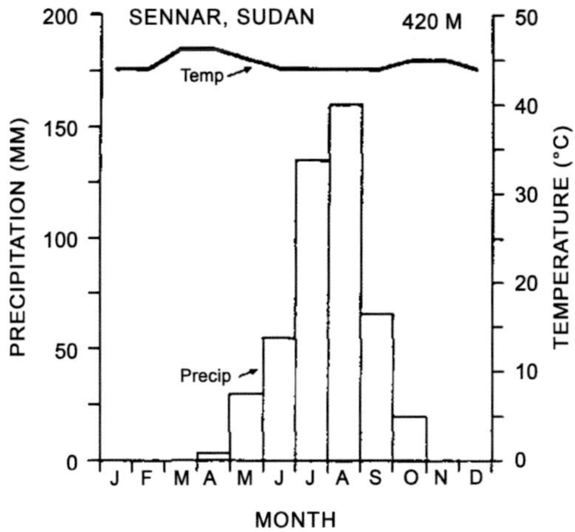

Figure 1.2. *Annual precipitation and temperature in Sennar, Sudan.*

Source: https://www.fao.org/4/t0122e/t0122e01.gif.

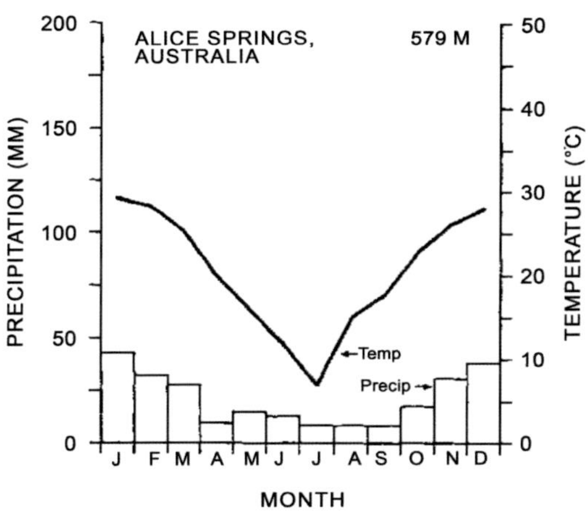

Figure 1.3. *Annual precipitation and temperature in Alice springs, Australia.*

Source: https://www.fao.org/4/t0122e/t0122e02.gif.

1.2.3. Rainfall

At a specific place, precipitation falls from the atmosphere and is either captured by vegetation such as trees and shrubs, or it hits the ground and becomes subsurface, groundwater, and overland flow. As the hydrologic cycle in Figure 1.4 illustrates, most rainfall eventually returns to the atmosphere, either through evaporation from streams and other bodies of water that are used by subsurface, groundwater, and overland flows, or through evapotranspiration processes from the vegetation. A region's soil and topographic features, vegetation, temperature and humidity regimes, rainfall patterns both spatially and temporally, and other factors all influence the relative dynamics of the hydrologic cycle in that area.

In contrast to temperate regions, arid zones have different summer and winter rainfall distributions. For instance, Rabat, Morocco, experiences precipitation during the chilly winter months, but virtually none during the hot summer months. Conversely, Sennar, Sudan experiences a protracted dry season in the winter, with summertime showers. Rainfall varies significantly between Rabat and Sennar, despite receiving roughly the same amount of precipitation in each location. While summer rains in Sennar fall on a hot soil surface and evaporate, especially when they come in the form of light showers, winter rains in Rabat can seep into the soil to subsurface storage. As a result, Rabat has more effective rainfall than Sennar does for plants.

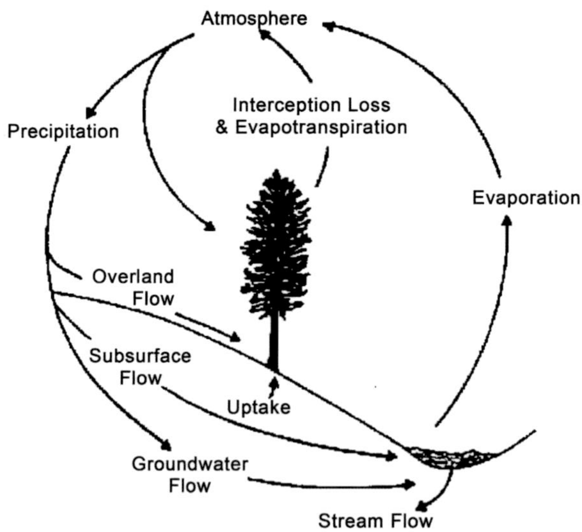

Figure 1.4. *The hydrologic cycle.*

Source: https://www.fao.org/4/t0122e/t0122e03.gif.

This example demonstrates that in order to provide the same amount of water available to the plants, summer rainfall regions need a greater annual precipitation total than do cooler winter rainfall climates. But in areas where plants hibernate in the winter, they are unable to utilize all of the water that is available to them. In dry regions, rainfall also fluctuates from year to year; this can be readily verified by examining rainfall data over an extended period of time for a specific location. Although it typically falls between ± 50% of the mean annual rainfall, there can be a significant difference between the lowest and highest rainfall recorded in different years. Even more variance exists in monthly rainfall. The average yearly rainfall observed over a period of years is typically different from the expected rainfall in a given location. Variability in rainfall is crucial to forestry operations because newly planted forests suffer when the rains are insufficient. For a forest plantation to be successful, the planting date must be chosen so that it coincides with rainfall.

The amount of rainfall is yet another factor that needs to be taken into account. Water loss through runoff may occur during a heavy rainfall because the soil may not be able to hold onto all of the water. Similarly, evaporation can cause a low-intensity rain's water to evaporate, especially if it falls on a dry surface. The quantity of rain that falls on a given day or, better yet, the number of rainy days can be used to gauge the intensity of the rainfall. Soil erosion risk is also correlated with rainfall intensity. It is well known that the energy contained in individual raindrops is strong enough to remove soil, especially topsoil. The soil structure can also be destroyed or deteriorated by splash erosion, which is erosion brought on by falling water droplets. It has been observed that the erosive power of the rain increases sharply as the intensity of the rainfall gets closer to 35 millimeters per hour. In the tropics, a significant portion of rainfall happens above this point, also known as the "erosion threshold."

1.2.4. Temperature

There is usually a relatively cool dry season, a relatively hot dry season, and finally a moderate rainy season that make up the climatic pattern in the arid zones. These seasons typically experience large daily temperature variations. Highs of 35 to 45°C are common during the cool, dry season, with nighttime lows of 10 to 15°C. During the scorching dry season, daytime highs can reach 45°C and nighttime lows of 15°C. Daytime highs of 35°C and nighttime lows of 20°C are common during the rainy season. The growth of plant species is often restricted by these diurnal temperature fluctuations.

Plants can only grow in specific temperature ranges, between maximum and minimum. Plants may suffer from extremely high or low temperatures. As long as they can use transpiration to cool down, plants may be able to withstand high temperatures, but growth will suffer as a result. Elevated temperatures in the soil's top layer cause a swift depletion of soil moisture as a result of increased transpiration and evaporation. Although low temperature issues are generally less frequent in arid zones, they can nonetheless limit plant growth when they do occur for extended periods of time. At temperatures below zero centigrade, plants can even perish.

1.2.5. Atmospheric Humidity

While temperature and rainfall are the main determinants of aridity, other factors also play a role. The air's moisture content affects the soil's water balance. Water has a propensity to evaporate into the air when the moisture content of the soil is higher than that of the air. Water will condense into the soil when the opposite is true. Arid zones typically have low humidity levels. In many places, the presence of mist and dew is essential to plant survival. Mist is a suspension of tiny water droplets in the air, whereas dew is the product of water vapor from the air condensing onto surfaces during the night. Plants can occasionally absorb water through their open stomata in the form of dew or mist, or they can let the water drop to the ground and add to the moisture content of the soil. Higher air humidity brought on by the presence of dew and mist reduces evapotranspiration and preserves soil moisture.

1.2.6. Wind

In general, arid regions are windy because they lack vegetation that can slow down air movements. Increased evapotranspiration results from winds removing the moist air surrounding plants and soil. Wind-induced soil erosion happens in areas where the soil, vegetation, and climate are suitable for this type of erosion. The following are common in arid zones: smooth ground surface, sparse vegetative cover, loose, dry, or fine soil, and wind that is strong enough to start soil movement. Soil erosion by wind is primarily caused by the loss of vegetation on the land. Wind erosion gradually removes silt, clay, and organic matter from the surface soil, causing the most significant harm from wind-blown soil particles: sorting of soil material. Sand and infertile materials could make up the remaining components. Sand frequently accumulates in dunes, endangering nearby areas. The primary conveyance of moisture from the air's water vapor to the ground is represented by precipitation. Evaporation is the process by which this hydrologic cycle is concluded. When thinking about effective rainfall, evaporation-related soil water loss is crucial. Strong winds, hot weather, and low humidity all cause more evaporation.

Plants need to transpire in order to adjust to high temperatures, as was previously mentioned. Significant moisture losses from the soil are attributed to transpiration. The plant itself, the wind, the temperature, and the humidity all affect how much transpiration occurs. Certain plants transpire less than others because they are better suited to dry environments. Thus, the rate of transpiration is greatly influenced by the composition of the vegetation. The main aspect of the water cycle that land management can affect to increase water yield is evapotranspiration, which is the combination of transpiration and evaporation.

1.2.7. Arid Zone Soils and Importance of Soil Properties

As vegetation and climate interact with parent rock material over time, soils are created. Significant daily temperature variations that cause the mechanical or physical disintegration of rocks are important factors in the formation of soil in arid climates. - Sands carried by the wind that erode and splinter exposed rock surfaces. Rocks break down physically into relatively large fragments that can only be broken up chemically through weathering. Due to the typical water shortage, chemical weathering proceeds slowly in arid zones. Long-term water shortages are also crucial for the leaching or removal of soluble salts, whose buildup is accelerated by high evaporation. Salts can only travel short distances during brief intervals of water runoff, which prevents them from penetrating deeply and frequently leads to the buildup of salts in closed depressions.

The process of soil formation is greatly aided by vegetation, which breaks up rock fragments and enriches the soil with organic matter from both underground and above ground sources. However, because of the limited development of aerial parts and the sparse canopy cover, the role of vegetation is reduced in arid zones. However, the root systems frequently show remarkable growth and exert the most impact on the soil. A forester is typically more interested in the characteristics of the soil that are necessary for the growth of trees and shrubs than in the processes that have led to the formation of the soil profile or in the regional systems of soil classification. The ability of arid zone soils to retain water and provide nutrients is of utmost significance. The physical properties of a soil, such as its texture, structure, and depth, determine how much water it can hold. The relative distribution of the particles (clay, sand, and silt) is referred to as texture. Water retention generally increases with texture fineness. The amount of organic matter that binds the soil particles has an impact on the structure, or internal arrangement of the particles. Sandy soils lack structure, whereas clayey soils have various types of structure and allow water and air to circulate because of the gaps between the particles. The permeability increases with the size of these spaces.

The type of tree root system and the amount of soil moisture are determined by the depth of the soil. Alluvial and colluvial soils are typically deep; residual soils, on the other hand, vary greatly in depth based on factors such as the slope angle, the rate and duration of weathering, and biotic factors like livestock grazing and cultivation). While the mid-slopes and valleys have moderately to extremely deep soils, the ridges and upper slopes frequently have shallow soils. In arid areas, a hardpan layer frequently limits the depth of soils. These hardpans can be found 5 to 60 centimeters below the surface and are composed of consolidated calcite in the Mediterranean region and ironstone or laterite gravel in the tropical zone. The soil's organic matter content is low in arid zones because there is less organic litter deposition and accumulation there. This limited organic matter content is rapidly depleted when these soils are farmed. The availability of nutrients is determined by the chemical composition of the soil. Significant nutrient leaching and intense mineral weathering are features of arid soils; however, these processes slow down with decreasing rainfall. Natural fertility is frequently low and is mostly dependent on the amount of organic matter in the topsoil.

Due to the climate's aridity, edaphic traits that lessen water constraints will be advantageous for planting trees or shrubs. The existence of a water table at a depth that the roots can reach is one of these edaphic traits. - A sufficient thickness of soil to accommodate a water reserve. - A soil texture that holds onto the most water possible. It is important to remember that the terrain's topography can have a significant impact. For example, the lower portions of sand dunes and their shallow areas can retain a significant amount of water, which can be utilized by vegetation that has adapted. Lastly, soil conservation and fixation are critical because arid zone soils are susceptible to erosion from both water and wind.

1.2.8. Arid Zone Vegetation

The vegetation cover in arid zones is scarce. Nevertheless, three plant forms can be distinguished:

- Ephemeral annuals.
- Succulent perennials.
- Non-succulent perennials.

After the rains, ephemeral annuals emerge and spend a brief season (approximately 8 weeks) completing their life cycle. Their growth is limited to a brief period of precipitation. The xeromorphic characteristics of perennials are absent from ephemerals. Ephemerals are typically small-sized, shallow-rooted, and have an active growth pattern as a physiological adaptation. In the form of seeds, ephemerals survive the dry season, which can extend for several years. Ephemerals can occasionally create dense stands that offer some forage.

Due to their low transpiration rates and the growth and proliferation of the parenchymal tissue in their stems and leaves, succulent perennials have the capacity to store and accumulate water, which they may use during dry spells.

Succulent perennials like cacti are common. The vast majority of plants in the arid zone are non-succulent perennials. These plants are resilient to the harsh conditions of the arid zone and include woody herbs, trees, shrubs, and grasses. The seeds of a lot of non-succulent perennials are hard and don't germinate easily; these seeds usually need to be treated (soaked in water or acid) in order for them to do so. Three growth forms of non-succulent perennials can be distinguished:

- Evergreen - biologically active throughout the year.
- Drought-deciduous - biologically dormant during the dry season.
- Cold-deciduous - biologically dormant during the cold season.

In general, succulent and non-succulent perennials are drought-enduring and drought-resisting species and are regarded as true xerophytes; ephemerals are species that escape drought and are not considered so. Plants that have the ability to survive with minimal moisture content are said to exhibit xerophytism. The development of a vast root system is one characteristic of xerophytic plants. Depending on the site, the primary root growth can be either vertical or horizontal. It is common for roots to extend horizontally in shallow soils, and roots that delve 10 to 15 meters are not unusual. In reaction to mild rainfall or during the dew-forming season, certain xerophytic species develop rain roots beneath the soil's surface. The shoot-to-root ratios are often between 1:3 and 1:6, with the shoots being smaller than the roots. The reduction of transpiring surfaces occurs as a result of rolling and shedding of leaves. Reduced transpiring surface as a result of seasonal variation in the plant; this characteristic lower water loss in the dry season. Some evergreen species, known as sclerophylls, have unique adaptations that reduce transpiration. Their leaves are leathery and frequently heavily wax-coated.

Other distinguishing anatomical characteristics associated with xerophytism are:

- Cuticularization – the formation of a surface plaster-like layer of cutin.
- Cutinazation – the impregnation of the cell wall with cutin, which forms a watertight layer with abundant hairs.
- Special arrangements of the stomata in recesses and grooves which provide protection from the arid atmosphere.

1.2.9. Classification of Vegetation

In classifying the vegetation of the arid zones, the major "delineations" are usually characterized in terms of rainfall amount and the pattern of occurrence.

- **Desert:** The term "desert" is used in this manual to refer to areas of land that are essentially devoid of vegetation, save for watercourses. Infrequent rain showers can bring forth ephemeral grasses and herbs. Rainfall is typically less than 100 millimeters annually.
- **Semidesert:** Grass, herbs, and small, short trees and shrubs up to two meters in height are the predominant vegetation types in semidesert regions, with bare patches in between. The vegetation of semidesert grasslands is a uniform cover of mixed grasses and herbs; these areas are formed when geological erosion has not been as severe and the soils have the capacity to absorb the little rainfall that falls on them. In regions with an abundance of water, along drainage lines, and on catchment sites, there are sporadic trees and shrubs. Open plant communities with succulent plants predominating—grasses may or may not be present—make up succulent shrubs. A profusion of plants with drastically reduced leaf size, the formation of succulent stems from storage tissues, and the presence of spines and thorns are the typical characteristics of the vegetation found in semidesert regions. Between 100 and 300 millimeters of rain fall on these areas each year; however, the majority of this precipitation is erratic, occurring in isolated rain showers or localized storms that last for a few months.
- **Low Rainfall Woodland Savanna:** The amount of grass to shrubs or trees (or both) in this area's vegetation is influenced by the frequency and severity of fires, which results in a mixed type of grasses and herbs. Trees and shrubs frequently have flat crowns resembling umbrellas. Instead of creating a closed canopy, their crowns leave wide gaps that are occasionally covered in ephemerals after rains. These gaps are mostly filled with low shrubs, grasses, and herbs, but there are also bare spots. Trees and shrubs are no taller than six meters, and grasses hardly ever grow to two meters. Usually, the trees and shrubs don't provide enough shade to stop the growth of grasses. These plants become potentially dangerous fire hazards during the dry season. On the other hand, some woodland savanna species can withstand fire to some extent. An area is designated as open and wooded grassland when it is dominated by grasses and herbs and contains less than 50% shrubs and trees. With brief periods of rain followed by protracted, sweltering dry spells, this type of woodland savanna is typical of the dry tropics. There is a 300–600 mm range in rainfall.
- **Evergreen Scrub**: This kind of vegetation found in arid zones is

made up of climbers, small trees, and an enclosed scrub of evergreen or semievergreen shrubs. Occasionally, some large trees are present. Shrubs are 2 to 3 meters tall with glossy, leathery leaves or thorny, succulent leaves. More substantial trees are widely dispersed. It rains more than 500 millimeters a year.

1.2.10. Highlights of Section

An area is considered to be arid when its evapotranspiration loss is less than 50% of its total precipitation, as determined by dividing the total precipitation received by the area by its fraction. Three zones can be distinguished within the arid environment: hyper-arid, arid, and semi-arid. The zones are vegetated by drought-resistant species and have low annual precipitation (0 to 800 millimeters), which falls rarely and irregularly. Arid regions make up nearly one-third of the world's land area. There are three main climatic types found in arid zones: continental, tropical, and Mediterranean. The distribution of these kinds of rainfall varies from year to year and between winter and summer. Additionally, different climatic types can have different effects on the arid zones in terms of runoff and evaporation losses; additionally, the potential for soil erosion varies based on edaphic characteristics. Significant daily variations in temperature have an impact on plant growth. The generally low atmospheric humidity has an impact on plant growth as well. Wind increases evapotranspiration, which can lead to soil erosion. In arid zones, soils are created when large daily temperature variations cause rocks to mechanically crumble, wind-blown sand scratches the surface of rocks, or plant roots break up small pieces of rock. Since the soil's low water content slows down chemical weathering, it generally has little effect. The structure (internal arrangement of soil particles), texture (relative distribution of different-sized particles), and depth (often restricted by a hardpan layer) of arid zone soils determine their ability to hold water. A decrease in rainfall slows the weathering of minerals and the leaching of nutrients.

Planting sites with soils and edaphic traits that reduce water constraints are preferred. Ephemeral annuals, whose growth is limited to the brief wet season, succulent perennials, which retain water for use during dry spells, and non-succulent perennials, which comprise most plants capable of withstanding the harsh conditions of an arid environment, are the three types of vegetation that grow in arid zones. Xerophytic plants are those that are found in arid environments and are characterized by their large root systems, lower transpiring surfaces, cutin layer, and other anatomical features that shield the plant from the dry air. The vegetation found in arid zones is divided into four categories: low rainfall woodland savanna (300–600 millimeters per year), semidesert (100–300 millimeters per year), and evergreen scrub (more than 500 millimeters).

1.3. CHALLENGES OF WATER SCARCITY

Worldwide, water scarcity poses a serious risk. Water use has increased over the past 100 years, and it is currently growing at a rate of 1% annually. The relationship between the availability and demand of water resources is referred to as water scarcity. Therefore, we also talk about the uneven temporal and spatial variations of water when we talk about scarcity. A quarter of the world's population, or one-third of developing countries' populations, live in regions that are scarce. The issue of scarcity is made worse by changes in consumption patterns, economic growth, and population growth. The pollution of groundwater resulting from salinization, wastewater, and agricultural load is an additional factor. Furthermore, precipitation and demand may be impacted by climate change. The management, conservation, and monitoring of groundwater; wastewater treatment for pollution reduction, sustainable use, and human development; land use management; and technologies for capturing and reserving water are some of the solutions that are sought after in response to these issues in order to preserve the quality and quantity of water, particularly in semi-arid areas.

Water scarcity is made worse in semi-arid areas, endangering ecosystems, food production, and human health. It is anticipated that semi-arid regions of the southern African coast, the northern Mediterranean, and North and South America will spread quickly. Thus, developing development strategies requires an understanding of the effects of hot weather and semi-arid conditions. The variations in weather between the dry and rainy seasons are referred to as a semi-arid climate. The nations where access to water is impacted by climate change have the worst development issues. In order to use water sustainably, one must first ensure that people have access to it in order to meet their basic needs. However, water is becoming more and scarcer due to global water market trading, which has an impact on water distribution processes from a socio-hydrological standpoint. Therefore, it is crucial to comprehend how participatory and sustainable criteria help these water-human systems adapt to water scarcity.

1.3.1. Types

Water scarcity can be divided into two categories by experts. Physical water scarcity is one. Economic water scarcity is the other. The International Water Management Institute led a study in 2007 that provided the initial definitions of these terms. This looked at how much water has been used in agriculture over the last 50 years. Its goal was to determine whether the world's water resources would be adequate to support future food production for the world's expanding population.

1.3.1.1. Physical Water Scarcity

When there aren't enough naturally occurring water resources to meet demand, physical water scarcity happens. This includes the water required for ecosystems to sustain themselves. Physical water scarcity is a common problem in arid regions. In places where it was already an issue, water scarcity has gotten worse due to human influence on the climate. In situations where resources are overcommitted but water appears to be plentiful, it also happens. Hydraulic infrastructure overdevelopment is one instance. This can be used to produce energy or for irrigation. Physical scarcity of water manifests itself in various ways. A few of these are the drastic deterioration of the environment, the depletion of groundwater, and the distribution of water resources that benefits some groups more than others. Experts have suggested an additional indicator. It's known as ecological water scarcity. It takes into account the quantity, quality, and flow requirements of the environment's water.

In populated arid areas, water is scarce. It is anticipated that each person will have less than 1000 cubic meters available annually. Examples include North Africa and Central and West Asia. More than 1.2 billion people, according to a 2007 study, reside in regions with physical water scarcity. Water that is needed for food production is scarcer than that which is needed for drinking, which is much less. A third category known as ecological water scarcity is something that some academics support including. The water requirements of ecosystems would be the main topic. That would be the bare minimum of water discharge quantity and quality required to keep ecosystems healthy and sustainable. Certain publications contend that this is merely a component of the physical water scarcity definition.

1.3.1.2. Economic Water Scarcity

Lack of investment in technology or infrastructure to extract water from aquifers, rivers, or other water sources is the cause of economic water scarcity. It also illustrates how inadequate human capacity is to meet the world's water needs. It forces those without consistent access to water to travel great distances in search of water for domestic and agricultural purposes. Such water is frequently filthy. According to the United Nations Development Program, the most frequent cause of water scarcity is economic. This is so that water needs for agriculture, industry, households, and the environment can be met in the majority of nations or regions. However, they are unable to offer it in an easily accessible way. Currently, about 25% of people on Earth reside in areas where there is a physical shortage of water.

Scarcity of water due to economic reasons affects 25% of the global population. A large portion of Sub-Saharan Africa has it. Thus, improved water infrastructure could contribute to a decrease in poverty. Infrastructure for irrigation and water retention investments could contribute to a rise in food production. This is particularly true for underdeveloped nations that depend on low-yield farming. Ensuring the public has access to potable water would also be beneficial. This goes beyond just building new infrastructure. Addressing poverty and social inequality requires both political and economic intervention. Planning is required because of the lack of funding. Usually, the focus is on enhancing domestic and drinking water sources. However, other than drinking and cooking, more water is used for things like cleaning, laundry, bathing, and caring for animals. This implies that focusing too much on drinking water only takes care of a portion of the issue. Thus, it might reduce the variety of options.

1.3.2. Solutions for Addressing Water Shortages

Although certain human endeavors have intensified the water crisis, technology has also been developed by humans to more effectively obtain or preserve freshwater. Dams and reservoirs, rainwater collection, aqueducts, desalination, water reuse, and water conservation are some of the ways to deal with the water scarcity issue.

1.3.2.1. Dams and Reservoirs

In order to store water for use during dry spells, reservoirs, or artificial lakes, that develop behind dams in rivers can collect water during wet seasons (Figure 1.5). Urban water supplies can also be supplied by them. Recreation, hydroelectricity, and flood control are additional advantages of dams and reservoirs. A few disadvantages are river channel erosion downstream and evaporative water loss in arid climates. Dams also lessen the flow of water downstream, which can cause political unrest when rivers cross national borders. A significant disadvantage of dams is their detrimental effects on ecosystems. For instance, dams alter the environment of a river to a lake and obstruct fish migration and spawning. In addition, the warming of the reservoir's surface water affects the downstream water temperature, which in turn affects fish and aquatic invertebrates that are cold-water adapted. Additionally, dams create habitat and supply nutrients downstream by trapping sediments that would otherwise continue to flow down the river.

Figure 1.5. *The hoover dam on the Colorado river in Nevada. Behind the dam is Lake Mead, the largest reservoir in the United States.*

Source: https://bio.libretexts.org/@api/deki/files/30104/Hoover_Dam.jpg?revision=1&size=bestfit&width=695&height=521.

1.3.2.2. Rainwater Harvesting

The process of collecting and storing rainwater before it hits the ground is known as rainwater harvesting. Figure 1.6 depicts a sophisticated rainwater harvesting system (also known as a rain water capture system) that is suggested for use in federal buildings; however, individual homeowners can use smaller, more basic systems (also known as rain barrels; Figure 1.7).

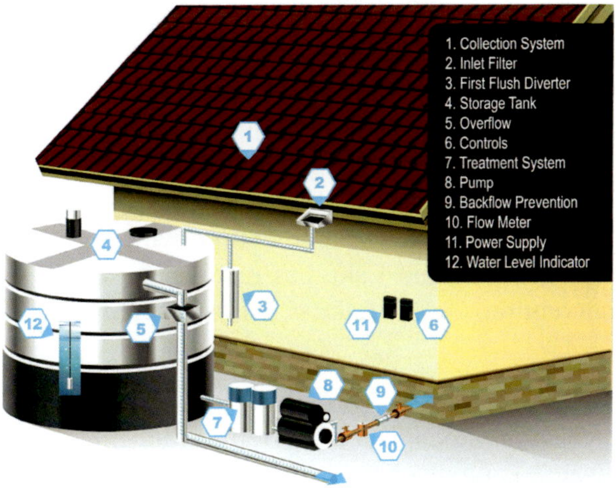

Figure 1.6. *The components of a rainwater harvesting system include.*

Source: https://bio.libretexts.org/@api/deki/files/30106/Rainwater_harvesting_graphic.png?revision=1&size=bestfit&width=837&height=647.

Figure 1.7. *A rain barrel is a simple rainwater harvesting system. Homeowners commonly use the stored water to irrigate their gardens.*

Source: https://bio.libretexts.org/@api/deki/files/30108/Rain_barrel.jpg?revision=1.

1.3.3. Aqueducts

Water can be transported via aqueducts from areas of abundance to areas of need. Aqueducts can be divisive and challenging from a political standpoint, particularly when it comes to long water transfer distances. One disadvantage is that the area where the water is drawn may experience a drought as a result of the water diversion. For instance, when the river flow from central California's Owens Lake and Mono Lake was redirected to the Los Angeles aqueduct, the lakes started to dry up. When Owens Lake ran out of water, it dried up and became a significant source of particulate matter, which contaminated the air during dust storms. While Mono Lake has recovered more significantly as a result of legal intervention, Owens Lake is still nearly entirely dry.

1.3.4. Desalination

Desalination, which entails taking out dissolved salt and minerals from seawater or saline groundwater, is one technique that can actually increase the amount of freshwater on Earth (Figure 1.8). This method has the benefit of having an almost limitless supply of saltwater. Boiling, filtering, electrodialysis (using an electric current to remove the ions that make up salts), and reverse osmosis are some methods for desalinating seawater (Figure 1.9). The water produced is far more expensive than freshwater from traditional sources because all of these processes range from moderately to extremely expensive and demand a significant amount of energy. Furthermore, the process generates highly salinized wastewater, which needs to be disposed of and has a big impact on the environment. The Middle East, where water is in short supply but oil energy is plentiful, is where desalination is most prevalent.

Figure 1.8. *Port Stanvac desalination plant under construction in South Australia.*

Source: https://bio.libretexts.org/@api/deki/files/30399/Port_Stanvac_Desalination_plant.jpg?revision=1.

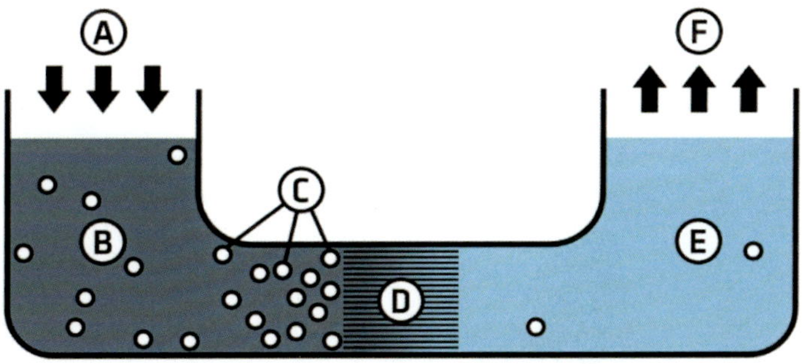

Figure 1.9. *Reverse osmosis is one method of desalination.*

Note: In this process, pressure (A) is applied to salt water, which contains ions (charged atoms or molecules) and other contaminants (C). The water is forced through a semipermeable membrane (D), through which the contaminants cannot pass. Clean freshwater that is safe to drink; and (E) is then distributed (F).

Source: https://bio.libretexts.org/@api/deki/files/30111/Reverse_osmosis.png?revision=1&size=bestfit&width=894&height=396.

1.3.5. Water Reuse (Water Recycling)

It supports green infrastructure and serves domestic, commercial, and industrial uses. Used water undergoes fit-for-purpose treatment, which is the minimum amount of treatment required. Environmental restoration, ground water recharge, agricultural reuse, surface water augmentation, portable reuse, irrigation reuse, home reuse, industrial reuse, and commercial reuse are examples of fit-for-purpose treatment and usage. Wastewater treatment is used to process discharge from residential, commercial, and industrial buildings (Figure 1.10).

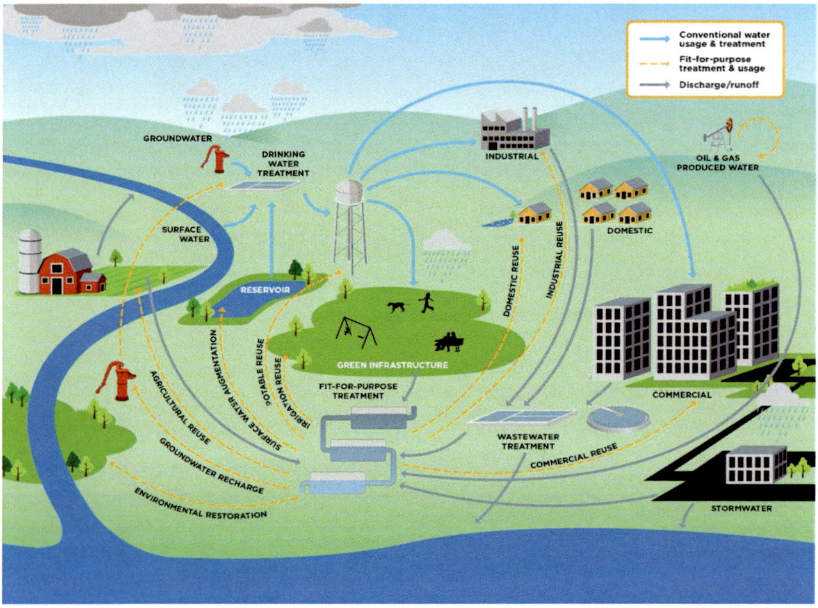

Figure 1.10. *Flowchart of water reuse. Surface water, groundwater, and water from reservoirs undergoes drinking water treatment.*

Source: https://bio.libretexts.org/@api/deki/files/30110/Water_Reuse.png?revision =1&size=bestfit&width=875&height=676.

1.3.6. Water Conservation

Conserving water means making better use of the water that is used. Water-saving devices and lifestyle choices can both contribute to conservation in the home. High-efficiency washing machines and low-flow showers and toilets are a couple of examples of water-saving technologies. Fixing leaky faucets, taking shorter showers and showering instead of bathing, and turning off the water while brushing your teeth are examples of water-conserving habits. Dishwashing by hand requires more water than using a dishwasher, since the dishwasher only runs when it is fully loaded. Comparatively speaking, doing fewer, bigger laundry loads saves water compared to doing more, smaller loads. Another way to save water is to choose foods like eggs that don't require as much water as ones that do, such as beef.

Water savings are numerous when gardening. Consider growing only native plants that can withstand drought and require little irrigation if you live in a dry climate (Figure 1.10). When watering your garden, make sure to do so early in the morning when evaporation is less likely to occur. Drip systems help to minimize evaporation by supplying only the necessary amount of water. These tactics can also be used extensively in agriculture, which is crucial given the substantial demands agriculture places on our water supply in comparison to municipal use. In agriculture, water conservation strategies include no-till farming, which lowers evaporative losses by covering the soil, growing crops in areas where natural rainfall can support them, and more effective irrigation systems like drip systems.

The water crisis cannot be resolved with bottled water. Bottled water is not always safer to drink than tap water. The U.S. public water supply is typically 700 times less expensive than bottled water, which uses about 200 billion plastic and glass bottles annually, with a comparatively low recycling rate. Bottled water also requires significantly more energy than tap water, mainly due to bottle production and long-distance transportation. If you do not like the taste of tap water, buying a water filter is a more environmentally friendly option than purchasing bottled water.

1.3.7. Sustainable Water Management in Arid Regions

Arid areas, which are marked by limited water resources, pose a significant challenge in light of the increasing scarcity of water worldwide. A sustainable approach to water management becomes critical as population growth and climate change intensify. Huge stretches of land are covered in arid regions, where millions of people live and rely on scarce water supplies. The World Health Organization (WHO) estimates that more than 2 billion people reside in nations with significant water stress. High evaporation rates and erratic precipitation make the situation worse in arid areas. Climate change is causing more frequent and severe droughts in arid regions, according to research by the Intergovernmental Panel on Climate Change (IPCC). Water supplies are further strained by changing precipitation patterns and rising temperatures. For example, researchers have noted a decrease in rainfall and an increase in the frequency of extreme weather events in the Middle East. Innovative technologies are essential to sustainable water management. For example, desalination has changed the game in dry coastal areas. The National Water

Company of the Kingdom of Saudi Arabia has made large investments in desalination plants, which supply a large amount of the country's water needs. In dry areas, agriculture is a significant user of water. The Food and Agriculture Organization (FAO) has conducted research that highlights the significance of implementing precision irrigation methods. Agricultural water use is maximized by the use of drip irrigation systems and soil moisture sensors, which drastically cut down on water waste.

Community involvement is essential for sustainable water management in arid regions. Studies carried out by regional non-governmental organizations in India demonstrate that teaching communities about water conservation results in a discernible decrease in water usage. Rainwater harvesting is one example of a community-driven project that has been successful in increasing water supplies. Water management policies are largely shaped by governments. The World Bank's research emphasizes the significance of integrated approaches to water resource management. In order to ensure a more sustainable water future, nations like Israel, which suffer from acute water scarcity, have put policies in place that encourage water recycling and reuse. Groundwater is frequently vital in arid areas. In order to stop depletion, groundwater management must be sustainable. The significance of artificial recharge techniques, like pumping treated wastewater into aquifers, is highlighted by hydrogeologist research. These methods stop seawater intrusion in coastal areas and replenish groundwater. Cooperation is crucial between countries that are experiencing a shortage of water. In order to guarantee that everyone has access to clean water by 2030, the UN has set forth Goal 6. Research collaborations that promote international cooperation, like the Joint FAO/IAEA Division, concentrate on exchanging knowledge about nuclear methods for water management.

1.4. WATER CONSERVATION IN ARID AND SEMI-ARID REGIONS

The type of the dominant hydroclimate largely influences the activities linked to effective water conservation management in arid and semi-arid regions. The comparatively tiny, unpredictable, and erratic rainfall events that are highly unevenly distributed spatially over the landscape and vary greatly in duration are what define the mean annual precipitation. For water suppliers and users alike, these circumstances present a number of intriguing challenges related to the location, volume, and accessibility of the relatively scarce water resources. Additional elements that contribute to water conservation include the potential for industrial and socioeconomic growth in some places that need infrastructure for water supply but are far from sustainable water sources.

Urbanization, population expansion, and the possibility of rising living standards, particularly in developing nations, increase demand for water resources, which are already in short supply. The population concentration in rural and urban areas with greater economic potential, such as urbanization, has a negative impact on the financial resources available to build new infrastructure for the supply of water. Increased water use from already limited water resources is necessary for both the expansion of industrial and mining activities and the improvement of food production. In order to minimize excessive demand and extend the amount of water available to satisfy the reasonable demand of all possible consumer groups, this situation necessitates resource management done with ingenuity. In this context, it should be noted that, compared to regions with higher rainfall, where it may be possible for contaminated water resources to be diluted and flushed, the pollution of water sources in arid areas poses a far greater, irreversible, and intolerable threat. Therefore, the best management practices—which include making the best use of all the institutional, technical, financial, environmental, and social opportunities available to water managers—should serve as the foundation for the conservation of water resources in arid areas.

1.4.1. Understanding an Arid Water Resource Base

A distinct rainy season always precedes a dry spell that lasts the entire year in arid countries. Convective rainstorms during the summer usually produce sparse, brief periods of precipitation, though they can occasionally produce extremely intense amounts. Shortly followed by high stormwater runoff, these downpours frequently result in catastrophic flood events. When rainfall occurs, runoff—which is typically referred to as a "flood"—occurs in the typically dry ephemeral streambeds or "rivers" directly. A further phenomenon that increases surface water availability is evaporation. Elevated temperatures and low humidity in the air mass above the land contribute to the high rates of evaporation encountered in arid regions. Since open water surfaces evaporatively lose a great deal of their accumulated runoff, the potential yearly evaporation is therefore far higher than the average annual precipitation. Up to 100 times less precipitation on average occurs each year than could be evaporating from the atmosphere. Aridity results from this water shortage, and evaporation is typically highest in regions with the least amount of rainfall. The watercourses in arid regions are therefore typically dry, but they may occasionally flow during the rainy season—that is, if and when it rains sufficiently.

Except for a small amount of perennial water flowing through the Border Rivers from neighboring countries like Botswana and Namibia have no permanent water resources at all. It can be inferred that the potential increase in internal water resources in those countries resulting from their shared resources could potentially escalate tensions if early mechanisms for collaboration and cooperative planning are not put in place. Significant variations from the average yearly rainfall are also observed in the rainfall received in arid regions. This complicates the planning, designing, and building phases of the development of water infrastructure as well as the management of the resulting water schemes. The variability of the rainfall is the mean annual rainfall divided by the deviation, expressed as a percentage of the mean annual rainfall. The statistical analysis of runoff, maximum probable runoff evens, and dam spillway design are all significantly impacted by this measure, which assesses the reliability that the expected mean annual rainfall will actually occur during a rainy season. This variation can vary from less than 20% in areas with higher rainfall to more than 80% in areas with lower rainfall in arid nations. Since the mean annual rainfall in arid regions is not a true representation of the rainfall that can actually be expected to precipitate, it should be used with caution due to the unpredictable and variable nature of the precipitation.

In general, the annual rainfall over a given duration is composed of many years with low, below-average rainfall and a small number of years with extremely high, above-average rainfall. Since the total volume of water produced will be less than expected based on the average rainfall over the observed period of time, the actual annual rainfall will typically be less than the calculated mean annual rainfall. One can compute an adjusted average known as the "median" if the few years with exceptionally high rainfall are disregarded or minimized. "In regions with heavier annual precipitation, the yearly precipitation may be relatively constant, making the mean more indicative of potential outcomes. This makes the reflection more accurate." Also, there will be little difference between the mean and the median.

The so-called "antecedent factor" is another significant hydrological phenomenon connected to runoff in arid areas. Because of the grazing of livestock and wildlife during the dry season, the vegetation cover gradually decreases. The vegetation cover may entirely vanish if there are several unfavorable rainy seasons followed by overgrazing. Upon the onset of the first good rains following a period of drought and overgrazing, there is an immediate beneficial runoff. Although the vegetation can contain or lessen runoff, it also begins to recover very quickly. The runoff produced may be less than the previous events with less rainfall because of the improved vegetation cover. This is often the case if heavier rains fall later in the same rainy season, or even better average showers during the next rainy season. It is commonly known that the high silt loads found in the runoff of ephemeral rivers are caused by the typically low levels of vegetation found in arid regions. Due to the absence of flora during dry spells, the silt loads in runoff are significantly higher right after and significantly lower after several wet seasons have helped the vegetation recover. Observing what transpires to the precipitated water in arid regions is also intriguing.

With an average yearly rainfall of 250 mm, it is estimated that in Namibia, only 2 percent of the rainfall is usable as runoff and that the remaining 1 percent eventually turns into groundwater. A little over 83% of the precipitation evaporates immediately after it falls, with the remaining 14% seeping into the dry soil and being used by the native flora. For dryland crop cultivation, livestock raising, and biodiversity preservation to continue, this amount of rainfall is vital. The economy of arid nations is highly vulnerable to fluctuations in the hydroclimate, so it's critical to comprehend the current situation as well as forecast potential outcomes in order to try and mitigate the effects of infrequent rainfall on human activity. Droughts in Namibia can be divided into two primary categories. They are known as an agricultural drought and a hydrological drought, respectively.

When the total precipitation during the rainy season is significantly less than the long-term mean or median annual precipitation, a rainfall drought occurs. One can distinguish between two types of droughts in the hydrological system: rainfall and runoff. Hydrological droughts are extremely important in arid regions, but people often don't realize that droughts are worse than normal dry conditions and that low rainfall is the typical occurrence in these regions. An excessive amount of precipitation, or none at all, or average rainfall that falls in the catchments but at such a low intensity that no runoff events happen are two scenarios that can result in a runoff drought. Consequently, the water supply dams cannot hold onto any surface runoff. When the rainy season begins, the same thing occurs: by the time heavy showers arrive, the established vegetation cover has greatly reduced runoff. This is because the vegetation has begun to recover significantly from the early little rains.

A rainfall drought happens when the total amount of precipitation that falls during the rainy season is significantly less than the mean or median annual precipitation over the long term. One can distinguish between a runoff drought and a rainfall drought within the hydrological drought. In dry regions, hydrological droughts are extremely rare. However, few people realize that low rainfall is the typical condition in these regions and that a drought is in fact worse than the usual dry conditions. A runoff drought can develop from either an excess of little (or no) precipitation or from average (or better) rainfall that falls in the catchments but at a level so low that no runoff events happen. Therefore, the water supply dams cannot accumulate any surface runoff. This is also true when the rainy season begins and the vegetation has begun to recover greatly from the early, light rains. By the time the heavy rains arrive, the established vegetation cover has greatly decreased runoff. When we talk about an agricultural drought, we're talking about the state of the grazing vegetation and the inability of rainfed crop cultivation as a result of insufficient rainfall during a rainy season. A drought in agriculture can be further classified as a drought for crops or for grazing. When the grass is either stunted or does not react well to rainfall, a grazing drought happens. This can occur in situations where there was a hydrological drought (low rainfall) or when there was average rainfall or higher, but the rain came either too early or too late to be useful during the ideal growing season. When rain falls during the rainy season at the incorrect time or does not occur at all, it results in a rainfed crop drought. Even though there has been more rain than usual, the crop will not succeed if the farmer plants after the rainy season's initial downpour and the subsequent downpours don't come. It can also occur that the planting season ends too late for a crop to be planted and harvested in the remaining time due to the first substantial rains falling so late. This is because countries that are arid may only receive more than 50 mm of rainfall in less than 4 months. Therefore, a key factor in crop productivity is the duration of the rainy season.

Because most arid countries only experience rain on 20 to 60 days a year, the frequency of rainfall is also crucial for crop productivity. These unpredictable weather patterns significantly impact crop yield forccasts, which in turn impact the production of staple foods under rainfed conditions. It is premature to determine whether or not there will be a good rainy season with sufficient crop yields, even in the event that good rains occur early in the rainy season. Using their institution and experience is probably the best course of action for farmers. Preparing the land and planting are preferable to waiting and watching to see what transpires. Still, a government cannot afford to take such a chance and must prepare backup plans to guarantee that, even well in advance of the start of the rainy season, a safe supply of staple foods will be accessible in the event of a crop failure. This also holds true for fodder availability. Rainfall that occurs either before or after the natural vegetation's growing season may cause a severe shortage of grazing land until the next rainy season. The productivity and economics of farming operations, in both the subsistence and commercial farming sectors, are obviously greatly impacted by the unpredictable rainfall.

One of the lucky things about living in an arid climate is that it is very difficult to forecast how a rainy season will turn out. If someone has good reason to think the worst, there's always a good chance that the reality of the rainfall will show them to be completely wrong. Anticipating the behavior of a rainy season in an arid region is solely dependent on waiting for the season to conclude and verifying the events that transpired.

1.4.2. Institutional Framework

In dry regions, effective management of water resources is essential, but without sufficient institutional frameworks, this cannot be achieved. A water conservation strategy cannot be implemented without the proper institutional framework and financial and human resource capacity. Water management in semiarid and arid nations such as Namibia, Botswana, and South Africa is overseen by Departments of Water Affairs that are suitably organized and skilled. The general management of the nation's water resources, including making sure that water is used sustainably and that the water resource base is thoroughly examined to assess its potential, should be one of the common responsibilities of these Departments of Water Affairs. Additional crucial roles include keeping water supplies free of pollution and distributing water to users in a way that maximizes the country's overall economic and social gains.

1.4.2.1. Water Legislation

There are several pillars that support the management of water affairs. These include the nation's constitution, the water law, the regulations enacted under the water law, and the policy for water supply and sanitation. The government may occasionally want to put additional policies into place in order to manage the nation's water resources in the most effective way possible while there is a drought. Adequate provisions in a nation's constitution facilitate the adoption of environmental, water, and sanitation policies as well as the enactment of relevant laws. Article 95 of the Constitution of the Republic of Namibia, for instance, addresses preserving the welfare of the populace. Sub-article 95(l) mandates that the Government establish policies for the sustainable use of living natural resources for the benefit of all Namibians, living and dead, as well as for the preservation of ecosystems, vital ecological processes, and the nation's biological diversity. Article 100 of the aforementioned Constitution addresses the sovereign ownership of natural resources in a similar manner. This Article states that, unless they are otherwise legally owned, the land, water, and natural resources of the nation shall belong to the State.

In light of competing demands, this gives the government the ability to control how limited water resources are allocated. A specialized Ministry or Department of Water Affairs and a representative advisory body to advise the Minister of Water Affairs on water matters should be established as minimum requirements of any water legislation drafted for arid countries. Rules governing water abstraction in subterranean water control areas should be established by legislation. Control over dam construction, water distribution, wise use of available resources, and fair distribution of those resources among various users in order of priority should also be made easier by the water law. Licenses and permits that are granted for the extraction and utilization of groundwater in underground water control areas or surface runoff in public streams need to be controlled.

A capable organization like the Department of Water Affairs should oversee the administration of water legislation on behalf of the government. Water legislation should also grant the Minister in charge of water affairs the authority to take any action he deems necessary for the exploration, development, management, and use of water resources in addition to putting a water Act's provisions into effect. When it comes to having the power to regulate water use in arid regions and accepting accountability for safeguarding and conserving water resources, the ownership of water is a crucial factor to consider. Namibia's Water Act distinguishes between public and private water sources. The Act governs the use of public water, which is defined as water that is typically found in a public stream, whether or not it is visible (groundwater in an ephemeral watercourse). Private water is defined as water that naturally exists on the land and may be used solely by the landowner for any purpose. Therefore, there is no property right in either private or public water. Therefore, the purpose of the Act's distinction between public and private water is to specify who has the right to use the water and how, rather than to establish who owns the water. The Act defines surplus water as any public water that does not flow normally, such as floodwater.

In dry regions such as Namibia, runoff is interpreted as floods, which can occasionally occur in those dry rivers since there are no perennial streams with regular surface flow. Though it cannot be used unless it is impounded, this excess water is available for additional beneficial uses. To safeguard the environment and the rights of riparian users downstream, water act regulations governing the construction of dams in public streams and their storage capacity must be followed. As a result, there will be a sufficient amount of runoff left over for use downriver. People who are not landowners or riparian landowners are permitted to use water under the Water Act. As with private citizens, local governments, or any other legal entity with the authority to provide public water to any consumer, the Minister may provide water from a State Water Scheme to any individual for any use on any land. Preventing the contamination of freshwater or marine waters from land-based effluent sources is crucial, and a water act should include provisions to regulate the purification and disposal of effluents.

It should be illegal to dispose of any wastewater in arid areas into ephemeral water sources unless authorized by a permit that includes requirements for purifying wastewater to a level where the finished product is deemed acceptable. In addition to being in charge of water affairs, a minister ought to have the authority to order the construction of any water works that he thinks appropriate or necessary to supply or conserve water. In addition, he needs to be able to control the abstraction of water from any surface or subsurface water source, store water, drain land, and stop water waste. Additionally, the Minister needs to have the authority to enact rules pertaining to the planning and administration of water projects. These could include provisions for financing the construction of water works, the imposition of fees and tariffs for the water supplied, and the regular adjustment of those fees. Additionally, the Minister must have the authority to grant permission for anyone to enter private property in order to monitor and inspect water supply schemes or to look into the viability of any water works.

REFERENCES

1. Hoelzmann, P., Keding, B., Berke, H., Kropelin, S., & Kruse, H. J. (2001). Environmental change and archaeology: Lake evolution and human occupation in the Eastern Sahara during the Holocene. *Palaeogeography, Palaeoclimatology, Palaeoecology*, *169*, 193–217.
2. Karimidastenaei, Z., Klöve, B., Sadegh, M., & Haghighi, A. T. (2021). Polar ice as an unconventional water resource: Opportunities and challenges. *Water*, *13*(22), 3220.
3. Kundzewicz, Z. W., Mata, L. J., Arnell, N. W., Döll, P., Kabat, P., Jiménez, B., Miller, K. A., Oki, T., Sen, Z., & Shiklomanov, I. A. (2007). Climate change and water. *Intergovernmental Panel on Climate Change* (IPCC).
4. Peel, M. C., Finlayson, B. L., & McMahon, T. A. (2007). Updated world map of the Köppen-Geiger climate classification. *Hydrology and Earth System Sciences*, *11*(5), 1633–1644.
5. Qiang, X. K., Li, Z. X., Powell, C. M., & Zheng, H. B. (2001). Magnetostratigraphic record of the Late Miocene onset of the East Asian monsoon, and Pliocene uplift of northern Tibet. *Earth and Planetary Science Letters*, *187*, 83–93.
6. Shi, N., Schneider, R., Beug, H. J., & Dupont, L. M. (2001). Southeast trade wind variations during the last 135 kyr: Evidence from pollen spectra in eastern South Atlantic sediments. *Earth and Planetary Science Letters*, *187*, 311–321.
7. Stadler, S. J. (1987). Aridity indexes. In J. E. Oliver & R. W. Fairbridge (Eds.), *The Encyclopedia of Climatology* (pp. 102–107). Van Nostrand Reinhold.

Chapter 2

Principles of Water Conservation in Irrigation

Contents

INTRODUCTION

A key component of sustainable agriculture and environmental stewardship is water conservation in irrigation. Optimizing water use efficiency, cutting waste, and making sure that every drop of water contributes to crop growth rather than evaporating or running off are the main goals of water conservation in irrigation.

First and foremost, it's critical to comprehend how much water crops require. Depending on their growth stage, the type of soil they are growing in, and the climate, different crops have different water requirements. Farmers can avoid both overwatering and underwatering by scheduling and supplying irrigation according to these particular requirements.

Soil moisture sensors and weather-based irrigation controllers are common tools used to put this idea into practice because they offer real-time data that helps irrigation decisions be made more precisely. Second, it's imperative to implement effective irrigation systems. Due to evaporation and runoff, traditional surface irrigation techniques like flood or furrow irrigation can result in significant water loss. Water delivery that is more precisely targeted is provided by modern irrigation technologies like sprinkler and drip irrigation. For example, drip irrigation minimizes evaporation and runoff by delivering water straight to the roots of the plants. Similar to this, by better focusing on the root zone, sprinkler systems can be adjusted to waste less water. The application of water-saving techniques and technologies is another fundamental idea. This involves using methods for conserving soil, such as mulching, which helps hold onto soil moisture and lower evaporation. Reliance on conventional water sources can also be decreased by installing rainwater harvesting and storage systems, which can collect and use precipitation for irrigation.

Furthermore, preserving and enhancing the health of the soil is crucial to water conservation. Water infiltration and retention can be improved by having healthy soils with a high organic content and good structure. Reduced tillage and cover cropping are two techniques that support soil integrity and lower erosion, both of which improve water management. Finally, it's critical to teach and train farmers in water-efficient techniques. Farmers can be effectively empowered to adopt conservation techniques and technologies by means of extension services, workshops, and demonstrations. In order to achieve sustainable water use, the agricultural community can cooperate by cultivating a culture of water stewardship and granting resources. Water conservation in irrigation refers to the following principles: matching crop requirements with water application, using effective irrigation systems, putting water-saving techniques into practice, preserving soil health, and encouraging education. Farmers can improve crop productivity, preserve water, and contribute to a more sustainable agricultural system by incorporating these ideas.

2.1. OVERVIEW OF WATER CONSERVATION

The goals of water conservation are to meet present and future human demand while safeguarding the hydrosphere and the natural resource of fresh water. It is possible to prevent water scarcity through water conservation. It includes every strategy, policy, and action to accomplish these goals. The size and growth of households, wealth, and population all influence the amount of water used.

The stress on natural water resources has increased due to climate change and other factors. This is particularly true for irrigation in agriculture and manufacturing. Numerous nations have effectively instituted measures to preserve water resources. To conserve water, there are a number of essential actions. One is the advantageous decrease in resource waste, loss, and usage of water. Preserving the quality of the water is another. Enhancing water management techniques that either decrease or increase the useful use of water is the third.

Technology solutions exist for households, commercial and agricultural applications to reduce the. Water conservation programs involved in social solutions are typically initiated at the local level, by either municipal water utilities or regional governments.

2.1.1. The Importance of Water Conservation in Irrigation

Since water is the most valuable resource, protecting it is essential to the long-term viability of our ecosystem and the welfare of future generations. Conventional landscape irrigation techniques frequently result in overwatering, evaporation, and runoff, which waste a large amount of water. In order to solve these problems, modern irrigation techniques that conserve water maximize water usage efficiency and make sure plants get the proper amount of water at the right time.

For a number of reasons that have a significant impact on agricultural productivity, environmental health, and long-term economic viability, water conservation in irrigation is essential. The efficient use of water in irrigation becomes more and more important as the world's population increases and climate patterns change.

- **Sustainable Resource Management**: Water is a finite resource, and many regions face significant water scarcity. Effective water conservation in irrigation helps ensure that available water supplies are used judiciously, preserving this critical resource for future generations. By optimizing water use, we can mitigate the risk of over-extraction from rivers, lakes, and aquifers, which can lead to ecological degradation and depletion of water sources.

- **Enhanced Agricultural Productivity**: Efficient irrigation practices directly contribute to better crop yields. When water is applied accurately and in the right amounts, crops receive the optimal conditions needed for growth. This leads to healthier plants, reduced stress, and improved quality and quantity of produce. Water conservation techniques, such as drip irrigation and soil moisture monitoring, help farmers achieve more reliable and consistent yields while using less water.
- **Economic Benefits**: Conserving water in irrigation can result in substantial economic savings for farmers. Efficient irrigation systems reduce water and energy costs, as less water is wasted and energy required for pumping is minimized. Moreover, improved crop productivity and reduced water stress can lead to higher profits and greater economic stability for farming operations. Additionally, efficient water use can prevent costly problems associated with waterlogging, salinization, and soil erosion.
- **Environmental Protection**: Proper water conservation practices in irrigation help protect ecosystems and biodiversity. Excessive water use can lead to the depletion of wetlands, reduced river flows, and harm to aquatic habitats. By conserving water, we reduce the impact on natural habitats and maintain the ecological balance necessary for wildlife and plant species to thrive. Furthermore, efficient irrigation helps prevent soil degradation and salinization, which can occur when water is applied excessively or unevenly.
- **Climate Resilience**: As climate change leads to more erratic weather patterns, including droughts and heavy rainfall, water conservation in irrigation becomes crucial for adapting to these changes. Efficient irrigation systems can help farmers manage water more effectively during periods of scarcity and adapt to varying climatic conditions. This resilience is essential for maintaining agricultural productivity and food security in an unpredictable climate.
- **Community and Social Impact**: Water conservation in irrigation supports the well-being of communities by ensuring a more stable and sustainable food supply. When water is used efficiently, it can help stabilize local economies, provide employment opportunities, and enhance food security. Furthermore, by reducing the strain on local water sources, communities can benefit from improved access to clean water for drinking and sanitation.

The importance of water conservation in irrigation extends beyond merely optimizing agricultural practices. It encompasses the need for sustainable resource management, economic efficiency, environmental protection, climate resilience, and community well-being. Embracing water-saving techniques and technologies is essential for addressing the challenges of water scarcity and ensuring a sustainable future for agriculture and ecosystems.

2.1.2. Strategies

The key activities to conserve water are as follows:

- Any beneficial reduction in water loss, use and waste of resources.
- Avoiding any damage to water quality.
- Improving water management practices that reduce the use or enhance the beneficial use of water.

One of the strategies in water conservation is rainwater harvesting. Digging ponds, lakes, canals, expanding the water reservoir, and installing rain water catching ducts and filtration systems on homes are different methods of harvesting rain water. Many people in many countries keep clean containers so they can boil it and drink it, which is useful to supply water to the needy. Harvested and filtered rain water can be used for toilets, home gardening, lawn irrigation, and small-scale agriculture.

Preserving groundwater resources is another tactic in water conservation. Some precipitation seeps into the ground and becomes subterranean. Groundwater is the term for the water in this saturation zone. The groundwater supply becomes unusable as a source of clean drinking water when it becomes contaminated, and it may take years for the contaminated groundwater to naturally regenerate. The following are a few instances of possible groundwater contamination sources: chemicals, road salts, landfills, uncontrolled hazardous waste, septic systems, storage tanks, and landfills. Preventive measures, such as shielding groundwater resources from contamination, are crucial to water conservation because contaminated groundwater reduces the amount of freshwater that is replenished. Adopting sustainable practices for using groundwater resources is another way to save water.

Gravity causes groundwater to flow, eventually discharging into streams. Groundwater levels drop as a result of excessive groundwater pumping, and if it persists, the resource may run out. Since surface and groundwater are interconnected, excessive groundwater use can limit and, in severe cases, completely eliminate the water supply in lakes, rivers, and streams. Over-pumping groundwater in coastal areas can lead to an increase in saltwater intrusion and subsequent contamination of the groundwater supply. Water conservation requires sustainable groundwater use. Outreach and education regarding various water programs is a vital part of any strategy for conserving water. Creating science-based communication for farmers, policy makers, land managers, and the general public is another crucial tactic used in water conservation. When developing a management plan to conserve a water system, communication of the science involved is crucial and is frequently used to ensure that the appropriate management plan is implemented. Water conservation is critical to maintaining the habitats of wildlife.

In temperate regions, water scarcity affects a wide variety of organisms. Additionally, as water pollution disturbs the ecosystem, a large number of freshwater organisms are becoming more and more affected.

2.1.3. Social Solutions

Programs for water conservation that are part of social solutions are usually started at the local level by regional or municipal water utilities. Public awareness campaigns, tiered water rates (increasing prices as water use increases), and prohibitions on outdoor water use, such as washing cars and lawns, are examples of common strategies. To cut down on the amount of water used outdoors, cities with arid climates frequently mandate or promote the installation of xeriscaping or natural landscaping in new construction. In California, residential use accounts for the majority of urban outdoor water use, highlighting the need for outreach to both homes and businesses.

Universal water metering is one of the main objectives of conservation. Globally, residential water metering is not as common as it once was. Less than 30% of households in the UK are thought to have metered water supplies, according to recent studies. In multifamily buildings or homes with private wells, individual water meters have historically been seen as impractical, but according to the US Environmental Protection Agency, metering alone can cut consumption by 20 to 40%. Metering is a crucial tool for locating and identifying water leaks in addition to making consumers more conscious of how much water they use.

metering could be advantageous to society since it offers a financial incentive to reduce water waste. Given that 70% of the fresh water used worldwide is used for crop irrigation, some researchers have proposed that farmers be the main target of water conservation initiatives. Most nations place a high value on agriculture, both politically and economically, and they frequently provide water subsidies. Environmentalists have called for the elimination of all subsidies in order to compel farmers to use less water when growing crops and to use less inefficient irrigation methods.

Customers now have more options thanks to new technology. Features like the ability to flush a toilet completely or partially aim to reduce waste and water usage. Additionally, the water can be used and contaminated incrementally, saving its use to flush toilets for last. This allows for more uses of the water within a single cycle before it needs to be purified again, which can also be done in-situ. Such a configuration is common in Earthships. Modern shower heads that minimize water waste are also available. According to reports, older fixtures used 5 to 10 gallons of water per minute, whereas the newer models offer equal water coverage and use 2.5 to 5 gallons per minute. Another approach is to directly recycle shower water using a semi-closed system that has a pump and filter. An identical configuration, known as a water recycling shower, has also been used at the VIRTUe LINQ house. In addition to recycling water, it also recovers the heat that would otherwise be lost in the process.

Contrary to the popular view that the most effective way to save water is to curtail water-using behavior (e.g., by taking shorter showers), experts suggest the most efficient way is replacing toilets and retrofitting washers; as demonstrated by two household end use logging studies in the US.

Water-saving technology for the home includes:

- Low-flow shower heads sometimes called energy-efficient shower heads as they also use less energy
- Low-flush toilets, composting toilets, and incinerating toilets. Composting toilets have a dramatic impact in the developed world, as conventional Western flush toilets use large volumes of water
- Dual flush toilets include two buttons or handles to flush different levels of water. Dual flush toilets use up to 67% less water than conventional toilets
- Faucet aerators, which break water flow into fine droplets to maintain "wetting effectiveness" while using less water. An additional benefit is that they reduce splashing while washing hands and dishes
- Raw water flushing where toilets use sea water or non-purified water (i.e., greywater)
- Wastewater reuse or recycling systems, allowing:
  - Reuse of greywater for flushing toilets or watering gardens
  - Recycling of wastewater through purification at a water treatment plant.
- Rainwater harvesting
- High-efficiency clothes washers
- Weather-based irrigation controllers
- Garden hose nozzles that shut off the water when it is not being used, instead of letting a hose run.

- Low flow taps in wash basins
- Swimming pool covers that reduce evaporation and can warm pool water to reduce water, energy and chemical costs.
- Automatic faucet is a water conservation faucet that eliminates water waste at the faucet. It automates the use of faucets without the use of hands.

Smart water meters are also a promising technology for reducing household water usage. A study conducted in Valencia, Spain, shows the potential that smart meter-based water consumption feedback has for conserving water in households. The findings showed that households that were equipped with smart water meters increased their water savings. This technology works to show people how much water they were using in their household, suggest ways they can reduce water usage, and incentivize water savings with physical rewards.

2.1.3.1. Applications

Many water-saving devices (such as low-flush toilets) that are useful in homes can also be useful for business water saving. Other water-saving technology for businesses includes:

- Waterless urinals (also can be installed in schools).
- Waterless car washes.
- Infrared or foot-operated taps, which can save water by using short bursts of water for rinsing in a kitchen or bathroom.
- Pressurized water brooms, which can be used instead of a hose to clean sidewalks.
- X-ray film processor re-circulation systems.
- Cooling tower conductivity controllers.
- Water-saving steam sterilizers, for use in hospitals and health care facilities.
- Rainwater harvesting.
- Water-to-water heat exchangers.

It is imperative to contemplate executing water-conserving modifications in industrial and commercial settings. It was discovered that whereas low-income countries use 8% of their water for industrial purposes, high-income countries use about 59% of it. Increasing the evaluation and upkeep of water systems is a significant improvement that commercial and industrial organizations can make. Water-efficient applications are simple to add, but long-term changes require regular maintenance and inspection. It is possible to create a plan for water conservation that includes benchmarks and goals for the company and its employees. Another adjustment that commercial and industrial businesses can make is to routinely inspect water-consuming systems for leaks or other issues. By doing this, it will be ensured that neither water loss nor unnecessary expenditure on utility bills occur. An additional modification that businesses and industrial organizations can make is the installation of a rain sensor. When precipitation is detected, this sensor ought to be able to recognize it and halt the irrigation program that would otherwise be running. The sensor should restart the program and go back to the regular watering cycle after the rain stops.

2.1.3.2. Agricultural Applications

An essential component of irrigation is water. Since plants constantly draw a large amount of ground water, it is necessary to replenish it. The goal of optimal water efficiency for crop irrigation is to maximize production while minimizing losses from evaporation, runoff, or subsurface drainage. The amount of water required to meet plant requirements can be ascertained using an evaporation pan and particular crop correction factors. The earliest and most popular kind of irrigation, flood irrigation, frequently distributes water extremely unevenly because some fields may need to receive more water than others in order to have enough for everyone. Utilizing center-pivot or lateral-moving sprinklers for overhead irrigation can result in a distribution pattern that is far more uniform and regulated. The most costly and infrequent kind of irrigation is drip irrigation, which has the advantage of minimal water loss when supplying water to plant roots. But with rising water prices, drip irrigation is becoming more and more affordable, especially for home gardeners. When drip irrigation replaces all-spraying irrigation systems, annual water savings of up to 30,000 gallons can be achieved. Similar to drip irrigation, there are also less expensive, more efficient techniques like using soaking hoses, which can be completely submerged in the growing medium to prevent evaporation.

Since it can be expensive to replace irrigation systems, conservation initiatives frequently focus on increasing the effectiveness of the current system. This could entail chiseling compacted soils, constructing furrow dikes to stop runoff, and optimizing irrigation schedules with soil moisture and rainfall sensors. Large efficiency gains are typically achievable through improved management of the current irrigation system and measurement. In order to maximize the use of rainfall and irrigation during dry seasons, the 2011 UNEP Green Economy Report notes that improved soil organic matter from the use of green manures, mulching, and recycling of crop residues and animal manure increases the water holding capacity of soils and their ability to absorb water during torrential rains.

Plastic mulch has the potential to help save water in agricultural practices, as demonstrated in China. Actually, the mulch is just a thin layer of plastic covered in soil. The plastic has holes in it that the plants can grow through. Plastic mulch conserves water by slowing down the evaporation of soil moisture, according to some studies. However, not enough practical research has been done to estimate the overall water savings that this practice may result in.

2.1.3.3. Water Reuse

The lack of water is becoming a more challenging issue to handle. Over 40% of people on the planet reside in areas where there is a greater demand than supply of water. Due to the mismatch between supply and demand, as well as ongoing problems like population growth and climate change, reusing water has become an essential part of water conservation. Waste water is treated using a range of techniques to make sure it's safe to use for drinking water or irrigation of food crops. Desalinating seawater uses more energy than desalinating freshwater.

In spite of this, a large number of seawater desalination plants have been constructed in response to global water scarcity. Because of this, it's essential to assess the effects of seawater desalination and look for methods to advance the technology. Experiments are being used in current research to identify the desalination techniques that are least energy intensive and most effective.

An additional technique for treating water is sand filtration. While there is still room for improvement, recent research indicates that sand filtration is getting close to optimal in terms of its ability to eliminate pathogens from water. Sand filtration works well at eliminating bacteria and protozoa, but it has trouble getting rid of viruses. Large surface areas are also needed to accommodate large-scale sand filtration facilities.

The elimination of pathogens from recycled water is crucial since wastewater invariably contains pathogens that can infect humans. A certain threshold must be reached for recycled water to be free of harmful virus concentrations that could endanger human health. To find more precise ways of estimating the quantity of pathogenic viruses in treated wastewater, more investigation is required.

2.1.3.3.1. Wasting of Water

The opposite of water conservation is wastage, which in domestic settings refers to allowing or causing water to be released into the atmosphere for no legitimate reason. Using water inefficiently is also regarded as wasteful. According to EPA estimates, leaks in US homes can waste up to 900 billion gallons (344 billion cubic meters) of water a year nationwide. Water management organizations are typically hesitant or unwilling to provide a precise definition for the somewhat nebulous concept of water waste.

However, local drought emergency ordinances frequently define water waste. One example is any deliberate or inadvertent acts or omissions that allow water from any tap, hose, faucet, pipe, sprinkler, pond, pool, waterway, fountain, or nozzle to leak, discharge, flow, or run to waste into any gutter, sanitary sewer, watercourse, public or private storm drain, or to any adjacent property. In this instance, the city code further specifies that "disposal," "flow," or "run to waste" refers to the fact that more water flows to waste when washing, wetting, or cleaning a dirty or dusty object like a car, sidewalk, or parking lot. Water utilities (as well as other media outlets) frequently list inefficient water use behaviors and forbid inefficient uses. San Antonio, Texas utilities are one example. Cities of San Diego, California; Las Vegas, Nevada; and California Water Service Company, California. Water-wasting practices like leaks, runoff, irrigating during and right after rainfall, and using potable water when non-potable water is available are all permanently restricted by the City of Palo Alto in California. The Australian State of Victoria has similar regulations in place. In England, Scotland, Wales, and Northern Ireland, there are temporary water use restrictions, commonly referred to as hosepipe bans.

Water that is released into the environment or into the sewer is not, strictly speaking, lost or wasted. It stays in the hydrologic cycle and returns as precipitation to the surface of the land and surface water bodies. But frequently, the water's source is located in a different catchment and is quite far from the point of return. In the riparian zone and watercourse, the distance between the extraction point and return point may indicate substantial environmental deterioration. The water supply for the community, which was collected, kept, moved, and treated to meet drinking quality requirements, is wasted. Using water resources wisely reduces the cost of providing a water supply while preserving more freshwater for ecosystem support and other users in lakes, rivers, and aquifers. For instance, we shouldn't use the toilet as a rubbish disposal. We waste gallons of water if we flush tissues or cigarette butts through it. because it is impossible to complete the water recycling process. Water-use efficiency is a concept that is closely associated with water waste. If less water can be used to achieve the same goal, then the use of water is deemed inefficient.

Technical efficiency is derived from engineering practice, where it is commonly employed to characterize the output to input ratio and is helpful in contrasting different processes and products. For instance, two showerheads with the same functionality could be deemed more efficient than one another i.e., of taking a shower) by utilizing fewer resources (e.g., reduced water pressure). When deciding how much money (or resources) to invest in water conservation measures, the technical efficiency concept is useless unless the inputs and outputs are valued. The idea of water conservation includes this type of efficiency, which is known as economic efficiency.

2.2. FUNDAMENTALS OF WATER-SAVING IRRIGATION

Water-saving irrigation is a critical component of modern agricultural practices aimed at addressing the growing concerns of water scarcity and ensuring sustainable crop production. The fundamentals of water-saving irrigation involve a combination of advanced technologies, strategic management practices, and a thorough understanding of crop and soil needs to optimize water use and minimize waste.

The precision principle is the foundation of water-saving irrigation. By applying water to crops precisely when needed, precision irrigation techniques minimize both overuse and underuse of water. Drip irrigation is among the most efficient techniques in this respect. Through a system of tubes, emitters, and valves, water is directly delivered to the roots of the plants by this system. Drip irrigation ensures that water is used efficiently and reduces water loss from evaporation and runoff by precisely targeting the root zone. Additionally, drip systems enable the fertigation application of pesticides and fertilizers, further improving the efficiency of resource use.

Another popular water-saving technique that, with careful management, can drastically cut down on water waste is sprinkler irrigation. The sophisticated features found in modern sprinkler systems include pressure regulators and programmable controllers, which alter the timing and volume of water applied in response to changes in the weather and soil moisture content. These systems can be set up to spray water in a consistent pattern, making sure that there are no excessive overlaps or gaps and that every part of the field gets the right amount of water.

Utilizing instruments for measuring soil moisture is essential to irrigation that uses less water. Farmers can make well-informed decisions about when and how much to irrigate their land thanks to the real-time data that soil moisture sensors provide on the soil's moisture content. To maintain ideal soil conditions and reduce water waste, this data-driven approach helps prevent overwatering and makes sure that water is applied only when necessary.

Effective water-saving irrigation not only uses cutting-edge irrigation technologies but also incorporates water-saving techniques into the broader farm management plan. By decreasing evaporation and enhancing soil structure, practices like mulching aid in the retention of soil moisture. Grown in between primary crops, cover crops improve soil health and lessen erosion, which further aids in water conservation. Reduced tillage techniques, for example, lessen soil disturbance, which can enhance water infiltration and retention.

A thorough irrigation strategy that conserves water must also include water harvesting and storage. Rainwater or runoff can be collected and stored to create an additional water source and lessen reliance on traditional water sources. By installing rainwater harvesting systems and building storage structures like ponds or tanks, one can lessen their dependency on surface or groundwater sources and increase the amount of water available during dry spells.

Water-saving irrigation techniques must be implemented successfully, and this requires education and training. Giving farmers the skills and resources they need to implement cutting-edge irrigation technologies and best practices guarantees that they can efficiently manage water resources and raise irrigation efficiency as a whole.

Soil moisture monitoring, integrated conservation practices, and precision irrigation are all part of the foundations of water-saving irrigation. Farmers can optimize water use, minimize waste, and support the sustainable management of water resources in agriculture by implementing these strategies. This all-encompassing strategy promotes environmental sustainability and resilience in the face of challenges related to water scarcity in addition to increasing crop yields and productivity.

2.2.1. Key Practices in Water-Saving Irrigation

Knowing when to stop irrigation in case of a malfunction is equally crucial. To irrigate effectively and professionally, your irrigation system must cut off the water when a problem is detected. With real-time water meter monitoring, your irrigation platform should be immediately shut off if the flow exceeds what is anticipated.

The most important thing about having this level of control is that it ensures that water will not be wasted in the event that a sprinkler breaks or a pipe bursts. Additionally, it drastically lowers runoff and any possible complaints from homes. Remember not to forget.

To achieve effective and professional irrigation, the precise amount of water irrigated still needs to be adjusted. The amount of water that needs to be adjusted for the plant as a result of evaporation from the previous day is typically the basis for this adjustment.

- **Soil Moisture Monitoring:** One of the most effective water-saving irrigation practices is to understand and monitor the moisture content of the soil. This can be achieved through the use of soil moisture sensors which provide accurate data on the soil's water content, allowing for more informed irrigation decisions. By understanding the specific water needs of the landscape, overwatering can be avoided, thus conserving water.

- **Drip Irrigation Systems:** Drip irrigation is a highly efficient method of watering plants. It involves the slow and precise delivery of water directly to the plant's roots, minimizing evaporation and runoff. This method is particularly effective in water conservation as it reduces the amount of water lost to non-target areas.
- **Timed and Automated Irrigation Systems:** Implementing timed or automated irrigation systems can significantly enhance water efficiency. These systems can be programmed to water during the cooler parts of the day, typically early morning or late evening, to reduce water loss due to evaporation. Additionally, automating the irrigation process ensures that the landscape is only watered when necessary and with the appropriate amount of water.
- **Mulch:** Applying mulch around plants is another effective water-saving practice. Mulch helps to retain soil moisture, reducing the need for frequent watering. It also aids in preventing weed growth, which can compete with plants for water.
- **Selection of Drought-Tolerant Plants:** Choosing plants that are naturally drought-resistant can significantly reduce the need for irrigation. These plants are adapted to survive in conditions with limited water supply, making them ideal for water-efficient landscapes.

2.2.2. Regular Maintenance of Irrigation Systems

Regular maintenance of irrigation systems is crucial to ensure their efficiency. This includes checking for leaks, ensuring that sprinkler heads are correctly aligned, and replacing any broken or worn-out parts. A well-maintained system helps to prevent water wastage.

Smart irrigation systems can help with regular ongoing maintenance, are equipped with sensors and connected to weather forecasting services and can adjust watering schedules based on real-time weather conditions and soil moisture levels. This technology ensures that water is used only when necessary and in the right quantities, thereby maximizing water conservation.

- **Challenges and Solutions in Implementing Water-Saving Irrigation:** While the benefits of water-saving irrigation are clear, there are still several challenges in its implementation. One of the primary challenges is the initial cost of setting up efficient irrigation systems. However, this can be mitigated by the long-term savings in water usage and the potential for reduced water bills. Additionally, educating landscapers and homeowners about the importance and techniques of water-saving irrigation is crucial for its widespread adoption.

- **The Future of Landscape Irrigation:** The future of landscape irrigation lies in the continued innovation and adoption of water-saving practices. As awareness of water scarcity grows, and as technology advances, it is likely that more efficient and cost-effective solutions will emerge. These advancements will not only contribute to water conservation but also promote the health and beauty of landscapes in a sustainable manner.

2.2.3. Water Balance in Agricultural Systems

Water balance in agricultural systems is a crucial concept for managing the availability and use of water resources efficiently. It refers to the equilibrium between the inflows and outflows of water in an agricultural system, ensuring that crops receive adequate moisture while minimizing waste and environmental impact. Understanding and managing water balance involves several key components:

2.2.3.1. Components of Water Balance

- **Precipitation**: This is the primary source of water for agricultural systems, including rain and snow. Effective management starts with understanding the amount and timing of precipitation, as this impacts irrigation needs.
- **Evapotranspiration**: This term combines evaporation from soil and water bodies with transpiration from plants. It represents the total loss of water from the agricultural system into the atmosphere. Monitoring evapotranspiration helps determine how much water is being lost and what is needed to compensate through irrigation.
- **Soil Moisture**: Soil moisture refers to the amount of water held in the soil, which is essential for plant growth. Managing soil moisture involves understanding its capacity to retain water and how it changes with rainfall and irrigation.
- **Runoff**: Runoff is the portion of water that flows over the soil surface and is not absorbed by the soil. Excessive runoff can lead to soil

erosion and nutrient loss, while insufficient runoff may indicate that the soil is not retaining enough water.

- **Infiltration**: Infiltration is the process by which water enters the soil from the surface. The rate of infiltration depends on soil texture, structure, and moisture content. Effective infiltration ensures that water reaches the root zone where it is needed.
- **Groundwater Recharge**: Some of the water that infiltrates the soil replenishes groundwater reserves. Groundwater is an important resource for irrigation, especially in areas with irregular precipitation.

2.2.3.2. Balancing Water Inputs and Outputs

Managing water balance involves matching water inputs (e.g., precipitation and irrigation) with outputs (e.g., evapotranspiration and runoff). This balance is crucial for optimizing water use and ensuring that crops receive the necessary moisture without excess or deficiency. Techniques such as rainwater harvesting can enhance water availability, while efficient irrigation methods can minimize waste.

2.2.3.3. Soil and Water Conservation

Maintaining soil health is integral to managing water balance. Practices such as mulching, cover cropping, and reduced tillage improve soil structure, enhance moisture retention, and reduce runoff. These practices help in maintaining a stable water balance by improving the soil's ability to absorb and retain water.

Water Management Technologies

Advanced technologies like soil moisture sensors, weather-based irrigation controllers, and remote sensing tools assist in monitoring and managing water balance. These technologies provide real-time data on soil moisture, weather conditions, and crop water needs, enabling precise and efficient water management.

Adaptation to Climate Variability

Climate change and variability can significantly impact water balance. Changes in precipitation patterns, increased frequency of droughts, and altered evapotranspiration rates require adaptive management strategies. Techniques such as adjusting irrigation schedules, using drought-resistant crops, and enhancing water storage capacities help adapt to these changes.

Sustainability and Environmental Impact

effects on the environment. Water body deterioration, soil erosion, and waterlogging are all less likely when there is a proper water balance. It guarantees that farming methods support environmental health and long-term sustainability.

Understanding and maximizing the interactions between precipitation, soil moisture, evapotranspiration, runoff, and infiltration are key components of managing water balance in agricultural systems. Farmers are able to address the needs of agriculture and the environment by achieving a sustainable and productive water balance through the integration of advanced technologies, efficient irrigation practices, and soil conservation methods.

2.2.4. Strategies for Reducing Water Losses

Improving irrigation water efficiency and sustainability in agriculture require reducing water losses. Adopting cutting-edge irrigation technologies like sprinkler and drip systems, which minimize runoff and evaporation by delivering water directly to plant roots or canopy, is one effective strategy. By using soil moisture sensors, irrigation schedules can be precisely planned so that water is only applied when needed and in the best possible quantities. While terracing and contour farming can aid in managing runoff and erosion, techniques like mulching help preserve soil moisture and minimize surface evaporation. Furthermore, leaks and other system problems that result in water waste are avoided by performing routine maintenance and repairs on irrigation infrastructure. Reducing reliance on traditional sources of water is possible by using rainwater harvesting techniques to collect and store precipitation. Water conservation initiatives are further strengthened by teaching farmers about water-efficient techniques and incorporating these strategies into farm management plans. Combining these techniques can greatly reduce water loss, which promotes more resource-conscious farming methods and sustainable agriculture.

2.3. MULCHING AND ITS IMPACT ON CROP PRODUCTION

Mulching is a soil improvement technique that includes applying different materials to the soil's surface. To shield the roots from the negative effects of the microclimate, it entails covering the soil surrounding the root zone of the plants. Mulching has gained popularity in agriculture due to its enhanced soil microbial performance as well as its immediate economic benefits, which include higher yields, earlier harvests, better fruit quality, and less water usage. Mulching improves soil temperature, retains soil moisture, lessens weed pressure and some insect pests, and maximizes the use of soil nutrients, all of which help plants grow in optimal conditions. By lessening the effect of raindrops on the soil's surface, mulching improves the soil's hydrothermal regime and its physical characteristics, including texture, porosity, and infiltration rate.

2.3.1. Types of Mulching

The mulching materials can either be organic or synthetic (Figure 2.1):

- **Organic Mulches**: These are mulches that can be easily degraded. They are usually available on the farm. The gradual decomposition of the organic mulches increases organic matter and soil fertility (Figure 2.1A). They host and nurture various beneficial soil organisms such as bacteria, fungi, insects, and worms and their remains in the soil do not create post utilization disposal problems. For example, leaves, paddy straw, grasses, sawdust, sugarcane trash, etc.
- **Paddy Straw:** It has a unique property of not absorbing water and makes water available to plants. Among all the organic mulches, paddy straw has the longest life span. It also serves as a nutrient reservoir and gradually releases them into the soil.
- **Sawdust:** It is a small granular chip wood that is obtained as the finished product in the sawmills. Easy to apply and inexpensive with high C/N ratio. It retains moisture for longer periods.
- **Sugarcane Trash:** It is the residue gotten from sugar cane after the removal of its juice. Helps conserve moisture and reduces weed growth and should be avoided in an area where there is an incidence of termites.
- **Synthetic Mulches**: These are mulches that cannot be easily degraded (Figure 2.1B). They synthesized materials that need prior work before using in the field. They are referred to as non-biodegradable as a result of the natural decomposition of organic mulches. They are available in different colors and thicknesses. Much expensive when compared to organic mulches and should be disposed of at the end of the growing season, for example, plastic films
- **White Plastic Mulch:** It is good for establishing crops under hot summer conditions and has little effect on soil temperature. It repels some insects and reflects more light on the plant as compared to black mulch.

- **Black Plastic Mulch:** It is the most predominant colored mulch and acts as an opaque black body absorber and radiator. It does not allow sunlight to reach the soil; therefore, it suppresses the growth of the weed. It helps to increase soil temperature and improves mineralization and nutrient absorption. It encourages plant growth through the warming of the soil during the winter season.
- **Transparent Plastic Mulch:** It is also known as clear plastic mulch. It absorbs little solar radiation with a transmission of 85% - 95%. It raises soil temperature drastically and affects the plant's growth adversely.
- **Degradable Plastic Mulch:** This can either be biodegradable or photo-degradable. It can be degraded by microorganisms or by sunlight.
- **Biodegradable Plastic Mulch:** It is made from plant starches such as corn, wheat, and potatoes which can be broken down by microbes. It can be easily plowed into the ground after harvest.
- **Photo Degradable Plastic Mulch:** It is formulated to break down after a certain period of exposure to sunlight. It has similar qualities to black or clear plastic film. Examples are plastigone and biolane.

Figure 2.1. *Types of mulch: Organic mulch, for example, rice straw (A); and plastic-film mulch (B).*

Source: https://cdnintech.com/media/chapter/82224/1512345123/media/F5.png.

2.3.2. Benefits of Mulching

- **Conservation of Soil Moisture:** The soil's loss of moisture could be caused by a number of abiotic factors. These factors include strong winds, warm temperatures, unfavorable weather patterns, etc. Mulching aids in lowering evaporative water loss and weed infestation. It has been demonstrated that straw mulch can reduce evaporation by up to 35%. Mulching, however, increases the amount of water available for transpiration by reducing direct soil water evaporation. This method of conserving water aids in the maintenance of the water balance in the plants, particularly in areas with low annual precipitation. Additionally, mulching protects the soil's surface, which lowers erosion and nutrient loss.

- **Minimize Soil Compaction and Erosion:** On many agricultural lands, compaction brought on by large machinery or equipment is starting to become a significant issue. The compaction issue may be mitigated by adding organic mulch materials. These materials stop erosion from wind and water, as well as compaction brought on by large tools or machinery. Additionally, it can lessen soil compaction, which can harm crops' roots and hinder their growth and development. As the best example of living mulch on slopes, organic mulch made of certain grasses and legumes aggregates soil particles into a complex unit, thereby reducing soil erosion.
- **Regulation of Soil Temperature:** Mulching contributes to the stability of soil temperature, which is advantageous for crop development and growth. Research has indicated that mulch can maintain soil temperature both in normal or warm weather conditions and in extremely hot weather. Severe heat limits the amount of nutrients and water that newly emerging plant roots can absorb. The extreme temperatures that plants grow in may cause stress to the plants, and newly established roots may not be able to take up enough water and vital nutrients. The way mulch is applied affects soil temperature differently. Because some mulches absorb solar radiation, they raise the temperature of the soil relative to bare soil. Furthermore, it has been noted that, in comparison to other mulch materials, organic and plastic-film mulches are superior at preserving a healthy soil temperature.
- **Reduces Infiltration Rate:** Organic mulch helps to retain water at the soil surface allowing water to slowly penetrate thereby minimizes surface runoff. The soil surface reduced the amount of irrigation water required by pepper and onion crops by 14–29% and 70%, respectively.
- **Reduced Fertilizer Leaching:** In sandy soil, in particular, excessive rainfall drains around the root zone, reducing leaching-related fertilizer loss. Additionally, applying organic mulch raises the organic carbon content of the soil, enhancing its ability to retain water and nutrients. The amount of N, P, and K in chili leaves was

increased by mulching with coconut fronds. Research has indicated that applying mulch to crops results in faster plant growth, early fruiting, decreased P concentration, and increased N concentration in the leaves and fruits.

- **Reduces Weed Infestation:** Mulching creates a physical barrier between the soil and the atmosphere, which inhibits the germination and nourishment of many weeds. The process of mulching keeps weeds under control by encouraging a decrease in weed growth and seed germination. The soil surface can be covered or mulched to physically suppress or prevent weed seed germination. You can use materials like rice and wheat straws to control weeds. Covering the soil's surface can physically stop seedling emergence or stop weed seeds from germinating. Rice straw and sugarcane bark are examples of organic mulches that work well to suppress weed growth.
- **Organic Matter Improvement:** Mulches break down and replenish the soil with organic matter and plant nutrients. enhancing the biological and physicochemical characteristics of the soil, which raises crop productivity. By adding organic matter to the soil, organic mulches not only improve soil nutrients but also aid in preserving soil moisture. A reduction in bulk density of 1.42 g cm^{-3} under straw mulch as opposed to 1.50 g cm^{-3} under bare soil. They came to the conclusion that adding more mulch resulted in noticeably higher organic matter levels.
- **Reduces Harvesting Period:** Mulching usually results in early maturity and higher yields for vegetables like cucumbers, muskmelons, watermelons, eggplants, and peppers. Tomato crops with organic mulches experience fewer days between fruit set and harvest because they bloom earlier than control plants. Mulch made of polyethylene shortened the growing season and improved the productivity and earliness of a variety of vegetable crops.
- **Improves Quality and Yield:** In many cases, mulch prevents soil rot, fruit cracking, and blossom end rot while keeping fruits away from the ground. Fruits have less scars and are smoother. Grading time is decreased when plastic mulch is properly laid because it keeps dirt from splashing onto the plants during rainfall. Additionally, the yield and quality of early potatoes, cabbage, and other vegetables can be enhanced by straw mulch.
- **Reduction of Diseases:** Mulch can lessen the impact of raindrops beating against one another or the force of irrigation water, both of which can spread disease spores. These spores adhere to the

leaves and branches of susceptible plants. Mulches feed a range of beneficial soil organisms that suppress disease-causing spores or compete with them for entry. They reduce the likelihood that plants will become ill. Through a variety of enzymatic processes, organic mulches compete with or digest pathogenic organisms, thereby inhibiting a large number of soil bacteria. An essential component of integrated pest management (IPM) is mulches.

- **Remediation of Heavy Metals:** Both humans and animals should not be exposed to heavy metals. Mulches are a great way to get rid of heavy metals from soil. It's common practice to extract heavy metals from soil solutions using eucalyptus leaves. Compost and woodchips can combine with copper metal in forest environments to create complexes that change the metal into a non-toxic form that is good for crop plant growth.

2.4. IRRIGATION WATER MANAGEMENT

Monitoring the application of water for gardens or crops is part of water irrigation management. It is typically applied to larger properties that require a system to help control the amount, rate, and timing of water application to match soil intake and water holding capacities. Monitoring soil moisture without runoff or deep percolation losses is crucial for promoting optimal crop yields. You can control your irrigation effectively by using tools that you can later modify to guarantee correctly produced results. Flow meters are excellent for capturing both the total volume of water used and the instantaneous flow rates. Soil water deficit can be tracked with soil moisture sensors and meters. In turn, a checkbook method that monitors the leveling of irrigated cropland may be able to balance soil moisture. Lastly, you can record the history of soil moisture in a field or area throughout the growing season by using data loggers.

2.4.1. Principles of Irrigation Water Heeds

The deliberate application of water to meet crop needs is known as irrigation. In addition to conventional sources like treated wastewater, desalinated water, or drainage water, irrigation water can also come from nearby lakes, reservoirs, rivers, or wells (groundwater). Cultivated land receives irrigation water via ditches, hoses, or pipes.

Farmers that use irrigation are less reliant on sporadic showers for their crops. Farmers can respond to the amount of water required by their crops by adding irrigation in addition to rainfall. Furthermore, managing water resources enhances the effectiveness of additional inputs that raise yield, like fertilizers and plant protection products. As a result, the producer has more control over the yields. Stable productivity and market integration depend on this.

The plant root zone's soil water storage is refilled through irrigation. Through photosynthesis and the sun, plants use their roots to draw moisture from the soil. This causes a flow of nutrients to reach the leaves via the stem, where they transpire water back into the atmosphere. Therefore, the only water that aids in the growth of plants and fruits is water absorbed through the root system.

2.4.2. Water Requirements

Crop Water Requirement (CWR) is the term used to describe how much water a plant needs. It is dependent upon the crop, the climate, the management style, and the surrounding circumstances. Places that are sunny, hot, dry, and windy have the highest CWR. The amount of water required by the roots for optimal plant growth depends on the type, variety, and stage of growth of the crop. Farmers can lessen the amount of water needed for their crops through applying various irrigation techniques, adjusting plant density, and mulching.

Because of this, a crop's CWR is highly site-specific and changes daily. It necessitates gathering data locally and on-site, e.g., using assistance from neighborhood extension services, to determine the CWR.

The most common units of measurement for CWR are millimeters (mm) and cubic meters per hectare (m^3/ha). Crops typically need between 2 and 10 m^3 per hectare each day.

It is crucial to remember that water is a scarce resource that is subject to price competition from other users (such as industry, energy, household consumption, etc., and that it is a commodity). Evaluations of the ecological and financial viability should take this into account.

2.4.3. Irrigation Efficiency

Applying the appropriate amount of water at the appropriate time is crucial when irrigating. Wilting and a decrease in yield will result from insufficient watering. An excessive amount of water can cause salinization, runoff, erosion, and nutrient leaching from the soil, all of which can lower yields. Every plant in the field would receive the same amount of water from an irrigation system that is 100% efficient. But e.g., Plants at the far end of the field typically receive less water than those near the water source.

As a result, certain plants receive too much water while others receive insufficient water, which compromises the farm's yields. This can have detrimental effects on large farms' operational costs and water resource management. The irrigation technology selected will decide how evenly the water is distributed across the field. These are the three most widely used irrigation techniques.

- surface irrigation;
- sprinkler irrigation; and
- drip irrigation.

Surface irrigation methods such as basin, furrow and border irrigation do not achieve more than 60% field application efficiency as only a limited amount of water actually reaches the root zone.

Sprinkler and drip irrigation systems rely on a network of pipes to distribute the water to the field, thereby reducing conveyance losses common for surface irrigation methods. Sprinkler systems achieve an average field application efficiency of 75%, while drip irrigation systems can go up to 95% uniformity. Drip irrigation allows a slow release of water at the plant's root zone. As a result, soil moisture conditions are good, and no water is "lost" between plants, or on the plants. Crop yields can be doubled, and significant savings can be made in water, energy and labor.

2.4.4. Types of Irrigation

Irrigation is essential to support agricultural crop growth. Relying on rain is not practical, especially in Australia where rainfall can be both unreliable and insufficient. However, overwatering can be just as detrimental to crop growth, so choosing the right irrigation system is one of the most important steps in setting up a successful farming operation.

The goal of irrigation is to encourage plant growth while minimizing soil erosion and water loss. To choose the right irrigation system you will need to have a knowledge of soil, equipment, plant species and land formation.

These are five of the most common types of irrigation systems:

2.4.4.1. Drip Irrigation Systems

Commonly used in orchards, vineyards and high-value vegetable crops, drip irrigation systems consist of a network of tubes that have small holes or emitters. They can be placed above or below the soil's surface and slowly drip water into the soil over long periods (Figure 2.2).

Figure 2.2. Drip irrigation systems.

Source: https://www.gvsprinklers.com.au/wp-content/uploads/2017/10/dripirrigation-1.jpg.

Advantages include:

- Uses 30–50% less water than other systems
- Prevents soil erosion and nutrient runoff
- Continuous flow allows water to penetrate deep into the soil and down to the roots
- Controls fungal growth
- Easy to modify

2.4.4.2. Sprinkler Irrigation

In sprinkler irrigation systems, water flows through a series of pipes and is delivered in a fine spray to specific areas. Micro sprinklers are particularly effective for tree crops. They also use less water and are cheaper to run (Figure 2.3).

Advantages include:

- Affordable and easy to set up
- Allows even distribution of water
- Easy to covers large areas
- Water at your chosen time of day to minimize evaporation

Figure 2.3. *Sprinkler irrigation.*

Source: https://www.gvsprinklers.com.au/wp-content/uploads/2017/10/sprinklerirrigation.jpg.

2.4.4.3. Centre Pivot Irrigation

A center pivot irrigation system is self-propelled and works with the use of a central pipe with outlets rotating around a central pivot point. It works like the sprinkler irrigation system, but it is much bigger and is supported by steel or aluminum towers (Figure 2.4).

Advantages include:

- Water is distributed evenly.
- Covers large areas in a short period of time.
- Prevents water runoff.
- Operates at a lower pressure saving energy.

Figure 2.4. *Centre pivot irrigation system.*

Source: https://www.gvsprinklers.com.au/wp-content/uploads/2017/10/centreirrigation.jpg.

2.4.4.4. Furrow Irrigation Systems

Furrow irrigation is a form of surface irrigation where small parallel channels are created and filled with water. Crops are grown on the ridges between the channels. Furrow irrigation is suitable for a wide range of crops, especially row crops like corn, sunflower, sugarcane and soybean and those that would be damaged by flooding like tomatoes and beans (Figure 2.5).

Advantages include:

- Usually results in a higher yield.
- Low installation cost.
- Saves time and labor.
- Saves money on water.

Figure 2.5. *Furrow irrigation.*

Source: https://www.gvsprinklers.com.au/wp-content/uploads/2017/10/furrowirrigation.jpg.

2.4.4.5. Terraced Irrigation

Terrace irrigation is an ancient agriculture practice that still exists today, generally in mountainous regions. A series of steps are cut into the sloped land so that when it rains, the water flows down from the top step down to the succeeding steps retaining the soil nutrients as it goes (Figure 2.6).

Advantages include:

- Manages water runoff.
- Prevents soil runoff.
- Improves soil fertility.

- Improves land productivity.

Figure 2.6. *Terraced irrigation.*

Source: https://www.gvsprinklers.com.au/wp-content/uploads/2017/10/terraced-irrigation.jpg.

2.4.5. Various Methods of Irrigation

- **Manual Irrigation:** This is when water is distributed across the land through manual labor and watering cans. It is an ancient method of watering but is still used, especially in developing countries. It is cheap but is very labor-intensive, and its efficiency is poor since there is an uneven distribution of water. Also, there are high chances of water loss.
- **The Sprinkler System:** It is a modern method of irrigation, and it uses sprinklers. A pump is connected to pipes that generate pressure and water is sprinkled through nozzles. They sprinkle water over the crops like raindrops, helping to achieve an even distribution of water. It is efficient and serves best in areas with unlimited water.

The sprinklers can be set permanently, temporarily or can be mounted on a movable platform. In Libya and Saudi Arabia, for example, sprinkler irrigation is by far the most predominant at 100% and 64%, respectively.

Unfortunately for the system, the water used has to be safe for animals and people, water distribution can be patchy, weeds get as much water as the crop, and can easily erode soils if the water pools.

- **The Drip or Trickle System:** In a drip or trickle system, water is supplied drop by drop, directly to the roots, using a hosepipe. Drip irrigation is considered one of the most efficient methods of irrigation as it focuses the water onto the plant itself.

- It is also best suited for areas where there is a shortage of water. Drip irrigation can use any type of water, including recycled and non-potable water, on any type of soil and any climate. It minimizes soil erosion, is energy- and cost-efficient and minimizes the growth of weeds as the water targets the crop itself.
- **Subsurface Irrigation:** This is where irrigation water is applied below the soil subsurface. It can occur naturally or artificially. For the former, water leaks from the water pipes or channels, moving through the subsoil and irrigating the crops situated nearby.

As for the latter, artificial water channels are provided underground and water is supplied to the crops through capillarity. It is a costly means of irrigation but provides high returns and higher profits.

- **Perennial Irrigation:** Irrigation water is diverted from a river to the main canal through the construction of a diversion weir. This is known as direct irrigation, as the water is going directly to the crops. If a dam is constructed and the water is collected and stored, this is referred to as storage irrigation. Direct irrigation is simple and the most economical form of perennial irrigation.

2.4.6. Importance of Irrigation

Main importances of irrigation supplies are discussed below:

- **Compensating for When There is no Rainfall:** Irrigation kicks in when there is insufficient rainfall or when there is uncertainty as to when the rain will fall. Without the rain or irrigation as the alternative, the crops are adversely affected, which can lead to a food shortage or crop/plant failure.
- **Increasing the Amount of Cultivatable or Agriculturally Productive Land:** Some areas of the world are dry, naturally. Irrigation has been responsible for turning such lands into cultivatable lands. Today, about 18% of the world's cropland is now irrigated. Irrigation is also responsible for bringing the most out of fallow lands, which are traditionally left idle after harvest, before the next cultivation season.
- **Helps Meet Food Demands:** In order to increase food demands without relying solely on rain, desert ecosystems, like those in Jordan and Israel, have adopted farming due to the expansion of irrigated land. By cultivating food crops like cereals, potatoes, and vegetables, these regions are able to meet the world's food demand by utilizing groundwater from wells and aquifers.

Additionally, nations and areas that use extensive and high-quality irrigation are well-known for their food exports, which benefits their economies.

- **Increased Productivity:** Irrigation, by default and in many instances, is employed when rainfall is insufficient. However, it can be applied at any time, even if there is sufficient rainfall to boost crop productivity.

According to a publication by Olayide, Tetteh, and Popoola (2016) on climate-smart agriculture, crop productivity on irrigated land is higher than in the un-irrigated areas, which primarily rely on rainfall.

- **Enables Multiple Cropping:** In most tropical countries, the rainy seasons are specific, making it impossible to grow multiple crops at the same time. Moreover, their water requirements are different, and over-irrigation spoils the crop production. Optimum irrigation makes it possible to grow more than one crop at a time in most areas of a country.
- **It is Effective:** Crops and plants only receive one pass of watering when it is done sporadically and with a simple hosepipe. This implies that regions that are difficult to access and those without consistent rainfall might not receive enough water. An effective irrigation system, like drip irrigation, makes sure that even the most difficult-to-reach parts of the land receive water.
- **Defeating and Efficient Nature:** When it will rain is out of our control. If it rains during the day, and it gets a bit warm, evaporation rates will be high, meaning the crops and the soil will not consume enough water.

It is possible to program irrigation systems to irrigate early in the morning or late at night, when evaporation is minimal. This optimizes the amount of moisture needed by the crops, flowers, or even the soil itself, while also using less water.

- **Contribution to Economic Growth:** No matter the season or weather, irrigation guarantees that food production continues. This implies that income and employment are constant, which lowers poverty. The economy can continue to grow thanks to the irrigation system's significant increase in revenue. Food exports to other nations or regions also help achieve this.

2.4.7. Environmental Effects of Irrigation

The quantity and quality of soil and water that are altered as a result of irrigation, as well as the ensuing effects on the natural and social conditions in river basins and downstream of an irrigation scheme, are all considered aspects of irrigation's environmental effects.

The effects are a result of the irrigation scheme's installation and operation changing the hydrological conditions. Over drafting is one of the issues contributing to the depletion of subterranean aquifers. Due to uneven distribution or poor management, soil can be over-irrigated, wasting water and chemicals and possibly contaminating water supplies. Over-irrigation can result in deep drainage from rising water tables, which can cause salinity issues with irrigation and necessitate the use of subsurface land drainage to control the water table. On the other hand, inadequate irrigation results in inadequate control of soil salinity, raising soil salinity and causing hazardous salts to accumulate on the soil's surface in regions with high evaporation. This calls for either leaching to extract these salts or some sort of drainage to remove the salts. Because alkaline soil is created when saline or high-sodium water is irrigated, soil structure may be harmed.

2.4.7.1. Direct Effects

An irrigation system distributes water over a certain area by drawing it from rivers, lakes, groundwater, or overland flow. This will have direct, or hydrological, effects on the area, such as decreased river flow downstream, increased evaporation in the irrigated area, and an increase in the water table due to the area's increased groundwater recharge and irrigated area flow. Similar to this, irrigation directly affects the atmosphere's ability to hold on to moisture. It can cause atmospheric instability and increase rainfall in the downwind direction, or it can alter atmospheric circulation to direct rain in different directions in the downwind direction. Precipitation shed studies, which look at how major changes to the delivery of evaporation to the atmosphere can affect downwind rainfall, focus a lot on changes in irrigation.

2.4.7.2. Indirect Effects

Indirect effects are those that have consequences that take longer to develop and may also be longer-lasting. The indirect effects of irrigation include the following:

- Waterlogging.
- Soil salination.
- Ecological damage.
- Socioeconomic impacts.

2.4.7.3. Waterlogging

Soil saturation with water is referred to as waterlogging. When soil is almost always nearly saturated with water, limiting its air phase and creating anaerobic conditions, the soil is said to be waterlogged. Severe prolonged waterlogging causes anaerobiosis, damages mesophyte roots, and initiates processes like denitrification, methanogenesis, and the reduction of iron and manganese oxides due to the subsurface reducing atmosphere. Different crops in agriculture require air—more especially, oxygen—to varying degrees of soil depth.

Air cannot enter the soil due to waterlogging. Depending on the intended use, there are different requirements for how close the water table is to the surface before the ground is considered wet. As with rice cultivation (*Oryza sativa*), a crop's need for freedom from waterlogging can change with the seasons.

Because wet soils inhibit the leaching of the salts brought in by irrigation water, waterlogging and soil salinity are frequently found together in agricultural land that receives irrigation.

When it comes to gardening, waterlogging is the process by which the soil becomes so hard that neither water nor air can pass through.

2.4.7.4. Soil Salinity

The amount of salt in the soil is called salinity, and salinization is the process of adding more salt. Salts are found in water and soils by nature. Natural occurrences like mineral weathering or the slow retreat of an ocean can result in salinization. Additionally, man-made methods like road salt and irrigation can cause it.

2.4.7.4.1. Natural Occurrence

Salts are a natural component in soils and water. The ions responsible for salination are: Na^+, K^+, Ca^{2+}, Mg^{2+} and Cl^-.

Over long periods of time, as soil minerals weather and release salts, these salts are flushed or leached out of the soil by drainage water in areas with sufficient precipitation. In addition to mineral weathering, salts are also deposited via dust and precipitation. Salts may accumulate in dry regions, leading to naturally saline soils.

Due to irrigation water containing salts, human activities have the potential to make soils more saline. By supplying enough drainage water to remove extra salt from the soil, effective irrigation management helps avoid salt buildup. Salt buildup can also occur when drainage patterns that allow for leaching are disturbed. When the Aswan High Dam was constructed in Egypt in 1970, it served as an illustration of this. Prior to construction, the groundwater level had changed, allowing for soil erosion and a high concentration of salts in the water table. The arable land became salinized as a result of the construction's ongoing high-water table.

2.4.7.4.2. Salinity Due to Irrigation

Salinity from irrigation can occur over time wherever irrigation occurs, since almost all water (even natural rainfall) contains some dissolved salts. When the plants use the water, the salts are left behind in the soil and eventually begin to accumulate. This water in excess of plant needs is called the leaching fraction. Salination from irrigation water is also greatly increased by poor drainage and use of saline water for irrigating agricultural crops.

Salinity in urban areas often results from the combination of irrigation and groundwater processes. Irrigation is also now common in cities (gardens and recreation areas).

2.4.7.4.3. Consequences of Soil Salinity

The consequences of salinity are:

- Detrimental effects on plant growth and yield
- Damage to infrastructure (roads, bricks, corrosion of pipes and cables)
- Reduction of water quality for users, sedimentation problems, increased leaching of metals, especially copper, cadmium, manganese and zinc.
- Soil erosion ultimately, when crops are too strongly affected by the amounts of salts.
- More energy required to desalinate

Salinity is an important land degradation problem. Soil salinity can be reduced by leaching soluble salts out of soil with excess irrigation water. Soil salinity control involves watertable control and flushing in combination with tile drainage or another form of subsurface drainage.

The indirect effects of waterlogging and soil salination occur directly on the land being irrigated. The ecological and socioeconomic consequences take longer to happen but can be more far-reaching.

Some irrigation schemes use water wells for irrigation. As a result, the overall water level decreases. This may cause water mining, land/soil subsidence, and, along the coast, saltwater intrusion.

Irrigated land area worldwide occupies about 16% of the total agricultural area and the crop yield of irrigated land is roughly 40% of the total yield. In other words, irrigated land produces 2.5 times more product than non-irrigated land.

2.4.7.5. Adverse Impacts

2.4.7.5.1. Reduced River Flow

The reduced downstream river flow may cause:

- reduced downstream flooding
- disappearance of ecologically and economically important wetlands or flood forests
- reduced availability of industrial, municipal, household, and drinking water
- reduced shipping routes. Water withdrawal poses a serious threat to the Ganges. In India, barrages control all of the tributaries to the Ganges and divert roughly 60% of river flow to irrigation
- reduced fishing opportunities. The Indus River in Pakistan faces scarcity due to thc over-extraction of water for agriculture. The Indus is inhabited by 25 amphibian species and 147 fish species of which 22 are found nowhere else in the world. It harbors the endangered Indus river dolphin, one of the world's rarest mammals. Fish populations, the main source of protein and overall life support systems for many communities, are also being threatened
- reduced discharge into the sea, which may have various consequences like coastal erosion and saltwater intrusion in deltas and estuaries. Current water withdrawal from the river Nile for irrigation is so high that, despite its size, in dry periods the river does not reach the sea. The Aral Sea has suffered an "environmental catastrophe" due to the interception of river water for irrigation purposes.

2.4.7.6. Increased Groundwater Recharge, Waterlogging, Soil Salinity

Increased groundwater recharge stems from the unavoidable deep percolation losses occurring in the irrigation scheme. The lower the irrigation efficiency, the higher the losses. Although fairly high irrigation efficiencies of 70% or more (i.e., losses of 30% or less) can occur with sophisticated techniques like sprinkler irrigation and drip irrigation, or by well managed surface irrigation, in practice the losses are commonly in the order of 40% to 60%.

This may cause the following issues:

- Rising water tables
- Increased storage of groundwater that may be used for irrigation, municipal, household, and drinking water by pumping from wells
- Waterlogging and drainage problems in villages, agricultural lands, and along roads - with mostly negative consequences. The increased level of the water table can lead to reduced agricultural production.
- Shallow water tables - a sign that the aquifer is unable to cope with the groundwater recharge stemming from the deep percolation losses.
- Where water tables are shallow, the irrigation applications are reduced. As a result, the soil is no longer leached and soil salinity problems develop.
- Stagnant water tables at the soil surface are known to increase the incidence of water-borne diseases like malaria, filariasis, yellow fever, dengue, and schistosomiasis (Bilharzia) in many areas. Health costs, appraisals of health impacts and mitigation measures are rarely part of irrigation projects, if at all.
- To mitigate the adverse effects of shallow water tables and soil salinization, some form of watertable control, soil salinity control, drainage and drainage system is needed
- As drainage water moves through the soil profile it may dissolve nutrients (either fertilizer-based or naturally occurring) such as nitrates, leading to a buildup of those nutrients in the ground-water aquifer. High nitrate levels in drinking water can be harmful to humans, particularly infants under 6 months, where it is linked to "blue-baby syndrome."

2.4.7.7. Reduced Downstream River Water Quality

The quality of the river water below the project area can deteriorate due to drainage of surface and groundwater in the area, which makes the water less suitable for use in industry, municipalities, and homes. These waters may also be salinized and contaminated by agricultural chemicals like fertilizers and biocides. It might result in worse public health. The ecosystem near the coast may be negatively impacted by contaminated river water that enters the ocean. By keeping sediments behind dams that are essential to surface water irrigation diversions, it is possible to completely eliminate the natural contribution of sediments.

An integral component of the ecosystem, sedimentation depends on the regular flow of rivers. The plants and animals that depend on the sediments carried downstream will depend on the natural cycle of sediment dispersion to replenish the nutrients in the soil. Big rivers like the Nile River benefit from significant sedimentation deposits. During flood season, the sediment from the delta has accumulated to create a massive aquifer that keeps water in the wetlands. Many bird species find a home in the wetlands that are formed and maintained by the sedimentation of the river basin. On the other hand, excessive sedimentation can worsen upstream floods and lower the quality of river water downstream. This has been documented in China's Sanmenxia reservoir. The Three Gorge Project, a larger man-made network of hydroelectric dams, includes the Sanmenxia reservoir. In 1998, the reservoir's ability to effectively carry out its flood-control function was significantly impacted by imprecise calculations and a large amount of sediment. Additionally, this lowers the quality of river water downstream. Utilizing water pragmatistically—using it where it is found—instead of going toward mass irrigation installations in an attempt to satisfy more socioeconomic demands would be against the natural balance of the environment.

2.4.7.8. Affected Downstream Water Users

Downstream water users often have no legal water rights and may fall victim to the development of irrigation.

Pastoralists and nomadic tribes may find their land and water resources blocked by new irrigation developments without having legal recourse.

Flood-recession cropping may be seriously affected by the upstream interception of river water for irrigation purposes.

2.4.7.9. Lost Land Use Opportunities

The original population's fishing and cattle grazing opportunities may be diminished by irrigation projects. Due to the forced relocation of the traditional pastoralist tribes, there may be a significant increase in the pressure from livestock on the remaining lands. This could lead to overgrazing, severe soil erosion, and the depletion of natural resources. 43000 hectares of savannah were destroyed by the Manantali dam in Mali, creating the Manatali reservoir, which crosses nomadic pastoralists' migration routes and may contribute to erosion and overgrazing in other areas. In addition, 120 km^2 of forest were destroyed by the reservoir. The forests downstream of the dam are being harmed by the depletion of groundwater aquifers, which is brought on by the suppression of the seasonal flood cycle.

2.4.7.10. Groundwater Mining with Wells, Land Subsidence

The aquifer's water storage is being mined when wells extract more groundwater than is replenished, making the use of that water unsustainable. Pumps will find it more difficult to maintain the intended flow rate and may need more energy per unit of water as levels fall, making the extraction of water more challenging. Groundwater extraction might eventually become so challenging that farmers are compelled to give up on irrigated farming.

2.5. MANAGEMENT MEASURE FOR IRRIGATION WATER

To reduce nonpoint source pollution of ground and surface waters caused by irrigation:

1. Operate the irrigation system so that the timing and amount of irrigation water applied match crop water needs. This will require, as a minimum:
   - the accurate measurement of soil-water depletion volume and the volume of irrigation water applied; and
   - uniform application of water.
2. When chemigation is used, include backflow preventers for wells, minimize the harmful amounts of chemigated waters that discharge from the edge of the field, and control deep percolation. In cases where chemigation is performed with furrow irrigation systems, a tailwater management system may be needed.

The following limitations and special conditions apply:

- In some locations, irrigation return flows are subject to other water rights or are required to maintain stream flow. In these special cases, on-site reuse could be precluded and would not be considered part of the management measure for such locations. In these locations, improvements to irrigation systems and their management should still occur.
- By increasing the water use efficiency, the discharge volume from the system will usually be reduced. While the total pollutant load may be reduced somewhat, there is the potential for an increase in the concentration of pollutants in the discharge. In these special cases, where living resources or human health may be adversely affected and where other management measures (nutrients and pesticides) do not reduce concentrations in the discharge, increasing water use

efficiency would not be considered part of the management measure.

- In some irrigation districts, the time interval between the order for and the delivery of irrigation water to the farm may limit the irrigator's ability to achieve the maximum on-farm application efficiencies that are otherwise possible.
- In some locations, leaching is necessary to control salt in the soil profile. Leaching for salt control should be limited to the leaching requirement for the root zone.
- Where leakage from delivery systems or return flows supports wetlands or wildlife refuges, it may be preferable to modify the system to achieve a high level of efficiency and then divert the "saved water" to the wetland or wildlife refuge. This will improve the quality of water delivered to wetlands or wildlife refuges by preventing the introduction of pollutants from irrigated lands to such diverted water.
- In some locations, sprinkler irrigation is used for frost or freeze protection, or for crop cooling. In these special cases, applications should be limited to the amount necessary for crop protection, and applied water should remain on-site.

2.5.1. Soil-Water-Plant Relationships

Understanding the fundamental interactions between soil, water, and plants is the first step towards effective and efficient irrigation. The soil can receive water from irrigation, precipitation, or groundwater (e.g., the water table rising as a result of improved drainage). Soil water, or water held in the soil, is absorbed by plants and used for growth (e.g., cooling, photosynthesis, nutrient uptake, and so forth.

The most significant element of the hydrologic cycle on farms is transpiration, with the majority of transpiration going toward cooling. Additionally, water evaporates from soil and leaf surfaces. Evapotranspiration, or ET, is the result of both transpiration and evaporation working together. Plant temperature, air temperature, solar radiation, wind speed, relative humidity, and the availability of soil water are some of the variables that affect ET. The quantity of water lost through ET is equal to the amount of water required by the plant, or its consumptive use. Owing to inefficiencies in the irrigated water delivery system (e.g., evaporation, runoff, wind drift, and deep percolation losses), irrigation requires more water than is used consumptively. Salinity control may be important in arid and semi-arid areas, and more water or a "leaching requirement" may be required (Figure 2.7).

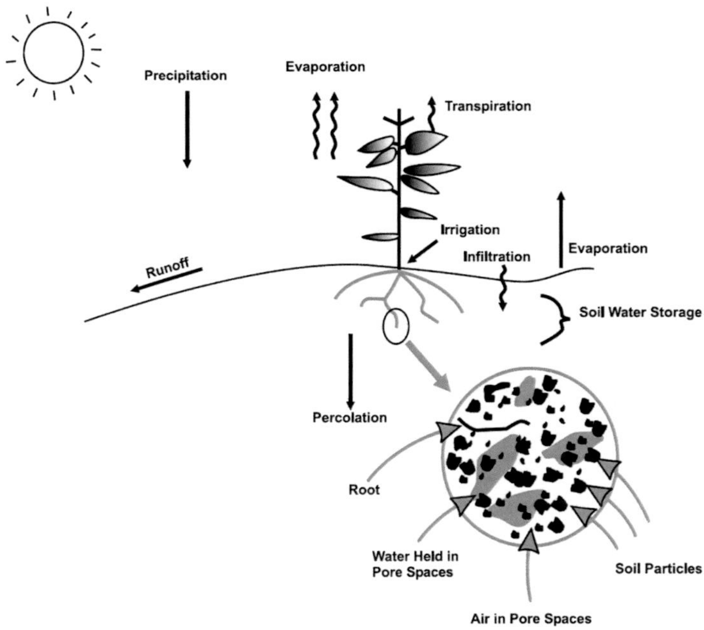

Figure 2.7. *On-farm hydrologic cycle for irrigated lands.*

Source: https://www.epa.gov/sites/default/files/2015-10/documents/chap4f.pdf.

Salt buildup usually happens in areas where evapotranspiration is higher than precipitation. Salts dissolved in the soil or contained in precipitation are carried to the surface by capillary action and evaporation, where they are deposited. In humid areas, where salts from rainfall naturally leach, salinity is typically not an issue.

Plant growth is negatively impacted by excessive salts in the soil. Plants find it more difficult to absorb water because of the osmotic force exerted by the total concentration of salts in the soil solution. Furthermore, at certain concentrations, certain ions—such as boron, sodium, chloride, and others—may be harmful to plants. Different crops react differently to specific and total salts; some are more sensitive than others.

A renewable supply of soil water is necessary for plant growth, and this supply is determined by the soil's water-holding capacity, water movement, amount of water that is easily accessible to plants, and rate of replenishment. Plants receive this replenishable supply of soil water through efficient irrigation, which uses the least amount of water, energy, and time wasted. Achieving optimal irrigation efficiency requires knowledge and comprehension of the variables that impact water movement in the soil, water storage in the soil, and water availability to plants.

2.5.1.1. Movement of Soil Water

When water is added to soil, it travels through evaporation, runoff, and infiltration. Capillary force and gravity are two of the forces that ultimately decide how applied water is transported. In soils with large pores, gravity pulls water downward freely, causing it to pass quickly through the root zone if it is not absorbed by the crop. Crop damage from too much water is avoided as the water moves through the soil and replenishes the air in the pores. Water is moved by capillary forces in soils with smaller pores. This "capillary water" tends to flow from wetter to drier regions and moves more slowly than gravitational water. Capillary water is more significant to an irrigated crop because of its lateral distribution, which increases soil wetting. The main force causing downward water movement in saturated conditions is gravity; in unsaturated soil, the main force is capillary action.

2.5.1.2. Storage and Availability of Soil Water

The soil serves as the plant's reservoir for water, so the amount of water it can hold—also known as its water holding capacity—is an important consideration in irrigation planning and management. Water is held in reserve in the pore spaces between soil particles as well as in the film surrounding each soil particle. Not all soil water is equally accessible for plants to draw from and utilize. Many elements, such as soil texture, soil structure, and soil layering, affect a plant's capacity to draw water from the soil. Based on the percentage of sand, silt, and clay particles in the soil, textures are categorized. Layering is the vertical distribution of soils in the soil profile, whereas structure is the arrangement of soil particles in groups or aggregates (e.g., clay soils sitting on top of a layer of sandy loam. The kind and degree of layering can affect how applied water percolates and is distributed laterally (Figure 2.8).

The size, form, and number of pores in the soil, as well as the amount of space that can hold water or air, are all influenced by the texture and structure of the soil. For instance, coarse sand can hold between 0 and 0.4 inches of water per foot of soil depth (in/ft), silt can hold between 1–2.2 in/ft and clay can hold between 1–3.9 in/ft. Certain volcanic ash soils have the structure to hold very high water contents at field capacity levels, but there may be some water trapped in the pumice and cinder fragments that plants cannot access. Greater soil water tension can prevent plants from using a significant amount of soil water in fine-textured soils and soils impacted by salinity, sodicity, or other chemicals.

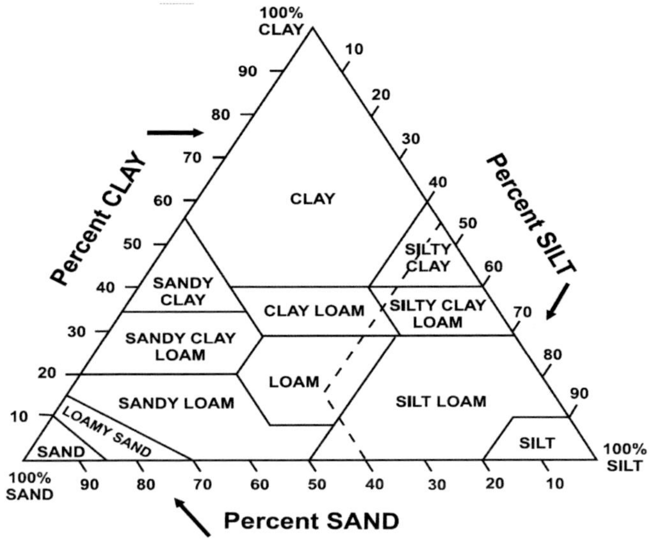

Figure 2.8. *Soil textural triangle for determining textural class.*

Source: https://www.epa.gov/sites/default/files/2015-10/documents/chap4f.pdf.

The amount of water that remains in a soil after "free" water has drained due to gravity is known as field capacity. Conceptually, "free" water and "gravitational" water are comparable. Coarse-textured (e.g., sandy) soils within a few hours following irrigation or rainfall, from medium-textured (e.g., In approximately a day, loamy) soils will become fine-textured (e.g., clay) soils over a few days. Texture, structure, bulk density, and water-restricting soil strata within the soil profile are all factors that influence field capacity. The difference between the amount of water held at field capacity and the amount held at the permanent wilting point is known as available water capacity.

2.5.1.3. Uptake of Soil Water by Plants

It is easiest for plants to retrieve water from soil pore spaces, but it is far more difficult for them to do so from the film that surrounds soil particles. The soil holds on to the water that remains more firmly as evapotranspiration takes water from the soil. As soils get drier, the soil moisture tension rises, making it harder for plants to take up water from the soil. Wilting is the result of the plant's inability to resist the forces securing the water in the soil particles (i.e., the water-soil tension). At this point, irrigation is necessary to preserve the plant. The soil-water content at which most plants are unable to get enough water to avoid irreversible tissue damage is known as the permanent wilting point. Before watering, growers determine how much water to let plants remove from the soil based on goals for yield and product quality. This quantity, known as the Management Allowable Depletion (MAD), varies depending on the crop and irrigation technique used. It is expressed as a percentage of the available water-holding capacity. MAD is 50% as a general rule of thumb. When using saline water, shallow root zones, micro-irrigation, and unpredictable water supplies, smaller MAD values—which lead to more frequent irrigations—may be preferable. When using hand-move and hose-pull sprinklers, in situations where soils are sandy and furrows are long, or for crops like certain cotton varieties that require stressing on heavy soil in order to produce an adequate number of cotton bolls, large MAD values may be preferred.

REFERENCES

1. Cominola, A., Giuliani, M., Castelletti, A., Fraternali, P., Gonzalez, S. L. H., Herrero, J. C. G., Novak, J., & Rizzoli, A. E. (2021). Long-term water conservation is fostered by smart meter-based feedback and digital user engagement. *npj Clean Water*, *4*(1), 29.
2. Delgado, J. A., Groffman, P. M., Nearing, M. A., Goddard, T., Reicosky, D., Lal, R., Kitchen, N. R., Rice, C. W., Towery, D., & Salon, P. (2011). Conservation practices to mitigate and adapt to climate change. *Journal of Soil and Water Conservation*, *66*(4), 118A–129A.
3. Han, S., Rhee, Y. W., & Kang, S. P. (2017). Investigation of salt removal using cyclopentane hydrate formation and washing treatment for seawater desalination. *Desalination*, *404*, 132–137.
4. Hermoso, V., Abell, R., Linke, S., & Boon, P. (2016). The role of protected areas for freshwater biodiversity conservation: Challenges and opportunities in a rapidly changing world. *Aquatic Conservation: Marine and Freshwater Ecosystems*, *26*, 3–11.
5. Moorberg, C. J. (2019). *Soil and Water Conservation: An Annotated Bibliography*. NPP eBooks.
6. Pimentel, B., et al. (2004). Water resources: Agricultural and environmental issues. *BioScience*, *54*(10), 909.

Chapter 3

Soil-Water-Plant Relationships

Contents

INTRODUCTION

The deliberate application of water to arable land for the purpose of providing crops with the water they need that isn't provided by natural precipitation is known as irrigation. In dry regions (Figure 3.1), without irrigation, sufficient food and fibers cannot be produced. Semiarid regions require irrigation most of the time due to the possibility of low crop yields and the risk of crop failure caused by variations in rainfall. Furthermore, as a precaution against crop losses, irrigation is preferred in humid and subhumid climates. Although summer rainfall is usually adequate for crop growth, droughts can happen at any time of the year. The main goal of agriculture is usually to produce a crop that is profitable. In semiarid, subhumid, and humid regions, irrigation ensures a profitable agricultural industry; in arid regions, it is an absolute necessity.

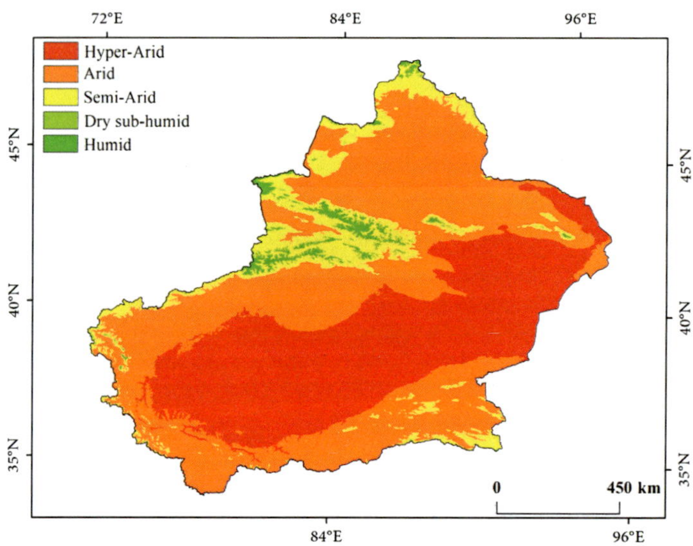

Figure 3.1. *Five climatic areas in the United States: hyper-arid, arid, semi-arid, dry sub-humid, and humid.*

Source: https://pub.mdpi-res.com/remotesensing/remotesensing-14-03911/article_deploy/html/images/remotesensing-14-03911-g003.png?1660298565.

An irrigation system, a controlled water table, or precipitation all contribute water to the soil. To meet plant evapotranspiration (ET) needs, it is stored in the soil matrix and then extracted by plant roots. This chapter discusses the physical characteristics of soils and plants that influence water movement, retention, and use. These characteristics are important to take into account when planning and implementing conservation irrigation systems. When planning and designing an irrigation system, the technician's main concerns are the soil's ability to hold water, especially in the plant's root zone; the soil's rate of water intake; the crop's root system; and the amount of water the crop requires. To effectively plan and manage irrigation for specific crops grown on specific soils and to modify the design to suit different conditions, one must also have a working understanding of all soil-plant water relationships.

3.1. SOIL

In addition to serving as a home for soil organisms and plant roots, soils also store plant nutrients and act as a water reservoir to meet the evapo-transpirational needs of plant communities. A soil's physical and chemical characteristics dictate how much water it can hold for plant use. The length of time a plant can be sufficiently maintained between irrigations or periods of rainfall is determined by this amount. The quantity also establishes the frequency, the amount to be applied, and the irrigation system's capacity required for continuous, ideal crop growth.

3.1.1. Soil Physical Properties

Mineral soils are porous mixtures of water, air, decomposing organic matter, and inorganic particles. They also have a variety of living things in them. Unconsolidated sediments or loose, unconsolidated pieces of weathered rock make up the parent material of mineral soils. An often-observed horizontal layering of the soil mass is caused by physical and chemical weathering, which results in the translocation and accumulation of different substances in trenches and road cuts. These layers or horizons taken together are referred to as the soil profile. Water transmission and retention in the soil, as well as root growth, are influenced by the properties of the profile's layers.

The structure and texture of soils are two crucial physical characteristics. The relative proportion of different sized groups of mineral particles in a particular soil or horizon is referred to as the soil texture. The way soil particles are arranged in groups or aggregates is referred to as soil structure. The supply of air and water in a soil is influenced by both soil structure and texture working together. Soil compaction can have negative effects on a soil's natural properties. Compaction has the ability to significantly alter a soil system's nutrient composition, root penetration, temperature relationships, water retention, and transmission characteristics.

3.1.2. Soil Texture

3.1.2.1. Mineral Soil

Separates are groups of mineral particles in a soil that vary in size. The U.S. uses classification and separation of soil. Table 3.1 displays the Department of Agriculture along with a range of diameter sizes. Parts that are coarse and have a diameter greater than two millimeters are excluded.

Table 3.1. *Range in Particle Size by Texture*

Soil Separate	Particle Diameter (millimeters)
Very coarse sand	2.0–1.0
Coarse sand	1.0–0.5
Medium sand	.5–0.25
Fine sand	.25–0.1
Very fine San	.1–.05
Silt	.05–0.002
Clay Less than 0.002	

Source: https://irrigationtoolbox.com/NEH/Part623_Irrigation/neh15-01.pdf.

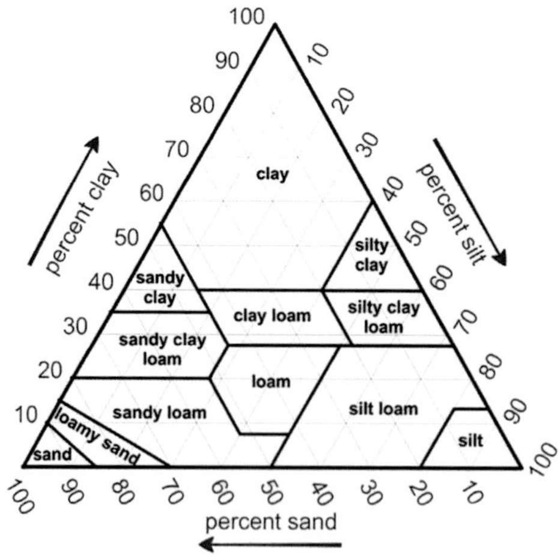

Figure 3.2. *Proportions of sand, silt, and clay in basic soil textural classes.*

Source: https://www.researchgate.net/profile/Ty-Ferre/publication/279631053/figure/fig2/AS:340509945024517@1458195305579/USDA-Soil-Texture-Triangle.png.

Different combinations of sand, silt, and clay are the basis for different soil textural classes. Fine texture distinctions are required for specific applications; Figure 3.2 illustrates the fundamental size distribution classes that were identified through laboratory mechanical analysis.

In some contexts, it is more convenient to discuss texture in general terms; Table 3.2 lists acceptable terms for various basic class groupings.

Table 3.2. *General Terms for Basic Soil Textural Classes*

General Terms	**Textural Classes**
Sandy soils:	
Coarse-texture	Sands (coarse sand, sand, fine sand, very fine sand), loamy sands (loamy coarse sand, loamy sand, loamy fine sand, loamy very fine sand)
Loamy soils:	
Moderately coarse-textured	Coarse sandy loam, sandy loam, fine sandy loam
Medium-textured	Very fine sandy loam, silt loam, silt
Moderately fine-textured	Clay loam, sandy clay loam, silty clay loam
Clayey soils:	
Fine-textured	Sandy clay, silty clay, clay

Source: https://irrigationtoolbox.com/NEH/Part623_Irrigation/neh15-01.pdf.

In the field, the texture of the soil can be assessed by feeling it with the fingers. This conclusion can be verified in the lab at a later time if needed. The general definitions of soil textural classes in terms of field experience are provided by the USDA Soil Survey Manual.

- **Sand:** The sand has a single grain and is loose. The individual grains are easily visible or tactile. Sand crumbles when squeezed in the hand and released from pressure when it gets dry. When wet, it forms a cast when squeezed, but crumbles when touched.
- **Sandy Loam:** A sandy loam is a soil type with a high proportion of sand and just enough silt and clay to provide some structure. The individual grains of sand are easily visible and tactile. A sandy loam that has been squeezed and dried produces a cast that easily breaks apart. Squeezing while damp will form a cast that won't break with careful handling.
- **Loam:** A loam is a soil type that has a fairly uniform distribution of clay, silt, and sand particles. It's smooth and slightly plasticky, yet it has a mellow, slightly gritty feel. Squeezing the moist soil when

it's dry creates a cast that can be handled freely without shattering. However, the cast must be handled carefully.

- **Silt Loam:** A silt loam is a soil with more than half of its particles being silt, with only a small amount of clay and a moderate amount of fine sand. A silt loam feels soft and floury when ground, but it looks cloddy and breaks easily into lumps when it's dry. The soil puddles and readily runs together when it's wet. Silt loam, whether it is wet or dry, creates a cast that is flexible enough to be handled without shattering; when wet, however, it does not ribbon and instead looks broken when squeezed between the thumb and finger.
- **Clay Loam:** A clay loam is a soil with a moderately fine texture that typically forms hard, dry clods or lumps. The damp soil barely holds its own weight when pinched between the thumb and finger, forming a thin ribbon that breaks easily. The wet soil is malleable and creates a cast that is resilient to repeated handling. Clay loam forms a dense, heavy mass when kneaded by hand; it does not crumble easily.
- **Clay:** A clay is a fine-textured soil that, when wet, becomes very plastic and typically sticky, and when dry, forms very hard lumps or clods. A long, flexible ribbon is formed when the moist soil is pinched out between the thumb and finger. Certain clays with extremely high colloidal content are friable and non-plastic under all moisture conditions.

3.1.2.2. Organic Soil

The percentage of organic matter in soils can range from 20 to 95%. They are typically categorized according to how much the organic deposits have broken down. Similar to how mineral textural terms are used, the terms peat, muck, and mucky peat are used for organic materials. Muck is organic soil material that has broken down well. Peat is made up almost entirely of the original fibers and is a mixture of raw, undecomposed organic materials. The material known as mucky peat lies in between muck and peat.

The texture of mineral soil can be changed with mucky. The word suggests that there is sufficient organic matter present to give the substance some characteristics of both organic soil and mineral material. But there isn't enough organic matter in the material for it to be considered "muck." Mucky material has a basic mineral composition and is typically dark, friable, and moisture-retentive. Typically, the organic matter content exceeds 10%.

3.1.3. Soil Structure

The way soil particles are arranged and grouped into what soil scientists refer to as peds—natural units of aggregation—is known as soil structure. Planes of weakness that endure through repeated cycles of wetting and drying in place divide pedals from one another. The majority of peds are big enough to view without a magnifier. In addition to affecting root penetration and the soil's nutrient supply, structure has an impact on how quickly air and water enter and move through the soil.

Structure type (Figure 3.3) refers to the particular kind of particle grouping that predominates in a soil horizon. Single grained and massive soils are structure less. In single-grained soils, such as loose sand, water percolates very rapidly. Water moves very slowly through massive soils such as some clays. The more favorable water relations are usually in soils that have prismatic, blocky, and granular structure; platy structure impedes the downward movement of water.

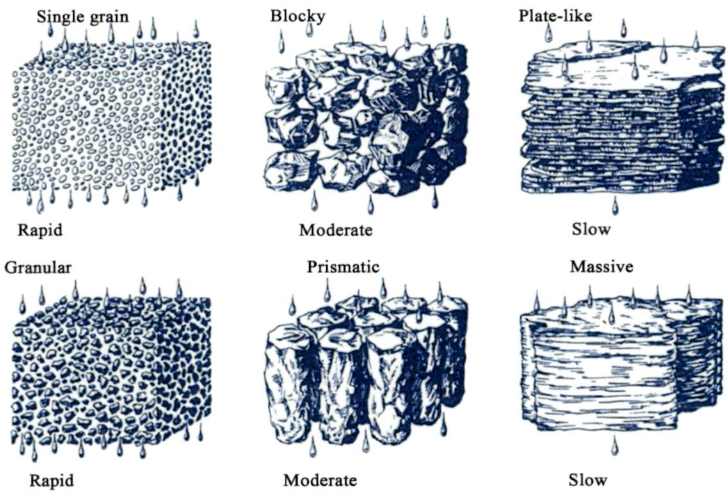

Figure 3.3. *Types of soil structure and their effect on downward movement of water.*

Source: https://slideplayer.com/slide/6677057/23/images/17/Soil+structure+types+and+their+effect+on+downward+movement+of+water+%28Source%3A+USDA%2C+1997%29.jpg.

In contrast to texture, the depth of tillage can alter the soil's structure. The top layer of soils with a high organic matter content, where perennial grass grows, develops excellent structure. The plow layer's structure is enhanced by cycles of freezing and thawing or wetting and drying. Conversely, growing on medium- or fine-textured soils with high moisture content tends to degrade structure. By dispersing the soil aggregates, irrigation water with high sodium content creates structure that is highly undesirable.

3.1.4. Tilth

Tillage-friendliness and plant growth are directly related to the physical state of the soil, which is known as tilth. It is dependent upon the soil aggregates' degree and stability. The usual terms used to describe tilth are good, fair, and poor. They speak about how easily soil can be tilled and how quickly it absorbs water. Most soils can have good soil health if appropriate soil management techniques are used.

3.1.5. Soil Porosity

In mineral soils, the volume of pore space typically makes up between 30 and 60% of the total volume, with an average of nearly half. The main factors influencing soil porosity are burrowing insects, worms, and other animals, as well as soil aggregation, texture, root activity, and trapped gases. While sandy soils typically have larger mean pore sizes, coarse-textured soils are generally less porous than fine-textured soils. Because clayey soils have a greater propensity to aggregate or disperse and can expand and contract during wetting-drying cycles, their porosity tends to be more variable. Soil pore space can be thought of as an enormous web of interconnected voids that extends in all directions. The majority of living things are found in the voids, which also act as entry points for roots to spread out and grow in. The voids also contain liquids and gases and control their movement. Total soil porosity can be determined for a soil sample from the equation:

Total porosity = 1 - (bulk density/average particle density)

Typically, a core sampler with a known volume that is intended to retrieve undisturbed samples from different profile depths is used to measure bulk density. Soil scientists occasionally use the water displacement technique to calculate bulk density from a sample of clods. A pre-saturated sample is subjected to a stepwise series of incremental suctions in the laboratory to measure the pore-size distribution. The equivalent pore-size distribution is then obtained using the capillary theory. It is possible to distinguish between macropores and micropores in the pore-size distribution where aggregation is quite distinct. The main channels for water infiltration, drainage, and aeration are the pore spaces between aggregates, which are known as macropores. The smaller pores found inside aggregates, known as micropores, are primarily used for solute and water retention. The distinction between macropores and micropores is mostly arbitrary, and the demarcation is rarely clear.

3.1.6. Soil Compaction

The term "compaction" pertains to the process of decreasing the porosity of agricultural soils by partially collapsing the pores and releasing the air that penetrates them. Soils are deemed compact in the agricultural sense when the air-filled porosity is low enough to limit aeration, which hinders drainage and root penetration.

It is possible for soils to naturally become compact due to their formation process, moisture content, or textural makeup. Agricultural soils often compact because of the mechanical pressure that cultural operations apply to the soil's surface. Although livestock trampling can result in soil compaction, wheels, tracks, and other soil-engaging tools are the most frequent causes of soil compaction in modern agriculture. The impact of higher compaction (high bulk density) on alfalfa root growth and proliferation in greenhouse pots after 80 and 110 days is shown in Figure 3.4. The root length density was not significantly altered by longer growth periods compared to the 110-day period.

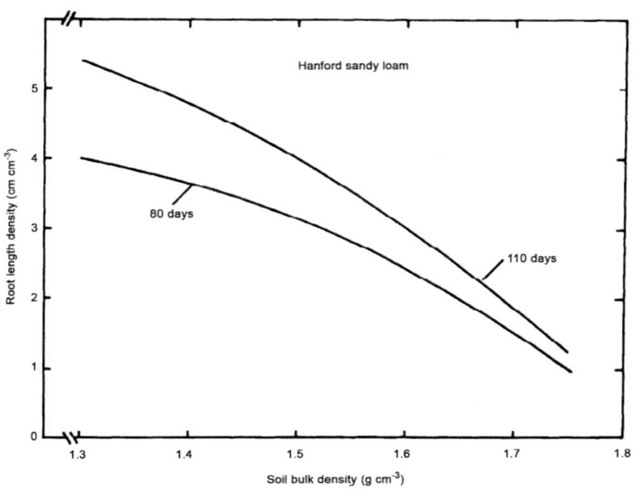

Figure 3.4. *Effect of soil bulk density on alfalfa root growth in greenhouse pots at the end of 80 and 110 days.*

Source: https://image.slidesharecdn.com/section15-240724070331-2142e2f0/85/Section-15-Irrigation-Chapter-1-Soil-Plant-Water-Relationships-pdf-17-320.jpg.

Deep root systems are not typically developed by crops in the soils of the southeast United States. A significant number of these soils belong to a textural class—sandy loams, for example—that is susceptible to soil compaction from traffic and excessive tillage during seed bed preparation. These types of soils could progressively hinder root growth.

3.1.7. Soil Salinity and Sodicity

Because rainfall in arid and semiarid regions is insufficient to meet plants' potential evapotranspiration requirements, saline and sodic soils are more prevalent in these environments. These soils develop when salts build up to levels that are harmful to plant growth and are not leached. Salt issues can arise in humid and subhumid areas, especially those close to the coast. An estimated 5 million hectares of American irrigated land, primarily in the 17 Western States, are impacted by salinity. Up to 70 million hectares, or one-third of all irrigated land worldwide, suffer from salinity issues.

Soil salinity originates from three primary natural sources: weathering of minerals, atmospheric precipitation, and fossil salts. Furthermore, irrigation and industrial and agricultural wastes enrich soils with salts. Salts are frequently moved from over-inundated areas to areas with inadequate drainage, where they build up. High concentrations of salts persist as irrigation return flows or drainage water evaporates. During a typical irrigation, water is applied to the soil's surface and any leftover water is raked through the soil in later irrigations. It is possible for some drainage water to eventually seep below the crop root zone. Water is lost by transpiration as well as evaporation at the soil's surface. The remaining concentration of dissolved salts is raised by both transpiration and evaporation. In well-drained soil, salt concentration typically rises with soil depth. Salt accumulation in the lower profile decreases as more irrigation water (the leaching fraction) passes through the root zone.

Exchangeable sodium levels can rise significantly when soils are irrigated with waters that are high in sodium. These kinds of soils usually crust badly, swell, or disperse, which significantly reduces the soils' hydraulic conductivity or permeability to water.

3.1.8. Categories of Salt-Affected Soils

Based on the amount of soluble salt found in soil solutions that have been extracted, salt-affected soils are categorized. The accepted indicator of salinity in a saturated extract is its electrical conductivity (EC_e). The salinity class and electrical conductivity of soil saturation extracts used by the Soil Conservation Service are listed in Table 3.3. The exchangeable sodium percentage was replaced by the sodium adsorption ratio (SAR), which is now the accepted method of determining a soil's sodicity. The following relationship is used to calculate the sodium adsorption ratio based on the concentrations of sodium, calcium, and magnesium ions in the saturation extract (measured in milliequivalents per liter).

$$SAR = Na/ [(Ca + Mg)/2]^{1/2}$$

Table 3.3. *Classification of Salt-Affected Soils*

	Electrical Conductivity (EC_e)	
Class	**Siemens Per Meter (S/M)**	**Decisiemens Per Meter (dS/m)**
Very slightly saline	0–0.4	0–4
Slightly saline	0.4–0.8	4–8
Moderately saline	0.8–1.6	8–16
Strongly saline	>1.6	>16

Source: https://irrigationtoolbox.com/NEH/Part623_Irrigation/neh15-01.pdf.

3.2. SOIL WATER

It is common practice to characterize soil water in terms of gravimetric percent content, volume-based percent content, or equivalent water depth per unit of soil depth. When the main concern is how much irrigation water is needed to return the soil to a specific water content, these descriptions are typically sufficient for irrigation considerations. However, a descriptive property is required to explain why similar treatments of soils result in different water contents; why plants react differently on different soils despite having the same water content; and why water will migrate from a sandy soil to a clay soil if they have the same water content and are placed in close proximity to each other. Such a phenomenon is described by the property known as soil water potential.

3.2.1. Soil Water Potential

The total water potential is the work/unit quantity of pure water that is needed to transport a small amount of water, reversibly and isothermally, from a pool of pure water at a given elevation and atmospheric pressure to the soil water at the point of interest. The tendency of water to flow within a soil system is caused by variations in the potential energy of water at different points in the system. Water always flows in the direction of decreasing potential energy in the soil.

Soil-water potential is a fundamentally important concept. It takes the place of the haphazard classifications (gravitational water, capillary water, hygroscopic water) that were applied when soil physics was first emerging. The water in the soil varies in potential energy rather than form from time to time and from place to place. Despite acknowledging the shortcomings of such nomenclature, it is convenient to keep the terms "field capacity" and "permanent wilting point" for very practical reasons.

Total water potential consists of several components. It is the sum of matric, solute, gravitational, and pressure potential:

Total = Matric + Solute + Gravitational + Pressure

The method used to specify a unit quantity of water determines the potential's units. This will be used in some illustrations since weight is one of the easiest ways to specify the unit of water and because conversions between English and SI (International System) are simple.

3.2.1.1. Gravitational Potential

Figure 3.5 shows how gravitational potential is determined. This portion of total potential is solely dependent on the vertical distance between the reference and the point in question, and it is unaffected by the characteristics of the soil. Gravitational potential is positive for points above the reference and negative for points below. Two soil points are situated in Figure 3.5 at a given distance from Reference Point Z. The difference in gravitational potential between the two points is 6 inches - (-4 inches) = 10 inches because point A's gravitational potential is 6 inches and point B's is 4 inches.

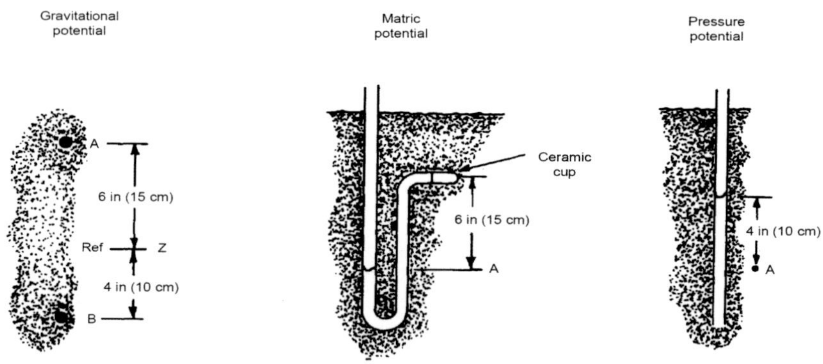

Figure 3.5. *Illustrations and example calculations for gravitational, matric, and pressure potential based on weight as specifying the unit of water. Increasing depth below the soil surface is considered to be negative in the illustrations.*

Source: https://irrigationtoolbox.com/NEH/Part623_Irrigation/neh15-01.pdf.

3.2.1.2. Mahic Potential

The vertical distance between the water surface of a water-filled manometer and the measured point of the soil (ceramic cup in Figure 3.5) is known as the matric potential when the unit quantity of water is expressed as weight. Matric potential is a dynamic soil property that, in the case of saturated soil, will be at zero. A soil system's matric potential is determined by the capillary and adsorptive forces arising from the soil matrix. These forces draw water into the soil, bind it, and reduce its potential energy below that of the water in bulk. The surface tension of water and its angle of contact with the solid soil particles cause capillarity. Previously, this potential was known as capillary water or capillary potential. In Figure 3.5, a water manometer and an unglazed ceramic cup buried in the ground are connected to create a tensiometer. The vertical distance, measured in inches, between the cup's center and the water level in the manometer represents the weight matric potential of the soil water at the cup.

3.2.1.3. Pressure Potential

Soils that are saturated are primarily affected by pressure potential. Pressure potential is the vertical distance between the water's surface and a given point when the quantity of water is expressed as a weight. This component is zero above and at the piezometer's water level in the field. It is consistently positive below the waterline. A piezometer tube, which is open on both ends, is inserted into the soil below the water table in Figure 3.5. The 4-inch gap between the point and the water level at point A is the pressure potential.

3.2.1.4. Solute Potential

The presence of a semipermeable membrane and soluble materials (usually salts) in the soil solution cause solute or osmotic potential to arise. The cell walls of plant roots and air-water interfaces are two known membranes in soil-water systems. The solute potential can be approximated from the relation:

Solute potential = RTC

Where C is the solute concentration (mol/cm^3), T is the absolute temperature (°k), and R is the universal gas constant (82 bars cm^3/mol °k). Utilizing SI units makes calculating solute potential much simpler due to the properties of the universal gas constant (R). With the units displayed, all units except bars cancel when temperature and solute concentration values are entered into the equation. As demonstrated in the discussion that follows, it is now simple to convert this unit (bar) to another unit.

3.2.1.5. Units

In the past, numerous units have served to measure suction, tension, stress, or potential. A number of these units include: bars, centimeters of water (cm), centimeters of mercury, inches of water, atmospheres, centibars, millibars, joules per kilogram, pounds per square inch, ergs per gram, and dynes per square centimeter. The bar unit is in extensive use; some conversions for this unit are:

- 1 bar = 1020 cm of water
- = 75.01 cm of mercury
- = 401.5 inches of water
- = 0.987 atmospheres
- = 100 centibars
- = 1000 millibars
- = 100 joules/kg
- = 10^6 ergs/g
- = 10^6 dynes/cm^2

3.2.2. Soil Water Characteristic Curves

Smaller pores drain as saturated soils are subjected to higher suction pressures; at very high suction pressures, only the narrowest pores hold water. Additionally, a decrease in the thickness of the water film covering the surface of soil particles is linked to an increase in soil-water suction.

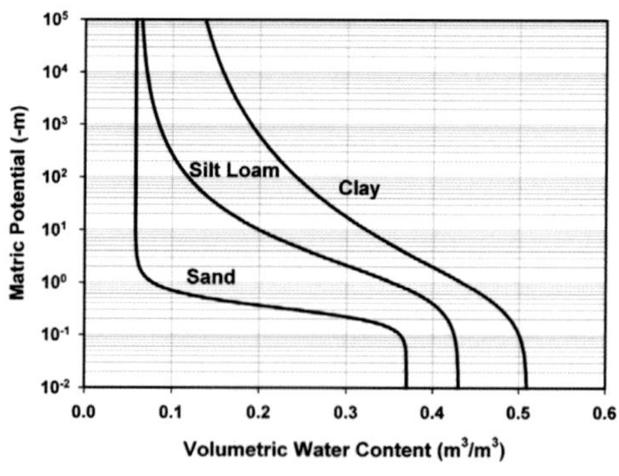

Figure 3.6. *Soil water characteristic curves.*

Source: https://www.researchgate.net/profile/Markus-Tuller/publication/268055840/figure/fig1/AS:295530669330432@1447471410205/Typical-soil-water-characteristic-curves-for-soils-of-different-texture.png.

As a result, the amount of water in the soil at various equilibrium points depends on the matric suction and is influenced by the size and volume of water-filled pores. Using the proper tools, pressure is used in place of suction in an experiment to create a water content versus soil moisture tension curve. Figure 3.6 displays illustrative curves for contrasting soil types. There exists a complex and nonlinear function that relates changes in water content to changes in matric potential. Usually, this characteristic curve is established for a specific soil.

3.2.3. Water Movement in Soil

A soil's actual geometry and flow pattern are very complex because, in saturated conditions, the pore space's properties determine how quickly water moves through a soil system. The flux density, or the volume of water flowing through a unit cross-sectional area per unit of time, is expressed by an equation known as Darcy's law. The equation is:

q = K (delta H)/L

where K is the proportionality factor commonly referred to as the hydraulic conductivity, q is the flux density, and (delta H)/L is the hydraulic gradient (head drop or change in head per unit distance in the direction of flow). Each soil system's complexity is reflected in the K factor.

According to this law, the direction and rate of water flow through the soil are determined by the driving force operating on the system.

In the crop root zone, a number of processes involving water flow or movement take place in unsaturated environments. Water movement under unsaturated conditions is much more complex and challenging to quantify than it is for saturated flow; however, for definitive information, a number of excellent quantitative reviews are available. Here, a summary of useful unsaturated flow properties will be examined.

Water in unsaturated soils experiences sub-atmospheric pressure, which is the same as a potential negative pressure. A driving force is a gradient of this potential. As was previously mentioned, the reason for matric suction is water's attraction to the surfaces of soil particles and capillary pores. Water is extracted from an area with thicker hydration films to one with thinner ones, as well as from an area with less curved capillary menisci to one with more highly curved ones. Stated differently, water moves from an area with a higher matric potential to one with a lower matric potential (low to high suctions).

The hydraulic conductivity (K) of a given soil will not change for saturated flow. Under unsaturated conditions, the water content has a significant impact on K. Since larger, more highly conductive pores drain first, K drastically decreases with soil drying. K values vary significantly amongst soils due to the different pore space characteristics; unsaturated hydraulic conductivity is strongly influenced by contrasting textural soil classes. In general, water will move through saturated sandy soil more quickly than through saturated clay soil. The exact opposite situation occurs when the soil drains; even at noticeably low potentials, the tiny pores of the clay soil will retain and conduct water. Fine-textured soils exhibit significantly more pronounced flow than sandy soils under unsaturated field conditions.

This talk has shown that, with respect to potential energy gradients, water in soils flows in any direction. The many intricate relationships that exist in soil systems control the rate and amount of movement. Figure 3.7 illustrates a few of the textural class's practical applications.

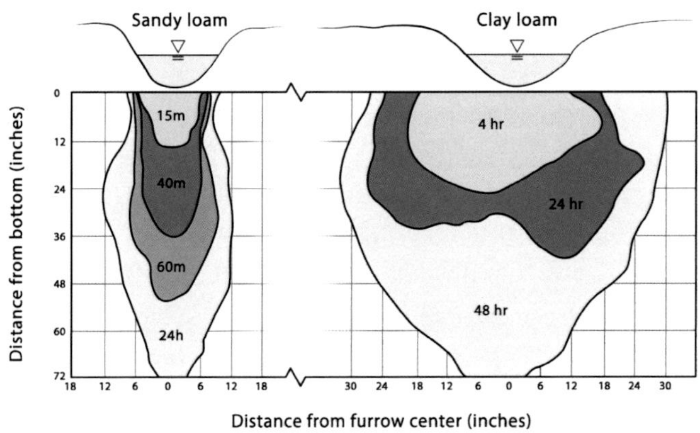

Figure 3.7. *Water penetration and movement in sandy and clay loam soils; to achieve complete wetting, furrows have to be closer together on sandy soils.*

Source: https://image.slidesharecdn.com/section15-240724070331-2142e2f0/85/Section-15-Irrigation-Chapter-1-Soil-Plant-Water-Relationships-pdf-22-320.jpg.

3.2.4. Plant Available Water

The amount of water that is available to plants in soils is important to know when designing an irrigation system because soils act as finite reservoirs. Plant available water has historically been defined as the soil's capacity to hold water between field capacity (FC) and permanent wilting point (PWP). Even though these two points only offer qualitative information on the qualities of soil water retention, people still use them, and these concepts can yield helpful planning information. Nonetheless, it's critical to comprehend the constraints put in place.

FC is, by definition, the volume of water retained in a well-drained soil following the drainage of "free" water. Because the pores in coarse-textured soils are relatively large, drain-off happens quickly after irrigation. Because the pores in fine-textured soils are so small, drainage takes a lot longer. Texture and water-restricting strata in the profile are two characteristics of the soil that have a material impact on field capacity. More water can be held by fine-textured soils than by coarse-textured soils. The definition of field capacity is -1/10 bar for sandy soils, -1/5 bar for silty soils, and -1/3 bar for clayey soils. In stratified soils, restricted flow slows redistribution but may cause a plant to consume more water. The broken line in Figure 3.8 near the left vertical axis shows how different soil textures affect the soil-water potential. Field capacity can be calculated using water-characteristic relations in a laboratory setting or in the field following a thorough irrigation or rainfall of the soil.

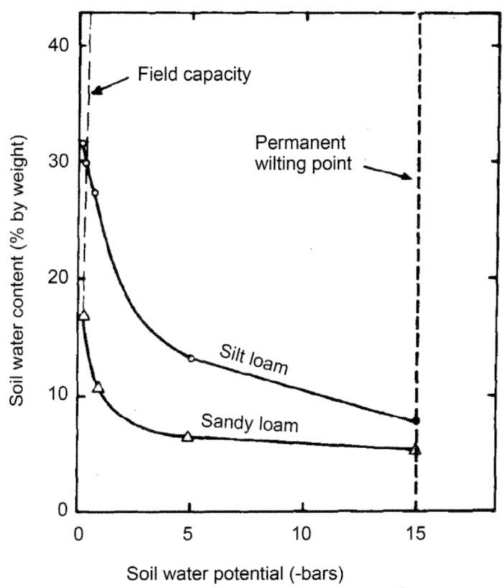

Figure 3.8. *Relation between the water content and the water tension of a soil.*

Source: https://irrigationtoolbox.com/NEH/Part623_Irrigation/neh15-01.pdf.

The soil-water content at which plants can no longer get enough water to meet their minimal transpiration requirements is known as the permanent wilting point (PWP). At this point, the plants wilt and, even if watered, will not recover. If soil water is not absorbed by plants quickly enough to meet the climatic ET demand, they will wilt. When wilted, plants still take in water, just not quickly enough to restore their turgor. The water potential of –15 bars, which was initially determined with sunflowers on a variety of soil types, is frequently utilized for PWP. The vertical line on the right in Figure 3.8 represents this parameter. Certain plants can draw water from the soil down to –15 bars before they wither, while other plants will wither at levels higher than –15 bars.

The potential energy level of soil water that is thought to be suitable for plant growth is situated between FC and PWP. It should be noted, though, that these calculated values only reflect the soil water system's matric potential. The total soil water potential may have a significant osmotic component due to the presence of salts. Plant availability of soil water is determined by the total potential. Examining the impact of soil texture on the water retained between FC and PWP is nevertheless beneficial. In soils with low levels of soluble salts, the available water capacity (AWC) increases with texture fineness.

The variation in FC and PWP water content by texture is displayed in Figure 3.9. For different soil bulk densities, the soil water content is displayed on the left margin in percentage terms based on the dry weight of the soil and on the right margin in inches of water per foot of soil. Until local curves can be created, the figure can be used as a general guide for estimating the AWC of soils based on texture. It is typically applicable to homogeneous soil profiles with low salinity levels.

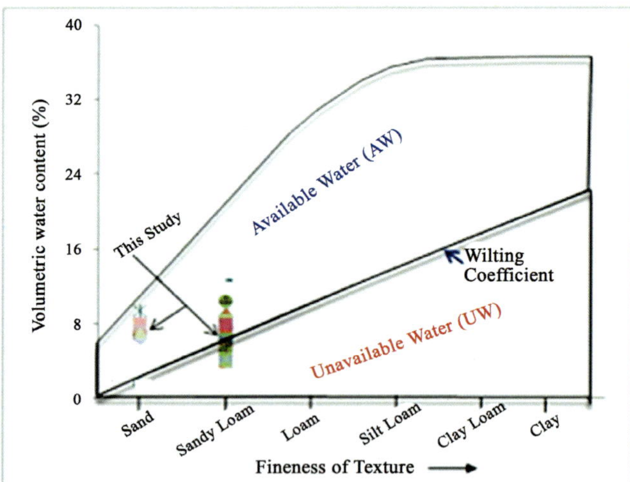

Figure 3.9. *General relationship between soil water characteristics and texture.*

Source: https://www.researchgate.net/profile/Akinola-Eluwole/publication/346956248/figure/fig3/AS:967662620381187@1607720155542/Relationship-of-soil-texture-with-soil-water-content-Adapted-from-Gurevitch-et-al-2002.png.

It should be noted that although FC is thought to be the maximum amount of water that is accessible, this is not entirely accurate. Upon rainfall or irrigation, water that moves downward in the soil can be efficiently utilized by developing plants. Although this water is temporary, it may have an impact on irrigation scheduling and is typically ignored in calculations to estimate a soil's capacity to retain water.

It is common to think of water held between FC and PWP as 100% of the available water supply. To demonstrate this idea, the water release characteristic curve from Figure 3.6 is replotted in Figure 3.10. Figure 3.10 makes it evident that when water has been utilized by plants to that degree, a certain level of permitted depletion, such as the 50% level, will represent very different soil water potentials. In the section on irrigation scheduling, this topic is covered in more detail. Under water balance, calculations pertaining to plant available water are shown.

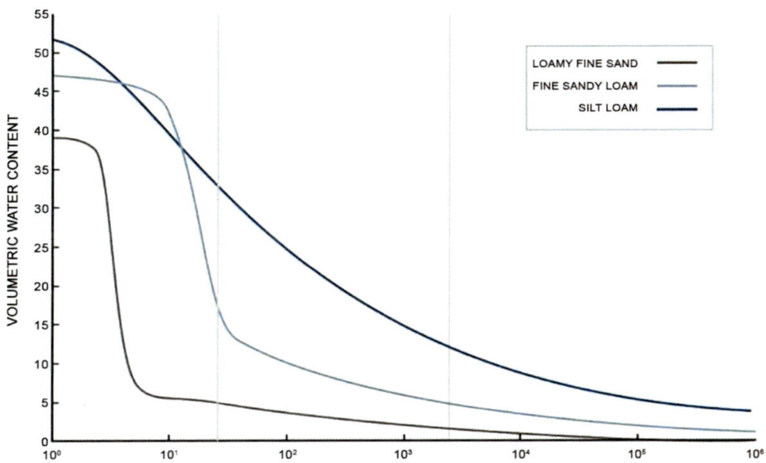

Figure 3.10. *Water release curve for three soils.*

Source: https://metergroup.com/measurement-insights/soil-moisture-release-curves-what-they-are-why-you-need-them-how-to-use-them/.

Figure 3.11 illustrates the dynamic process by which soil water is drawn up by plant roots. In the San Joaquin Valley of California, cotton was planted on April 16. The soil water content of the profile at that time can be assumed to represent FC. During the growing season, there was no rainfall and no irrigation of the crop. At depths shallower than 160 centimeters, some soil water was harvested at a total water potential lower than PWP. However, limited root growth in the lower portions of the profile meant that by harvest time, some water held at more than –15 bars was not needed.

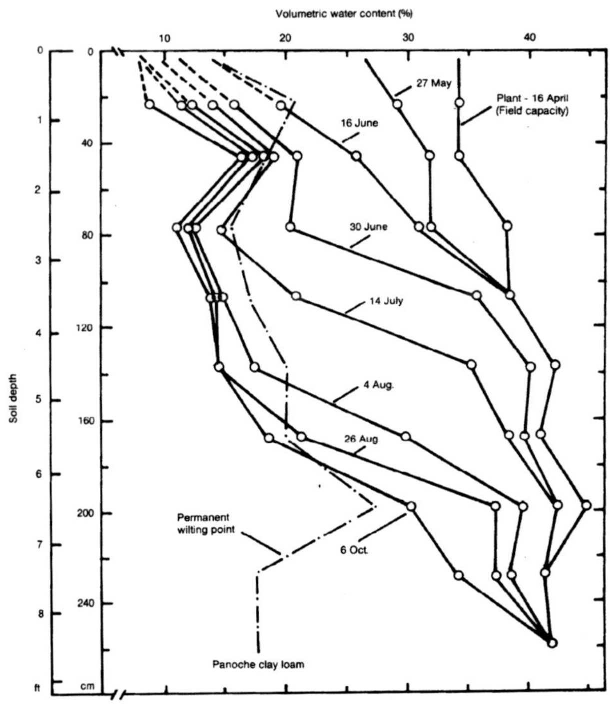

Figure 3.11. *Soil Water depletion by cotton on a panoche clay loam.*

Source: https://irrigationtoolbox.com/NEH/Part623_Irrigation/neh15-01.pdf.

3.2.5. Soil Water Measurement

It is crucial to measure the amount of water that soils can hold as well as their capacity to do so. The fact that certain humid soils yield crops even when there are days or even weeks without rain between rainstorms is proof of their ability to retain water, as all developing plants need water on a daily basis. Soil storage capacity plays a crucial role in determining the depth of water applied during irrigation and the time interval between irrigations in regions that receive irrigation. For this reason, soils that have the ability to store water for crop growth are highly valued. When there is a lack of irrigation water, productive crops can be grown on irrigated soils with a large water-storage capacity.

Understanding a soil's ability to hold onto available irrigation water is crucial for effective irrigation. The extra water applied by the irrigator is wasted if it exceeds the root-zone soil reservoir's capacity to hold it during a single irrigation. Plants may wilt from a lack of water before the next irrigation if less is applied than the soil can hold, unless water is applied more frequently. To maximize the use of rain, irrigations are planned in humid regions. It is impossible to see water losses from deep percolation beneath a crop's root zone. The amount of water applied in a single irrigation, less runoff, can be subtracted from the storage capacity of the different soils to determine or estimate losses.

3.2.6. Methods Used for Characterizing Soil Water

Measuring or estimating the water level in the soil is the best and most efficient method of deciding when to irrigate. An irrigator with knowledge of and experience with a specific crop on a specific soil type can accurately determine when irrigation is required by knowing the available water volume. Many of the various techniques available for estimating and measuring soil water are unsuitable for use in the field. However, irrigators are currently employing a number of promising techniques, and more are being developed, to help them decide when to irrigate.

3.2.6.1. Gravimetric

The recognized gold standard for measuring soil water is the gravimetric approach. For every type of soil, soil samples are collected from multiple locations within a field at a desired depth. Samples are weighed, dried at 105 to 110°C for 24 hours, and then weighed once more. The difference in weight is the amount of water, dry weight basis, in the soil, which can be converted to inches or centimeters of water remaining in the soil. Despite producing good results, growers rarely employ this technique. The quantity of samples collected and the level of skill required to collect and handle the samples determine how accurate it is. It involves a lot of time and labor and calls for the use of facilities that growers typically do not own. The technique is primarily applied in experimental work and serves as a benchmark for comparing various soil water determination techniques.

3.2.6.2. Feel and Appearance Method

Water content can be inferred from the feel and appearance of soil samples that were collected in the field at suitable depths and locations. Samples can be taken with a shovel, but for certain soils, a soil auger or sampling tube work better. Three field tests are conducted on the soil, and the results are recorded and compared to the water content and locally developed feel. These three tests are as follows: the ball test, in which soil is packed into a firm ball and dropped from various heights, recording the effects; the rod test, in which soil is rolled into a 0.10-inch (3 mm) diameter rod and held out vertically, recording the effects on length; and the ribbon test, in which soil is smeared between the thumb and first finger, recording the length of the resulting ribbon. The irrigator should be able to estimate the soil water content level within 10 to 15% with experience, even though feeling and looking at the water is not the most accurate way to gauge its conditions. Although this method is cheap, it takes a lot of work to obtain the soil samples.

3.2.6.3. Tensiometers

Tensiometers (Figure 3.12) operate on the premise that water passing through a porous ceramic cup and into the surrounding soil creates a partial vacuum in a closed chamber. A water manometer, a mercury manometer, or a vacuum gauge can all be used to measure tension. Typically, the scales are calibrated in centimeters of water or in hundredths of an atmosphere. Because they offer such high precision, mercury manometer-based tensiometers are typically chosen as research instruments. Tensiometers with Bourdon vacuum gauges are simpler and more appropriate for practical applications, such as irrigation control on specific soil types.

The tensiometer must be filled with water after the cup is inserted into the ground at the appropriate depth. Until the water in the cup and the water in the soil reach equilibrium, water flows through the porous cup. The above-ground vacuum-gauge reading increases with any increase in tension that happens as the soil dries. On the other hand, tension decreases and the gauge reading drops as soil-water content rises. The tensiometer records changes in the amount of soil to water until the tension reaches 0.85 atmosphere. At that point, air enters the system and the instrument stops working. If this happens, the device won't work after a rain or irrigation cycle until it is filled with water once more.

Using a tensiometer requires some experience. Measurements become unreliable if air seeps into the unit through any leaks at the rubber connections. In addition to defective cups, air leaks can also occur at the points where the setscrews holding the porous cup to the metal support come into contact with one another. A test pump that can be used to empty the instrument's air and test the gauge is offered by certain manufacturers.

Tensiometer readings only reveal the relative moisture of the soil surrounding the porous tip; they do not reflect the soil's water tension. They don't offer precise details about how much water is retained in the soil. Tension measurements are helpful in determining when irrigation is necessary, but they don't tell you how much water to use. In order to translate water-tension measurements into available water percentages, a unique water-characteristic curve specific to the given soil site is required.

Figure 3.12. *Tensiometers are used to measure soil water.*

Source: https://esi.com.my/wp-content/uploads/2019/10/IRROMETER-banner.jpg.

Tensiometers are not able to measure all soil types' available water to a satisfactory degree. However, in the wet range, they are arguably the best field instruments to use for determining the water conditions in medium to coarse-textured soil. Because a large portion of the water available to plants in sandy soils is held at a tension of less than one atmosphere, these soils are best suited for their use. In fine-textured soils, where only a small portion of the available water is held at a tension of less than one atmosphere, tensiometers are less useful. In finer textured soils, tensiometers are typically placed in the lower half of the root zone so that, even after significant water extraction, the readings remain within the gauge range.

3.2.6.4. Electrical-Resistance Instruments

These devices operate on the idea that variations in the water content of the soil or an instrument buried in the soil will result in variations in some electrical property of the soil (Figure 3.13). They are made up of two electrodes that are fixedly positioned in conductivity units, which are often blocks made of gypsum, nylon, fiber glass, plaster of Paris, or mixtures of these materials. A resistance or conductance meter that measures variations in the electrical resistance in the blocks is connected to the electrodes in the blocks via wires. The units react to variations in the soil's water content when they are buried in the ground because they practically blend in with the soil. Electrical resistance is determined by the water content of the blocks; therefore, if the block is calibrated for a specific soil, measuring any change in resistance is an indirect measure of soil water.

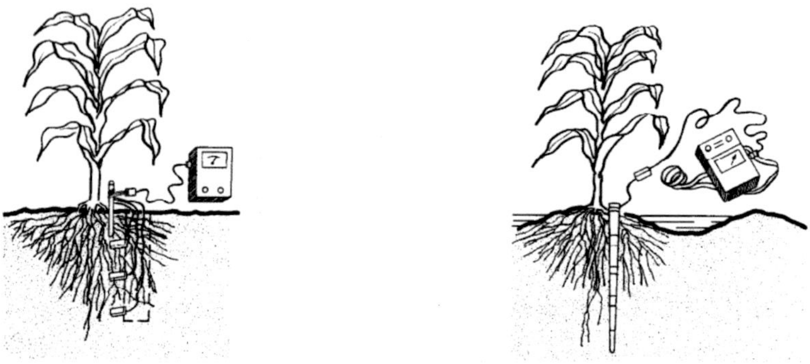

Figure 3.13. *Electrical resistance soil water meters.*

Source: https://irrigationtoolbox.com/NEH/Part623_Irrigation/neh15-01.pdf.

Plaster of Paris blocks are less sensitive to changes in soil moisture than nylon and fiber-glass blocks, but they frequently have poor contact with alternating wet and dry soil. The most sensitive tension for nylon blocks is less than two atmospheres. Fiberglass blocks perform satisfactorily across the whole range of available water, while plaster of Paris blocks work best at tensions between 1 and 15 atmospheres. Good soil-to-unit contact and sensitivity in both wet and dry ranges are achieved by combining fiber glass and plaster of Paris. Particularly in soils with a sandy texture, there can be a lag between the soil water change and that in the block. This is especially valid for blocks made of gypsum. One to three-day lag times have been recorded.

Salts in the soil can affect electrical resistance instruments; fiber-glass blocks are more sensitive than plaster of Paris. Fertilizer concentrations therefore have an impact on electrical resistance readings as well. Electrical-resistance instruments should be positioned well to one side of fertilizer bands when they are spread. Temperature has an impact on all unit readings.

Calibration drift has led to changes in tension in some units of up to one atmosphere in a single season. The number of drying intervals and the days between each determine how much of a change occurs. Soil type also affects readings. The instrument needs to be calibrated for the soil it will be used in because different soil textures may result in different amounts of water being available for the same reading. The blocks need to be installed correctly if readings are to be considered representative of a region. Since each block needs to be placed in a hole, the soil is disturbed. The root development and moisture pattern may not be representative if the soil is not replenished in the hole at the same density and in the same manner as the remainder of the profile. Forcing the blocks into the undisturbed soil along the sides of the hole you dug is a good technique. Refer to Figure 3.13 for the placement of the blocks. One kind uses tapered stakes for the casting of the blocks. Using a specialized auger, a tapering hole the same size as the stake is bored into the ground. After the stake is drenched in water, it is inserted into the hole so that it is in close contact with the ground.

In general, commercial instruments provide accurate estimates of soil water content when used in accordance with manufacturer instructions. However, the blocks need to be calibrated for each task in the field for optimal outcomes. To accurately estimate soil-water conditions, one must have experience and carefully interpret instrument readings.

3.2.6.5. Neutron Scattering

The method of estimating soil-water content using neutron scattering has become widely accepted. It has some advantages over the gravimetric method because it minimizes the impact of soil variability on subsequent measurements by allowing for repeated measurements at the same depth and location. Additionally, it calculates the water content based on volume, with the type of soil and the instrument's wetness influencing the volume of soil involved. The initial large equipment investment, Federal operating regulations, and the amount of time needed to install access tubes at each site are drawbacks.

A high-energy source, or fast neutron source, is lowered into an access tube that has already been installed to the required depth. An americium-beryllium or radium-beryllium source emits fast neutrons into the soil, which progressively lose energy upon collision with different atomic nuclei. The most efficient component of the soil to slow down neutrons is hydrogen, which is almost exclusively found in soil water. As a result, the amount of soil-water content can be determined by measuring how much neutrons slow down. An electrical pulse is produced on a charged wire when some of the thermalized, or slowed, neutrons randomly return to the detector, creating a cloud surrounding the source. A ratemeter can be used to measure the rate of pulsation, or a scalar can be used to count the number of such pulses over a specified period of time. The relationship between the count rate and the water content is roughly linear. The radiation source is kept safe when not in use in a shield made of high-hydrogen materials like paraffin wax or polyethylene. This material acts as a benchmark for confirming that the instrument is operating properly. Using the count ratio method instead of just a count is more satisfactory because it accounts for source decay and instrument variations. Plotted against water content is the ratio of sample count to standard count. By doing this, any systematic errors that the instruments might occasionally introduce are eliminated. The energy of the first fast neutrons and the soil's moisture content determine the measured volume of soil. When using the radium-beryllium source, the volume of soil is measured as a sphere with a diameter of up to 20 inches (50 centimeters) in dry soil and about 6 inches (15 centimeters) in wet soil. Neutrons may be lost through the surface, making measurements close to the surface inaccurate. Because the sphere of influence integrates individual layers, it is challenging to detect any abrupt change in soil water with depth caused by a wetting.

A calibration curve is typically provided by the manufacturer; however, it is advisable to confirm its suitability for a particular type of soil. SCS has established standard operating procedures for calibrating neutron gauges for a given soil site.

3.2.6.6. Heat Dissipation

Based on the idea that heat transfers much more quickly through water than through dry soil, heat conductivity can be used as a measure of water content. A heating element embedded in a porous block may be subjected to a continuous current for a predetermined amount of time. The heat that results is transferred out of the element, and the element's temperature is correlated with the porous block's water content. A linear diode temperature sensor can measure temperatures without requiring correction for variations in the surrounding air temperature. The goal of this system's irrigation control is to keep the soil matric potential within a specific range. A soil water potential sensor is a device that combines a temperature sensor and a heating element. Although they are occasionally utilized for field applications, these sensors have primarily served as research tools. They are being used in irrigation controllers and are available commercially.

3.2.7. Sampling Error

Irrigators have long been hampered by sampling errors when attempting to estimate the amount of water in the soil. Getting representative samples is a significant issue. Because they result in differences in the amount of water in the soil, uneven plant growth and uneven root penetration must be taken into account. Variations in soil texture and structure affect how moisture is absorbed, transferred, and retained. The opportunity for rainfall and irrigation water intake is influenced by variations in the land surface configuration, and the rate of irrigation water intake is influenced by the sizes and shapes of furrows. In a field, the water content varies depending on all of these factors. Unless the process for determining soil water naturally incorporates a large volume of soil, multiple samples must be taken in order to obtain a representative soil water sample. As the soil water variation increases, more samples are needed to obtain a representative sample. The fact that practically all techniques for determining soil water content rely on tiny samples is another element that makes the process more difficult to measure. It is reasonable to anticipate that individual samples will differ from the mean of a large number of samples by at least 20%, give or take.

3.2.8. Location of Soil Water Measurements

It is crucial to determine the precise location of any soil water measurements. Finding locations that will provide a reliable estimate of the soil water level over a field typically requires familiarity with the soil, past experience, and sound judgment. Finding the areas to be examined is easier in fields with the same type of soil than it is in fields with different types of soil. As a general guideline for initiating the irrigation cycle, one location should be close to the side of the field where irrigation is to begin. To ascertain whether the field is being covered quickly enough to maintain a sufficient level of soil water there, at least one location needs to be at the other end of the field. Measurements ought to be taken at additional spots that are indicated by any critical soil condition, like an area that dries out first or remains wet for the longest. It is best practice to have two or three measurement stations in areas that are representative of the majority of the field, and to have at least two stations in each critical area. An effective soil water measurement system gives the irrigator enough information to control the system and maintain a consistent soil water level throughout the field. This type of information acts as a guide when adjusting the frequency and volume of irrigation for various field locations or stages of the growing season.

Measuring stations for sprinkler irrigation should be positioned between sprinkler heads and 10 to 15 feet from the lateral line. Measurements for row crops should be taken within the row or close to the plant, not at the furrow's bottom. Measurements for trees are typically taken inside the drip line and four to six feet away from the trunk.

Measurements should be taken in the area of the soil where plant roots draw water, as well as in accordance with the crop's specific pattern of water extraction. One measurement should be taken in the upper quarter of the root zone and one to three additional measurements should be taken at lower levels in soils with uniform textures. Measurements should probably be taken at 6, 12, and 18 inches, for example, if the maximum water-extraction depth for a given crop is 24 inches. Measurements from the various textural strata should be obtained for stratified soils. For most crops, the 6-inch measurement is sufficient to determine when to irrigate during the early phases of root development. Measurements from all three sites are required as the root system reaches maturity in order to provide a clear picture of the water content throughout the water-extraction zone. To find the overall soil water content for the profile, add the water from each measurement.

3.3. WATER INTAKE

Infiltration is the natural flow of water from the surface into the soil. Through worm and decayed root holes, pores, cracks, and cavities created by tillage, water can enter the soil. The water that has infiltrated the soil can either evaporate from the surface, transpired by the plants, or seep below the roots of the plants and enrich the ground water table.

Water that has been added to the soil through precipitation—whether natural or artificial, like sprinkler systems—infiltrates the soil; if the soil is unable to absorb it, some of the water may be momentarily retained on the soil's surface. It follows that when the rate of application is higher than the rate of infiltration, water accumulates on the surface and either ponds or runs off. The amount of water that enters the soil and the amount that can be retained in the soil profile for crop use are determined by the infiltration rate. The amount of possible runoff and the corresponding risk of soil erosion are also determined by the infiltration rate. For instance, surface crusting or sealing can limit the amount of water available to plants, increase erosion, and reduce infiltration.

3.3.1. Percolation

The soil's capacity to transfer water through the soil profile and away from the soil surface when the surface becomes ponded limits the infiltration rate. Percolation is the term used to describe this flow of water through the soil profile. The permeability or hydraulic conductivity of the soil controls the rate of percolation. Both terms refer to how easily water moves through soil.

Percolation is dependent on the relative quantity and continuity of large pores in a soil, as these are the main channels through which water percolates. High hydraulic conductivity is found in soil that has a course, open texture and high porosity. Because small pores have a higher flow resistance than large pores, when two soils have the same total porosity, the soil with small pores has lower conductivity than the soil with large pores. When large pores create continuous pathways through the profile, soil with a wide range of pore sizes conducts water more quickly. Conductivity in fine-textured soils is primarily determined by the pore space that exists between structural units. Particles in certain soils are cemented together to create what are known as "hardpans," which are almost impermeable layers. In other soils, colloidal or extremely finely divided material expands upon absorbing water to create an impermeable gelatinous mass that impedes the flow of water. Hydraulic conductivity may be significantly impacted by the quality of the transmitted water, especially by its salinity and alkalinity. Water's viscosity changes have an impact. Hydraulic conductivity can be significantly impacted by chemical changes in water without affecting viscosity. Even minute amounts of sodium chloride added to the soil water—not enough to noticeably alter the viscosity—can alter the structure of the soil to such an extent that hydraulic conductivity is significantly decreased.

3.3.2. Factors Affecting Infiltration

3.3.2.1. Soil Water Content

The amount of water that seeps into the soil when there is a pond affects how quickly water enters the soil (Figure 3.14). Large variations in matric potential in dry soils force water into the soil profile, increasing the amount of water that the soil can hold than in a wet starting point. Whether the soil was initially wet or dry, as irrigation continues and the intake rate drops to the steady infiltration rate, the surface soil gradually becomes saturated.

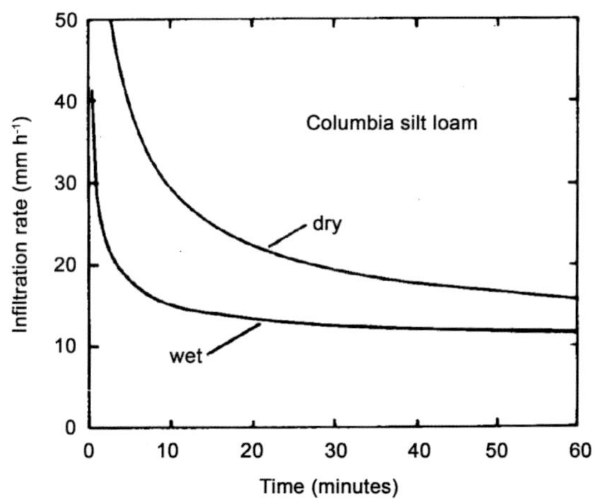

Figure 3.14. *Predkted infiltration rates for a deep Columbia silt loam with different initial water contents.*

Source: https://irrigationtoolbox.com/NEH/Part623_Irrigation/neh15-01.pdf.

3.3.2.2. Soil Sealing

Any flow restriction brought on by a shift in hydraulic conductivity or a barrier at the soil-water interface can restrict infiltration. The rate of water entry through the soil surface is quickly decreased by the formation of a thin, compact layer on the surface. This layer is the consequence of a breakdown in soil structure brought on by water flowing over the soil surface and raindrops or sprinkler system drops hitting the soil. Fitting smaller particles around bigger ones creates a seal that is comparatively impermeable. A little bit of cultivation prior to irrigation can aid in opening the seal and boosting infiltration. By covering the soil's surface with mulch or another permeable material, sealing can be partially avoided. Close-growing plants, such as grasses, absorb droplets, release their energy, and lessen surface sealing.

3.3.2.3. Surging

The cyclic flow that intermittently wets the soil surface during surge irrigation decreases infiltration. Numerous theories have been put forth to explain the decline in infiltration. Water can soak up and dissolve clods through wetting and drying, which helps the soil settle and compact after dewatering. As water moves across the surface, a seal could also form in addition to dissolving. Therefore, sealing and compaction both lessen infiltration. Reduced infiltration has also been attributed to air trapped during dewatering. Water holds air bubbles, preventing infiltration by obstructing the tiny pores in the soil's surface. Soil swelling has been proposed as a mechanism to reduce infiltration because of the hydration of clays and the decrease in hydraulic gradient as soil wetting proceeds.

3.3.2.4. Compaction

If soils are tilled too wetly, tillage operations may result in compaction and the formation of plowpans below cultivation depth. A plowpan slows down the flow of water, which lowers the rate of infiltration. The infiltration rate of certain soils is lowered in the furrows where tractor wheels pass. Deep tilling, also known as subsoiling, breaks up the impermeable sublayer, which temporarily improves water movement. Water flow is improved by the larger apertures. Compaction will return if cultural practices remain the same, with less tillage, crop residues added, fewer tillage operations performed, or appropriate timing of tillage operations.

3.3.2.5. Tillage

Plowing, cultivating, or doing any other type of stirring that causes the soil's pore size to increase can temporarily increase the infiltration rate. Cultivation only improves soil porosity and intake temporarily; after that, the soil returns to its original state due to compaction, flooding, or precipitation. Tillage operations are not likely to increase the rate of infiltration of loose, porous sand. Through pore space disruption and compaction, cultivation can lower intake.

3.3.2.6. Soil Cracking

In certain clay soils, when water is removed, cracks appear. Flood irrigation produces a high initial intake rate because cracks fill up quickly before the soil swells. As the soil gets wet, the cracks enlarge and finally close. Consequently, intake on these fine-textured soils is frequently very slow or nonexistent. As a result, the size and quantity of cracks should determine how much water is applied. If the rate at which water is applied during sprinkler irrigation is lower than the rate at which water infiltrates, the amount of water applied depends on the irrigation duration rather than the size and quantity of cracks.

3.3.2.7. Organic Material

When high residue crops are produced, organic material becomes available, and porosity stays high for relatively long periods of time. Using a cropping system that allows for high rates of crop residues in the top few inches of soil will maintain and even increase the infiltration rate. Legumes and grasses are two types of crops that help raise the soil's organic matter content. In order to produce larger pores and higher infiltration rates, the proportion of stable soil aggregates is increased. Alfalfa is one of the perennial crops that increases infiltration because it keeps the soil from sealing, keeps organic matter in the soil, and increases the amount of water-conducting pores created by decomposing roots.

3.3.2.8. Salts in Soil

Irrigated soils accumulate salts from irrigation water, which can alter the properties of the soil. In arid areas where irrigation provides the majority of the water, this accumulation is dangerous. In order to control, lessen, or eliminate soluble salts from the soil in the root zone area, it is frequently required to overirrigate (leach) on occasion. In humid regions, rainwater seeps through the soil and extracts the majority of soluble salts. Certain soluble salts found in irrigation water, like potassium nitrate, may have a direct positive impact on crops. Magnesium and calcium can improve the physical characteristics of soil in certain situations. On the other hand, high sodium sulfate or chloride concentrations are harmful. The breakdown of soil structure leads to the eventual dispersion of soil colloids, which lowers tilth and the rate of infiltration. This kind of sealing might be apparent on certain sandy soils as well.

Certain sodic soils can have their physical characteristics, like infiltration, enhanced by the addition of chemicals or soil amendments that replace exchangeable sodium with calcium. Gypsum, or calcium sulfate, is a reasonably priced amendment that is frequently used to improve aeration and infiltration, which in turn promotes root development and plant growth. If there is sufficient calcium present in the soil, additional substances like sulfur and aluminum sulfate are also applied.

3.3.2.9. Sediments in Irrigation Water

Irrigation water quality is impacted by suspended fine silt and clay particles. Depending on the amount of silt transported, how long the silty flow lasts, and the soil texture to which water is applied, this may or may not be advantageous. On coarse-textured soils, silty water deliveries can occasionally be advantageous because the sediments enhance the root zone's physical condition and slow down water movement. As silty water slows intake and makes the soil harder to cultivate, it typically exacerbates surface sealing issues when applied to fine-textured soil. Some plant nutrients, like calcium, phosphate, and potassium, are added to the soil by sediments.

3.3.2.10. Soil Erosion

Many soils experience a decrease in infiltration rate as erosion worsens due to the loss of organic matter and surface soil. This occurs as a result of either mixing finer-textured subsoil into the plow layer or leaving less permeable material, like dense clay subsoil, exposed. Certain soils may have coarse-textured layers, like sand and gravel, exposed by erosion, which promotes infiltration.

3.3.2.11. Land Leveling

During land leveling, soil movement and mixing may alter the characteristics of infiltration. When more or less permeable soil is exposed, the consequences resemble those of erosion. Compaction of the soil caused by earth-moving machinery used for land leveling can lower infiltration. To solve the issue, adding organic material and subsoiling are frequently * required. If a permeable layer is covered by a less permeable layer, better infiltration may occur by removing the top layer to reveal the permeable layer.

3.3.2.12. Temperature

Rainfall in the summer draws in more water than it does in the winter. It seems that as temperature rises, water's coefficient of viscosity rapidly drops, accelerating infiltration. However, the majority of authorities believe it has little bearing on infiltration.

3.3.2.13. Surface Storage

The amount of water that can collect on the surface and be set aside for infiltration is influenced by the slope and roughness of the soil's surface. When surface storage is filled and the application rate is higher than the infiltration rate, runoff starts. Smooth, steep, bare slopes tend to have less storage than flat, rough, vegetated slopes. As a result, surface storage influences the amount of water that seeps through.

3.3.3. Infiltration Stages

If the amount of precipitation or water applied by a sprinkler is less than the soil's capacity to absorb water, water does not accumulate on the soil's surface. The rate of water infiltration into the soil over time for both high and low steady application rates is depicted in Figure 3.15. The application rate determines the infiltration rate early in the process and is equal to it; the two curves are on the same horizontal line. Over time, the soil's capacity to hold water decreases and may even be surpassed by the rate at which water is applied; in such an event, ponding occurs and water builds up on the soil's surface. Surface storage that could be lost to runoff is represented by the shaded region between the horizontal steady water application line and the falling infiltration rate curve. With the high application rate, ponding occurs sooner and potential runoff increases.

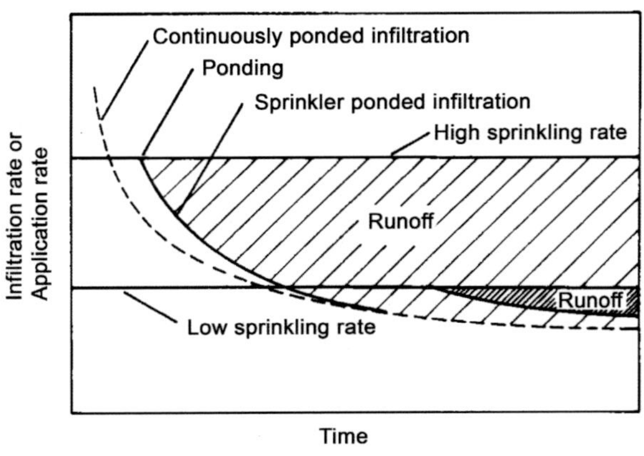

Figure 3.15. *Potential runoff for high and low steady water application rates similar to stationary sprinklers.*

Source: https://irrigationtoolbox.com/NEH/Part623_Irrigation/neh15-01.pdf.

The broken line in Figure 3.15 represents the decrease in infiltration rate over time under continuously ponded or flooded conditions. The infiltration rate is high when irrigation first begins but quickly drops. Because it fluctuates over time, the infiltration rate is known as transient. The rate becomes the steady infiltration rate when it reaches a very small change. As long as the rate of precipitation is lower than the constant rate of infiltration, water will not pool.

When the broken continuously ponded line meets the horizontal, steady precipitation rate line, ponding does not occur (Figure 3.15). Consequently, ponding time and possible runoff cannot be reliably predicted by overlaying an application rate line on a flooded infiltration test curve since it is later for both high and low application rates. As ponding occurs later than indicated by the above graphical method, runoff would be overestimated as a result. The three solid curves also have slightly different shapes. Thus, overlaying a precipitation rate line on top of a flooded infiltration test curve does not provide an accurate estimate of ponding time. Because ponding occurs later than anticipated by the aforementioned graphical method (Figure 3.15), runoff may be overestimated.

3.4. PLANTS

3.4.1. Plant Root Systems

Plant Root Systems Plant roots serve as a conduit for nutrients and water from the soil to the aboveground portions of plants. Fibrous roots and taproots are the two main categories of root systems that are known to exist. The fibrous root systems of cereal grains and other grasses (monocotyledons) are present. Other crops with taproot systems include sugar beets and alfalfa (dicotyledons). Figure 3.16 provides illustrations of the two types. Numerous thin roots with comparable lengths and diameters make up fibrous roots. A seminal or primary root is the first root to emerge from a seed that is germinating. The seminal root progressively lengthens and gets wider. From the main root, secondary roots emerge as lateral branches and subbranches. Nodal roots emerge from the subterranean stem nodes as they grow. Other aboveground nodes, like the brace roots of maize, can also produce roots. Some crop plants have an entire root system that is supported by a single taproot (taproot system), in contrast to the fibrous root system. Alfalfa and sugar beets are two examples of crops with this kind of root structure. First order laterals make up the majority of the entire root system, even though taproots can reach significant soil depths.

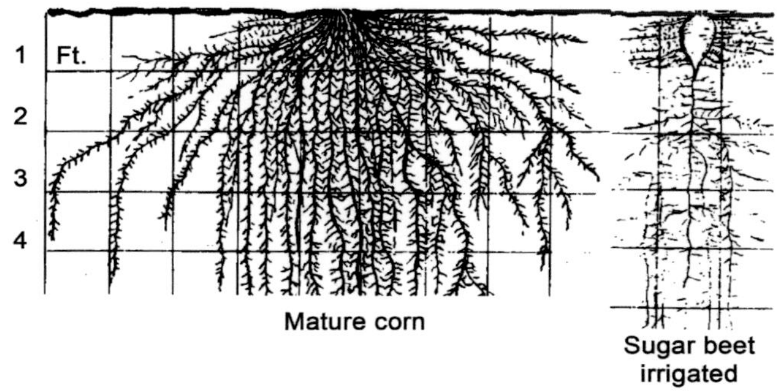

Figure 3.16. *Fibrous root system of corn (Left) and taproot I system of sugar beet (Right): both are approximately two months old.*

Source: https://irrigationtoolbox.com/NEH/Part623_Irrigation/neh15-01.pdf.

After a few weeks of growth, an intact root system has a significant length and complex branching, regardless of the fundamental rooting traits of dicotyledons and monocotyledons. Primary laterals give rise to secondary laterals, and so on, until an extensive network is formed under favorable conditions. Laterals are initiated in the parent root member. Typical growth rates for various root class members are:

root axes, 0.8 inch per day;

primary laterals, 0.2 inch per day; and

secondary laterals, 0.04 inch per day.

For maize, reports of root elongation up to 2 and a quarter inches per day have been made. On the other hand, adverse circumstances resulting from climate, soil chemistry, aeration, or composition may significantly impede the growth and multiplication of roots.

3.4.2. Crop Species Rooting Characteristics

Depth development and rooting density are two critical factors in crop rooting characteristics that must be well-understood for effective irrigation management. A comprehensive review of the literature led to the creation of a list containing the maximum rooting depth attained by about 55 plant species, as reported from 135 field observations. The range of depth attained within a crop species is most likely the result of unfavorable growth conditions and genetic varietal traits. Exclusion of data derived from less than ideal growth conditions was attempted.

3.4.3. Factors Affecting Root Growth

The rate of development and the maximum depth to which roots can grow can both be severely limited by a number of factors, even though root growth normally proceeds rapidly under ideal conditions. Crop root systems can support significant shoot growth even in the face of severe restrictions, provided that the effective root system receives proper aeration, water, and nutrients. Roots can be stressed by a variety of factors, which can be classified as follows: mechanical impedance, anaerobic conditions, lack of water, unfavorable temperatures, and plant pests and diseases; biological stresses; and nutrient deficiencies, imbalanced nutrient supply, and toxic substances.

Most root limiting conditions are caused by mechanical impedance considerations. For shallow soils that sit atop consolidated parent material or pans brought on by soil compaction connected to specific management practices, these might have genetic roots. When soil materials or pans are compacted, root limitations can occur suddenly, while in cases of soil compaction, they happen gradually. Reduced root growth rates are typically the result of compacted soil; complete root exclusion is only occasionally seen. The volume occupied by pores, particularly those with large sizes, is decreased when soil is compacted. This reduces the rate of gas exchange between the soil and atmosphere, alters the soil's capacity to retain and transmit water, and creates a mechanical impedance to root extension. Soil compaction affects all of these factors simultaneously and modifies root growth.

Soils with layers can provide significant obstacles for a growing root system. Given that sand and gravelly layers typically have high bulk density and strength, their decreased root proliferation can be significant. Normally well-drained, these layers become extremely rigid when a growing root expands and enters the matrix. Root growth through a clay layer that sits on top of a medium-textured or sandy soil zone may be restricted temporarily by a perched water condition and inadequate aeration.

3.4.4. Water Flow into Roots

Water in the soil, plant, and atmosphere moves in response to variations in the water's potential energy within the system. Soil water enters the root system along this potential gradient because transpiration lowers the water potential in the plant shoot and root system relative to the bulk soil. Water enters the root system via the epidermal cells that come into contact with the moist soil. It then passes through the pericycle cells, cortical cells, endodermis, and cortical cells before arriving at the xylem, which carries the water to the aerial plant parts. Important physical factors to take into account are the degree of root development and the physical contact between the root and soil. The uptake of water per unit volume of soil has been found to be proportionate to the rooting density when the upper portion of the root zone gets relatively dry and water is available in the lower zone. As a result, a key management concern is the distribution of roots, which varies depending on the crop species and physical characteristics of the soil. When assessing the amount of water that is available, salts in the soil water solution must be taken into account. A semipermeable membrane found in plant roots lets water through but most salt stays out. Thus, the primary impact of soluble salts on plants is osmotic, whereby elevated salinity levels impede the plant's ability to absorb sufficient water from the soil solution to fulfill its transpirational needs. Plants first show signs of cellular enlargement and the characteristic color changes linked to water stress.

3.4.5. Water Uptake-Root Profile Relations

The length of roots per unit volume of soil, or root length density, is usually greatest close to the soil's surface and decreases as one descends to the maximum depth at which a particular crop species' roots can be seen. The graphic representation of this overall trend is found in Figure 3.17. In the area with the highest concentration of roots and the best temperature and aeration, water is extracted most quickly. Water is quickly removed from the upper portion of the soil profile since it also evaporates from the top few inches of soil. At shallower depths, soils typically exhibit a faster rate of water loss until the potential drops to a point where it becomes rate-limiting.

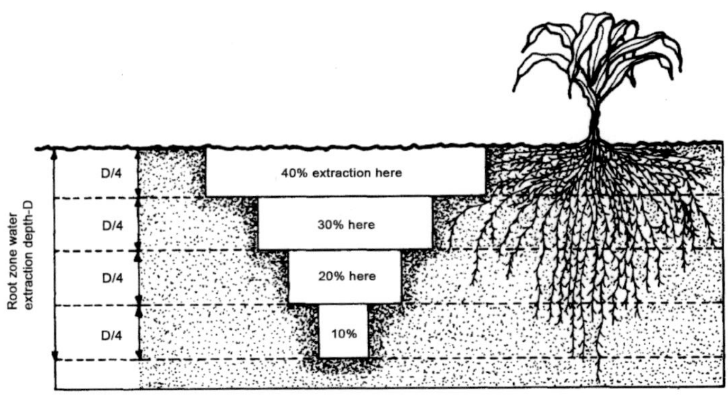

Figure 3.17. *Average water extraction pattern of plants growing in a soil without restrictive layers and with an adequate supply of available water throughout the root zone.*

Source: https://irrigationtoolbox.com/NEH/Part623_Irrigation/neh15-01.pdf.

The soil layer with the lowest soil water retention capacity may determine how much soil water is needed for crop growth in nonuniform soils. For instance, after irrigation or rain, a top soil layer with limited water retention capacity may quickly dry out. Water stress may occur in the early stages of plant development if the root system is not fully established, even though soil water may be sufficient at lower depths. Figure 3.18 shows some illustrations of limiting soil layers. When constrictive barriers are encountered, the typical extraction pattern for a given crop will alter. Additionally, the lower soil layers will supply more water than usual if the water level in the upper soil layers is allowed to remain abnormally dry.

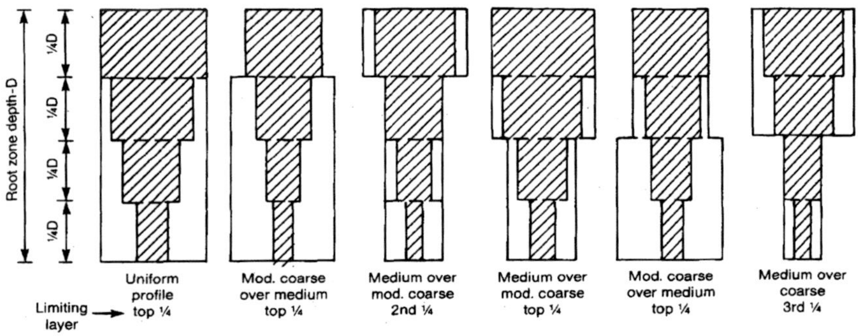

Figure 3.18. *Water extraction patterns as determined by available water content in various parts of soil profile. width of each profile represents total available water. gross area (height and width) of each profile represents total available water in profile; hatched area shows water extraction pattern for each profile.*

Source: https://irrigationtoolbox.com/NEH/Part623_Irrigation/neh15-01.pdf.

REFERENCES

1. Bristow, K. L., Campbell, G. S., & Calissendorf, K. (1993). Test of a heat-pulse probe for measuring changes in soil water content. *Soil Science Society of America Journal*, *57*(4), 930–934.
2. Broadbridge, P., & White, I. (1988). Constant rate rainfall infiltration: A versatile, nonlinear model. I. Analytic solution. *Water Resources Research*, *24*(1), 145–154.
3. Buckingham, E. R. (1907). *Studies on the Movement of Soil Moisture* (U.S. Department of Agriculture Bureau of Soils Bulletin 38). Washington, D.C.: USDA.
4. Cassel, D. K. (1974). *In situ* unsaturated hydraulic conductivity for selected North Dakota soils (Bulletin No. 494). Fargo, N.D.: Agricultural Experiment Station, North Dakota State University.
5. Chen, Z., Govindaraju, R. S., & Kavvas, M. L. (1994). Spatial averaging of unsaturated flow equations under infiltration conditions over areally heterogeneous fields: I. Development of models. *Water Resources Research*, *30*(2), 523–533.
6. Clothier, B., & Scotter, D. (2002). Unsaturated water transmission parameters obtained from infiltration. In J. H. Dane & G. C. Topp (Eds.), *Methods of Soil Analysis, Part 4: Physical Methods* (pp. 879–898). Madison, WI: Soil Science Society of America.
7. Dane, J. H., & Topp, G. C. (2002). *Methods of Soil Analysis, Part 4: Physical Methods*. Madison, WI: Soil Science Society of America.
8. Gardner, W. H., Hillel, D., & Benyamini, Y. (1970). Post-irrigation movement of soil water: I. Redistribution. *Water Resources Research*, *6*(3), 851–861.
9. Germann, P. F., & Beven, K. (1985). Kinematic wave approximation to infiltration into soils with sorbing macropores. *Water Resources Research*, *21*(7), 990–996.
10. Grossman, R. B., & Reinsch, T. G. (2002). Bulk density and linear extensibility. In J. H. Dane & G. C. Topp (Eds.), *Methods of Soil Analysis, Part 4: Physical Methods* (pp. 201–228). Madison, WI: Soil Science Society of America.
11. Hussen, A. A., & Warrick, A. W. (1994). Tension infiltrometers for measurement of vadose zone hydraulic properties. In L. G. Wilson, L. G. Everett, & S. J. Cullen (Eds.), *Handbook of Vadose Zone Characterization and Monitoring* (pp. 189–201). Boca Raton, FL: Lewis Publishers.

Chapter **4**

Traditional Irrigation Methods

Contents

INTRODUCTION

In arid and semi-arid regions, where water scarcity poses significant challenges to agriculture, traditional irrigation methods have been crucial in sustaining farming practices for centuries. These regions, characterized by low and unpredictable rainfall, require efficient water management strategies to optimize the use of available water resources. Traditional irrigation methods, developed through generations of local knowledge and experience, are tailored to the specific environmental conditions and cultural practices of these areas.

One of the most prominent traditional irrigation techniques is flood irrigation, where water is diverted from rivers or streams and allowed to flow over the fields. This method, although simple and widely used, can be wasteful due to evaporation and runoff, especially in hot and dry climates. However, it remains popular due to its low cost and ease of implementation.

In response to the limitations of flood irrigation, more efficient traditional systems have been developed, such as qanats and karez. Qanats, originating in ancient Persia and widely used in the Middle East and North Africa, involve underground channels that tap into water from aquifers or mountains and transport it to the surface for irrigation. This method minimizes water loss through evaporation and provides a reliable water supply even in the harshest conditions.

Another traditional technique is the terracing of fields, which is common in hilly or mountainous areas. Terraces reduce water runoff and erosion, allowing the soil to retain moisture for longer periods. Similarly, spate irrigation, practiced in parts of the Middle East and Africa, involves the diversion of seasonal floodwaters to irrigate fields, taking advantage of the brief but intense rains typical of these regions.

Traditional irrigation methods in arid and semi-arid regions reflect a deep understanding of local environments and the necessity for water conservation. While these techniques have been effective for centuries, modern challenges like climate change and population growth call for their integration with contemporary water management practices to ensure sustainable agriculture in these vulnerable regions.

4.1. TRADITIONAL IRRIGATION SYSTEMS

Water from various water sources (rivers, reservoirs, canals, channels, lakes, and wells) is drawn and directed to farms or fields by irrigation systems, which are designed to support farmers and supplement crop water requirements (Adamala et al., 2014). Many affordable, long-lasting, and indigenous irrigation systems

(TIS) have been created in India since prehistoric times. These systems rely on collecting and using surface and groundwater as well as rainwater. It has been discovered through various archeological excavations that TIS have been a part of Indian agriculture for at least 5000 years (Sengupta, 1985). Usually created, built, and overseen by nearby farmers, these TIS are intended for small-scale community applications (Srinivasan and Babu, 2000). As of right now, though, very few TIS still exist; these are located in places of worship and are owned by the community (Brebbia et al., 2010).

Since independence, major irrigation system construction has received the majority of attention. However, governments have focused their efforts on fixing the current TIS in recent decades. Because of this, a program called "Repair, Renovation and Restoration" of water bodies for agriculture was introduced in 2005 with the goal of enhancing and restoring the tanks that had been chosen in 26 districts across 15 States. Launched in March 2015, the Telangana state initiative "Mission Kakatiya," with the tagline "Mana Ooru Mana Cheruvu," aims to restore all of the identified tanks and lakes (more than 45000). Up to 250–270 TMC feet of water will be available for livestock, agriculture, irrigation, and drinking water needs once nearly all of the tanks have been restored. Furthermore, the government of Tamil Nadu's "Rain Water Harvesting (RWH)" campaign, which was started in 2001, had a significant influence on replenishing the groundwater table throughout the entire state. The Mukhya Mantri Ahar-Pyne Yojna was launched by the Department of Minor Water Resources, Bihar, in 2006. In accordance with this plan, 2237 of the 20938 Ahar-Pynes that currently exist—only 84% of which are operational—would be renovated in 17 state districts.

Notwithstanding TIS's enormous advantages, there are certain drawbacks that should be taken into account. For instance, the need for water for irrigation, drinking, domestic use, hydropower, ecology, industry, mining, leisure, tourism/culture, livestock, fisheries, and other purposes grows with population. Not all of these needs can be met by the TIS as it stands. Furthermore, by increasing the area under cultivation and introducing high-yielding varieties that demand more watering, the Green Revolution of the 1960s has intensified irrigation. Moreover, the water table in groundwater formations has been continuously lowered by the use of dug wells to levels that are inaccessible using conventional techniques. Farmers were compelled by the effects of these changes to adopt alternate irrigation methods in order to raise crops and provide for their communities (Sharma, 2006).

As a result, new technologies have been adopted, such as modern or micro irrigation (drip and sprinkler), in which emitters and a pipe network are used to apply water to the plant in the necessary quantity at regular intervals with a small flow (Adamala, 2016; Adamala, 2017). The WUE is enhanced as a result of this application technique (GOI, 2004). Additionally, drip irrigation reduces water losses through evaporation and distribution by delivering water directly to the crop's root zone (Dhawan, 2002).

Indian farmers have long learned to use communal water harvesting techniques to irrigate their agricultural fields despite unpredictable rainfall patterns and uneven land topography. Even though it's not as common as their other traditional methods, crop irrigation is still effective and in use.

4.1.1. Diversion Channels

Historically, people have diverted water from streams or water channels into agricultural fields using both simple engineering—made with boulders and tree branches—and complex engineering—built with Grand Anicut on the Cauvery River. With seasonal flows, diversion channels directed the water to a storage structure for later use in both semi-arid and arid regions. Farmers in the flood plains transport water from natural springs to fields for irrigation or drinking using bamboo pipes that are readily available in the area.

4.1.2. Tanks or Rain Water Harvesting Structures

Farmers in South India (Andhra Pradesh, Telangana, Karnataka, and Tamil Nadu) are big fans of tank-based TIS. These kinds of systems are required in areas with few rivers, sparse and irregular rainfall patterns, steep terrain, and low soil water retention capacities.

4.1.3. Wells

TIS lift water from wells for drinking or agricultural use using a "Rahat/Persian wheel." The inhabitants of arid regions of Rajasthan and central India built step wells and other obstructions across waterways to collect surface runoff and demonstrate cooperative water use. "Kundis" or "Kundds," which are covered underground tanks to store runoff from an artificially prepared catchment primarily to address drinking water problems, have even been developed in areas with an abundance of land resources (Ministry of Rural Development, 2004).

Farmers from various regions have constructed various TIS to fully satisfy their irrigation and drinking needs using fundamental engineering principles and distinctive modes. Table 4.1 displays a portion of the local TIS that is still in use in various Indian locations.

Table 4.1. Small Scale or Local Traditional Irrigation Practices in India

Region	System	Description	Practiced in
North	Zing	Tanks for collecting melted glacier water	Ladakh (Jammu)
	Naula	Small wells to collect water from nearby stream	Uttaranchal
	Kuhl	Diversion canals	Himachal Pradesh
	Guhl	Diversion channels leading directly to agricultural fields	
	Baudi	Tanks exclusively used in Hamirpur, Kangra, and Mandi districts	
	Nawn	Tank used for drinking, washing, and taking showers	
	Khatri	Pits made of rock to collect rainwater seeping through these rocks	
	Dighi	Small reservoirs or wells	Delhi
	Ahar-Pynes	Rainwater conservation, storage (Ahars) & distribution (Pynes) structures	Bihar
East	Apatani	Rice and fish farming with surface and groundwater	Arunachal Pradesh
	Zabo/Ruza	Impounding runoff	Nagaland
	Cheo-oziihi	Long water channel to divert to the terraces through bamboo pipes	
	Tuikhur/Jhoras	Springs	Manipur and Mizoram
	Bamboo drip	System of tapping of stream water to carry over long distances	Meghalaya
	Dongs	Ponds used by Bodo tribes	Barak & Brahmaputra (Assam)
	Garh and Dara	Big Naula	
	Dungs/Jampois	Irrigation canals from stream for cultivating paddy fields	Jalpaiguri (West Bengal)
	Inundation channels	Floodwaters to be diverted to agricultural lands	West Bengal
Central Highlands	Pat	Diversion bunds across the stream	Jhabua (Madhya Pradesh)
	Katas/Bandhas	Earthen dams	Odisha
	Haveli	Embankments/bunds in agricultural fields to store rainwater	Madhya Pradesh
West	Kuis/Beris	Deep pits to collect seepage	Rajasthan
	Baoris/Bers	Wells used mainly for drinking	
	Paar	Command area where percolated water is accessed using Beris	
	Kunds/Kundis	Circular underground well that has a saucer shaped catchment	Pali (Rajasthan)
	Talab/Nadi	Village ponds	Jodhpur (Rajasthan)
	Tankas	Underground tanks used only for drinking	Bikaner (Rajasthan)
	Khadins	Earthen embankments	Jaisalmer (Rajasthan)
	Baoli	Step well	Rajasthan
	Jhalaras	Human made tanks meant for community use and religious rites	
	Virdas	Shallow wells dug in low depressions called Jheels (tanks)	Rann of Kutch (Gujarat)
	Bandhis	Natural or man-made reservoirs	Bundelkhand & Udaipur

	Kuva/Saza	Open well	Mewar (Rajasthan)
	Johads	Stone barriers	Alwar (Rajasthan)
	Naada/Bandha	Stone check dam constructed across a stream to capture runoff	Mewar (Thar Desert)
	Vavs/Vavdi	Shallow wells	Rann of Kutch (Gujarat)
	Rapat	Tank	Rajasthan
	Chandela/Bundela	Tanks constructed by stopping the water flow between the hills	
	Bandharas	Diversion weirs across streams	
	Phad	Canals and series of check dams	
	Bandhini	Flooded embankments	
	Shilotris	Water courses	
	Ramtek	Surface and ground water bodies connected through canals	
South	Cheruvu	Reservoirs to store runoff	Telangana & Andhra Pradesh
	Bavi, Kunta, Revu	Wells, tanks, channels	
	Jarukattu	Tank	Sirsi (Karnataka)
	Kere/Talpariges	Channels branching off from Anicut built across streams	Central Karnataka
	Garanda	Small well (pilla bhavi) inside a large dug well	Kolar (Karnataka)
	Eri	Small tanks to cultivate the few acres of land in irregular slope	Tamil Nadu
	Surangam/Mala	Horizontal well	Kasargode (Kerala)
	Korambu	Temporary dam of brush wood & mud to raise water level in canal	Kerala
	Khazana	Regulatory systems to control ingress of saline river waters	Goa
Islands	Jack wells	Pits to collect rainwater through bamboos	Nicobar

Source: https://www.researchgate.net/publication/348540380_AN_OVERVIEW_OF_INDIAN_TRADITIONAL_IRRIGATION_SYSTEMS_FOR_SUSTAINABLE_AGRICULTURAL_PRACTICES/link/6002a92045851553a049485b/download?_tp=eyJjb250ZXh0Ijp7ImZpcnN0UGFnZSI6InB1YmxpY2F0aW9uIiwicGFnZSI6InB1YmxpY2F0aW9uIn19.

4.1.3.1. Northern Region

The most widely used techniques in this area are paddy farming and terraced agriculture. Small tanks called "zings" are used to collect melted glacier water for use in mountainous areas' irrigation systems. The person in charge of distributing water equally is known as "Chirpun." "Naulas" are drinking places with stone floors where people gather water from streams or springs. It is primarily done in Uttarakhand's hill country. Little waterways called "Kuhls" transport glacial water from "Khuds" (streams) into fields used for farming. It is made up of a headwall for storage and Moghas, or Kuchcha outlets, to divert the flow into a field close by. It is primarily funded by donations from the general public and is maintained by a village community or "Kohil," a designated local engineer. The Kohil is involving the village community in the repair and operationalization of the Kuhl during the start of the irrigation season. Kuhl's length ranges from 1 to 15 meters, and its trapezoidal cross-section has an area of 0 to 12 square meters. Its flow rate is between 15 and 100 liters per second (Centre for Science and Environment, 2011). One Khul can irrigate an area of about 20 hectares, providing support for 6–30 farmers. Natural contour water channels called "guhls" are primarily located in the Dehradun and Nainital regions. These are employed to move water downhill slopes by building barriers out of branches and trees that can be found nearby. "Baudi" is a small surface water tank intended for storage, while "Nawn" is a large irrigation system in the form of a tank that is used for showering, washing, and drinking. "Khatris" are rock-built pits measuring 3 m by 6 m by 1.8 m that are used to catch rainfall (Mohan, 2012).

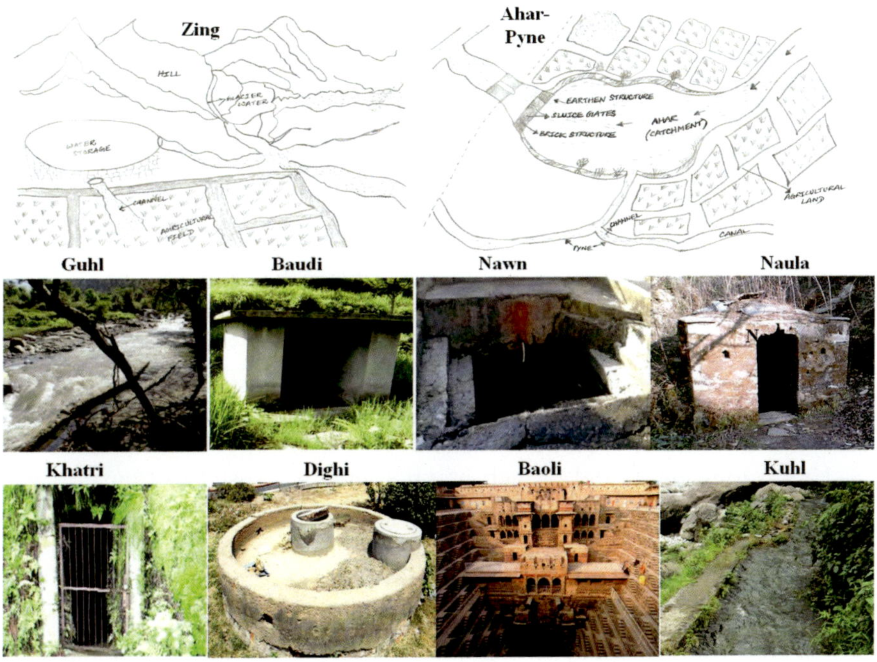

Figure 4.1. *Local irrigation systems in Northern region.*

Source: https://www.researchgate.net/publication/348540380_AN_OVERVIEW_OF_INDIAN_TRADITIONAL_IRRIGATION_SYSTEMS_FOR_SUSTAINABLE_AGRICULTURAL_PRACTICES/link/6002a92045851553a049485b/download?_tp=eyJjb250ZXh0Ijp7ImZpcnN0UGFnZSI6InB1YmxpY2F0aW9uIiwicGFnZSI6InB1YmxpY2F0aW9uIn19.

These are owned by private citizens, local governments, or Panchayats. In Khatris, rainwater collected from seepage through rocks is used for household purposes, while water collected straight from the roof through pipes is used for animals and washing. "Dighi" refers to a square or circular reservoir or well with a sluice gate that is primarily located in Delhi and its environs. Dighi water can be obtained for domestic use by hiring a person by the name of Kahar or Mashki. In Delhi, step wells known as "baolis" are used for bathing and washing. The most widely used baolis that are still in use are Dargah of Kaki Saheb baoli, Rajon baoli, Muradabad Pahad baoli, and Gandak baoli (the water contains gandak, or sulfur). Floodwater harvesting systems known as "Ahar-Pyne" (catchment-channel) are common in low-rainfall paddy-cultivated areas like South Bihar (Figure 4.1).

4.1.3.2. Eastern Region

The greatest potential for water resources, an abundance of groundwater resources, and paddy growing areas dominate this region. The Apatani tribes of Arunachal Pradesh use the most sophisticated and scientific method of rice-fish farming known as "Apatani" (Agarwal and Narain, 1997). Under this system, fields are irrigated using both surface and ground water, resembling contour bunds. Aji, a different system that is comparable to Apatani, also grows finger millet on the bunds that divide each plot. The term "Zabos or Ruzas" refers to the local integrated irrigation systems that include water storage structures, agriculture, forestry, fisheries, and animal care. widely used to grow the terraced fish cum paddy by Nagaland's Kikruma farmers. From the highest to the lowest slopes, it is composed of different components. agricultural lands and cattle sheds, catchment (forest land), and water harvesting systems (ponds, channels, and earthen embankments). "Cheo-oziihi" (person-water) refers to the 8–10 km long main and branch bamboo channels used to transport water to terraced fields (Agarwal and Narain, 1997). In the Manipur and Mizoram regions, "Tuikhur or Jhoras" are the traditional methods of storing water using numerous springs in the hills with the Jhum method of cultivation. The clever drip system known as "Bamboo Irrigation" is utilized by the tribal farmers in the Revoi and Jowai districts of Meghalaya, as well as in the Khasi and Jaintia hills of Cherrapunji, to grow black pepper and betel leaf. It is made of bamboo. Water from streams is diverted to lowland terrace fields in roughly four to five stages using bamboos of various sizes and shapes (Dhiman and Gupta, 2011; Borthakur, 2008). "Dongs" are ponds used to collect water for field irrigation using a device known as a Lahoni. "Garh and Dara" functions similarly to a large naula and is used to direct river water. Small channels called "dungs" or "jampois" are used for irrigation; these are created by cutting channels out of streams or nallahs. Broad, shallow channels known as "inundation canals" or "overflow irrigation systems" are used in South Bengal during the monsoon season to transport excess water (Chhabra and Gai, 1981). The most widely used TIS in this area, aside from inundation canals, are pukurs, which are open wells built downstream of ponds (Figure 4.2).

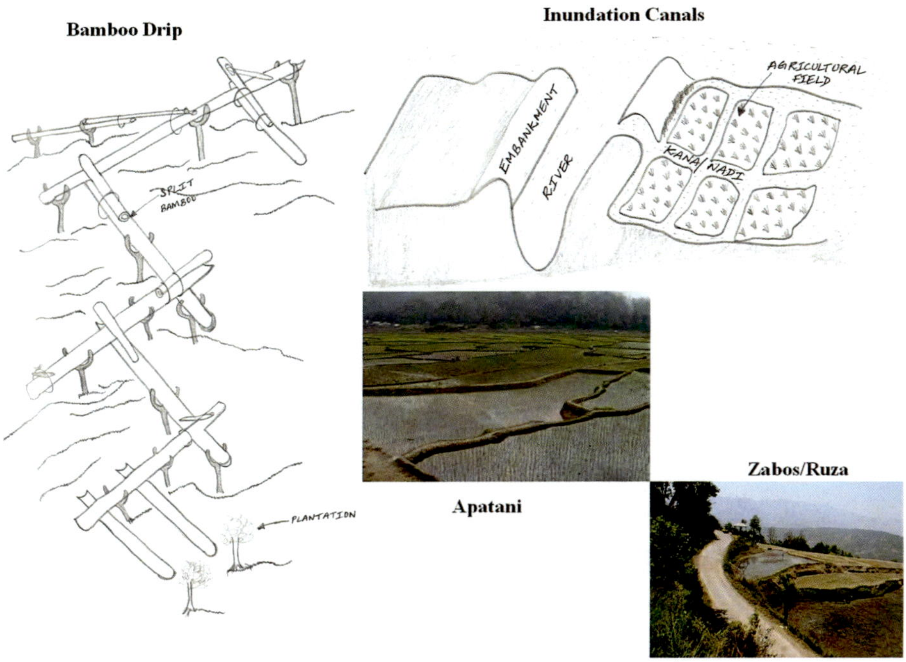

Figure 4.2. *Predominant traditional irrigation systems in Eastern region.*

Source: https://www.researchgate.net/publication/348540380_AN_OVERVIEW_OF_INDIAN_TRADITIONAL_IRRIGATION_SYSTEMS_FOR_SUSTAINABLE_AGRICULTURAL_PRACTICES/link/6002a92045851553a049485b/download?_tp=eyJjb250ZXh0Ijp7ImZpcnN0UGFnZSI6InB1YmxpY2F0aW9uIiwicGFnZSI6InB1YmxpY2F0aW9uIn19.

4.1.3.3. Central Highlands

Wells and tanks are typically the main sources of irrigation in these areas. The primary TIS that are used in this area are Pat, Haveli, and Bandhas. The "Haveli" system involves building enclosed embankments on four sides up until sowing, at which point the embankments are opened. A "path" system is a way to direct water from streams into irrigation channels by building various embankments throughout the landscape. "Kata, Munda, or Bandha" are unique water diversion systems that are built on uneven terrain, with an earthen embankment built across a drainage line (Figure 4.3).

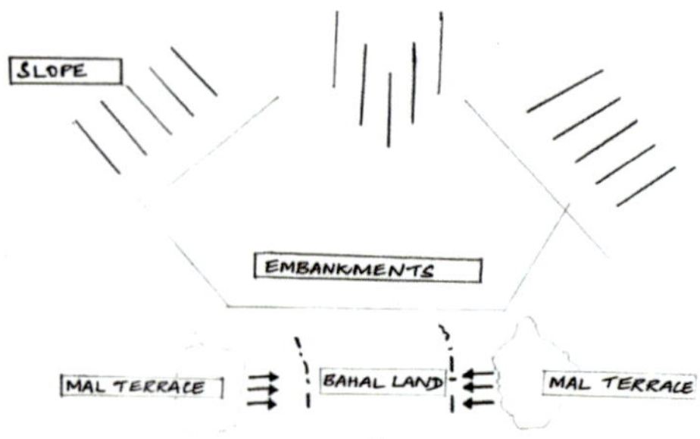

Figure 4.3. *Kata and pat irrigation systems in central highlands.*

Source: https://www.researchgate.net/publication/348540380_AN_OVERVIEW_OF_INDIAN_TRADITIONAL_IRRIGATION_SYSTEMS_FOR_SUSTAINABLE_AGRICULTURAL_PRACTICES/link/6002a92045851553a049485b/download?_tp=eyJjb250ZXh0Ijp7ImZpcnN0UGFnZSI6InB1YmxpY2F0aW9uIiwicGFnZSI6InB1YmxpY2F0aW9uIn19.

4.1.3.4. Western Region

Due to the region's general lack of water, wise techniques to preserve water for drinking and irrigation have been developed. These include the use of tanks, pits, dug wells, step wells, excavations, embankments, and reservoirs. Kuis, also known as Beris, are shallow pits with a broader base and a narrow mouth that are dug 10 to 12 meters deep to collect seepage water from nearby bodies of water (CE IIT Kharagpur, 2011). In areas where water is scarce, "Baoris or Bers" are the community step well structures that are common (Rajasthan). By replenishing the aquifers, the water table is rising as a result of the stored water in these wells seeping into the ground. In order to reduce evaporation losses, a sequence of steps is built next to bodies of water with the intention of deepening the wells and narrowing the width from above. "Paar" is a popular method of collecting water from catchments.

It involves digging kuis or Beris 5 to 12 meters deep to collect percolated water (Rajani Pani). This method of collecting rainwater is called Patali Paani. "Kunds," which resemble an overturned cup nestled in a saucer, are rainwater harvesting drinking structures. "Talabs or Nadis" are village ponds that are used to hold rainwater during the monsoon season from neighboring catchments. A "tanka" is a subterranean pit used to gather and hold rainwater. To ensure that the water is safe to drink, they are lined with finely polished lime.

In order to collect surface runoff for agriculture, "Khadin" refers to the locally constructed earthen embankment (Dhora) across streams (Mishra,

2011). Traditional step wells known as baoli are widely distributed in Rajasthan and Northern India. "Jhalaras" are man-made, rectangular step wells intended solely for religious ceremonies and communal bathing; they are not intended for drinking. There are currently eight Jhalaras in Jhodhpur city, with Mahamandir Jhalara being the most established and effective. "Virdas" are shallow wells or Jheels (tanks) that are excavated in the transient stream beds. Because of its higher density, the saline water in Virdas stays below the freshwater zone, while the sweet freshwater floats on the upper layer.

"Bandhis, also known as Talabs," are naturally occurring reservoirs that have valleys and depressions that are used for drinking and domestic use. Beris refers to the wells that are situated on the Talab beds. "Saza/Kuva" refers to a multi-owner circular open pit or well. "Johads" are improvised, transient barriers placed along the contours to catch and hold rainwater. It contributes to lowering surface runoff, allowing blocked water to seep into the ground, and raising the water table. "Naada or Bandha" is a check weir across a gully made of stones that stops runoff over a certain distance. Traditional step wells called "Vav/Vavdi" are used to irrigate fields.

They have a sluice and a trough. "Rapat" is a percolation tank with a waste weir that is used to recharge surrounding areas that are three to five kilometers away rather than directly irrigate agricultural lands. The "Chandela and Bundela" are earthen embankments and tanks built by blocking small streams' water flow. "Bandharas" are constructed check dams, either temporary or permanent, that span the rivers and streams in the western Deccan.

These dams ensured that the rivers flowed into channels by raising the water level of the rivers. The term "phad" describes an irrigated area of land (roughly 100–125 hectares) where elevated water from Bhanadaras is channeled into canals or kalvas. It is made up of earthen embankments that are placed over rivers in order to redirect water for use in agriculture (Bhattacharya, 2015). In Maharashtra's Konkan region, where the fields frequently flood during intense monsoons, "bandhini" is practiced.

Here, TIS such as Shilotris, Bandharas, and Phad are also adopted, in addition to Bandhini. Shilotris are small, narrow waterways used to reroute field embankments and small streams so that paddy fields can be watered. The Maharashtra "Ramtek" model is made up of a number of surface and ground water bodies, with wells, baoris, and kundis serving as storage areas for runoff (Figure 4.4).

Figure 4.4. *Traditional irrigation systems in Western region.*

Source: https://www.researchgate.net/publication/348540380_AN_OVERVIEW_OF_INDIAN_TRADITIONAL_IRRIGATION_SYSTEMS_FOR_SUSTAINABLE_AGRICULTURAL_PRACTICES/link/6002a92045851553a049485b/download?_tp=eyJjb250ZXh0Ijp7ImZpcnN0UGFnZSI6InB1YmxpY2F0aW9uIiwicGFnZSI6InB1YmxpY2F0aW9uIn19.

4.1.3.5. Southern Region

Using a Bavi (well), Kunta (percolation tank), and Revu or Cheruvu (diversion channel to collect stream water), irrigation is carried out in Andhra Pradesh and Telangana. "Cheruvu" are reservoirs designed to store runoff, featuring embankments, sluices, weirs, and canals. An embankment known as "Jarukattu" resembles a check dam. In some parts of Karnataka, there are small springs known as "talpariges," which are not the main source of irrigation. "Kere" refers to a group of tanks constructed in the Karnataka region that are fed by streams or Anicuts, a type of check weir. "Garanda" is the name given to a small well that is built inside a larger well to collect lower-level springs and percolation water. 'Eris' (Kinaru or Keni) is a system of cascading irrigation tanks with two sluices viz. one with full local community management at a higher elevation and another at a lower elevation. According to Bhagule and Bhavsar (2007), they aid in soil loss and flood prevention, runoff storage, and groundwater recharge. "Surangam/Mala" are man-made, vertical tunnel-like excavations that descend from hill slopes to lowlands. Kerala has a lot of horizontal wells, but the Kasaragad district of Northern Malabar uses a well-known local system called Thurangam/Thorapu/Mala/Tunnel for irrigation as well as domestic needs. These Surangams, which are 0.5–7 m wide, 3–300 m long, and 1.2–10 m high, have a constant water supply all year round with a discharge that varies between 2 and 60 lpm (Iyengar, 2006). The "Korambu/Chira" temporary dams in Kerala are man-made structures that span the mouths of channels. They aid in raising the canal's water level and directing water into channels. "Ooranis" are tiny irrigation tanks used in South Travancore to cultivate tiny agricultural plots. The Kuruma tribe in Wayanad, Kerala, uses a unique kind of well called a "Panam Keni" to collect and store rainwater. The "Khazana" irrigation system in Goa uses river water from the Zuari and Mandovi to irrigate large rice fields and low-lying agricultural fields (Figure 4.5).

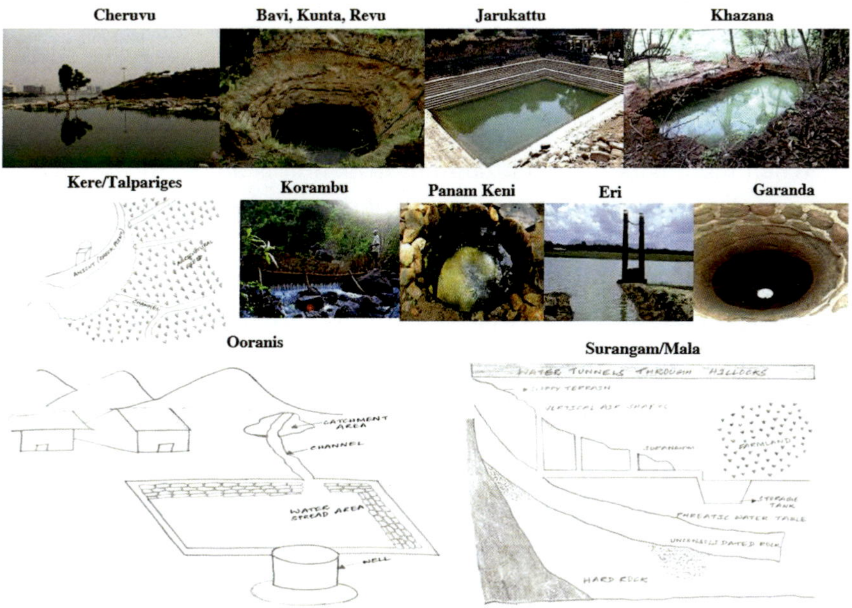

Figure 4.5. *South India's traditional irrigation systems.*

Source: https://www.researchgate.net/publication/348540380_AN_OVERVIEW_OF_INDIAN_TRADITIONAL_IRRIGATION_SYSTEMS_FOR_SUSTAINABLE_AGRICULTURAL_PRACTICES/link/6002a92045851553a049485b/download?_tp=eyJjb250ZXh0Ijp7ImZpcnN0UGFnZSI6InB1YmxpY2F0aW9uIiwicGFnZSI6InB1YmxpY2F0aW9uIn19.

4.2. TRADITIONAL METHOD OF IRRIGATION

For India, challenges such as low crop yields, unfavorable harvests, downpours, and other issues have always existed. The great majority of these issues have been resolved by using conventional irrigation techniques. In the conventional irrigation technique, farmers supply water to their fields using canals and groundwater supplies. These tried-and-true methods of irrigation are still in use today. These techniques are employed by Indian farmers to guarantee optimal seed germination, increased harvest yields, and faster growth in crop productivity.

The term "traditional irrigation method" describes an irrigation technique that has been in use for many years, if not centuries. Compared to contemporary irrigation techniques, these are the most economical, convenient, and environmentally friendly.

4.2.1. Importance of Irrigation

In contrast to relying solely on rainfall, irrigation entails supplying water to crops via canals, pipes, sprinklers, or other man-made methods. For farming and agriculture, irrigation is essential. The majority of India experiences no rainfall at all during the year. Nevertheless, even in the months without rain, plants require water. When there is not enough rainfall, irrigation is automatically activated. It can be applied whenever necessary to boost agricultural output, even in the event of insufficient rain. Thus, irrigation is essential to the development and maintenance of plants and crops.

In drought-prone areas such as Rajasthan and certain parts of Gujarat (Kutch and Bhuj), farming practices are supported by irrigation. Irrigation provides water supply to even the country's colder northern regions, like Jammu, where water is primarily found in the form of glaciers. Irrigation also makes plowing easier and guarantees the right water supply, both of which are necessary to maintain soil fertility. In addition, the nation needs a high crop yield in order to feed its expanding population. Together with the green revolution, irrigation serves the primary goal of farming in the nation.

4.2.2. Different Methods of Irrigation

There are four main types of irrigation techniques that are considered traditional. Every one of these methods works best with a particular combination of crops and types of land.

- **Moat (Pulley System):** In order to irrigate the soil with this method—also referred to as the pulley system—water must be drawn up from a well or some comparable source. To raise water, either human labor or livestock must be used. The user must devote a substantial amount of time and energy to this method. Moat systems are significantly less expensive to use and waste less water than other irrigation techniques.

- **Chain Pump:** Two enormous wheels connected to one another by a chain make up a chain pump. There are some connected buckets on the chain. There is a link in the chain that extends to the water supply. The wheel is turned to collect the water in the bucket. They will eventually be raised to the higher wheel by the chain, where the water is eventually emptied into a source. After that, the bucket is emptied and put back down.
- **Dhekli:** It is a method of extracting water from a well or other subterranean storage facility. A bucket and a rope are held atop a pole. A large object that can be used as a counterweight is fastened to the other end. You could use the pole to obtain water.
- **Rahat (Lever System):** The Rahat irrigation technique uses the labor of animals. A large wheel is fixed in position above the well. The wheel that was used to draw water from the reservoir was turned by an ox or a cow, and buckets were fastened to the belt and passed over the wheel to collect the water.

4.2.3. Traditional Method of Irrigation: Advantages

Traditional irrigation methods provide several advantages, some of which are region-specific and others general.

- They are the least complicated and least expensive approaches to the process of irrigation.
- They do not need any prior technical knowledge or expertise on the user's part.
- They may be used effectively to irrigate lands that have been levelled.
- Erosion is reduced with the amount of time that the water spends in the reservoir.
- Due to human supervision, water is utilized more efficiently, and the crop receives the appropriate amount of water.

4.2.4. Traditional Method of Irrigation: Disadvantages

In addition to the advantages, the traditional irrigation method includes several drawbacks that should be considered.

- They need a significant amount of effort and time to complete.
- The seepage that occurs in drains results in water being wasted.
- As a result of the development of the crops, an abnormally large amount of water flows into the basins, which results in a loss of usable water.
- If not done correctly, this may result in waterlogging.
- Too much water would flow off, which would be wasteful.
- The water can travel through the field if the soil has an appropriate slope and absorbs water easily. Plants are harmed by standing water, which also results in lower yields of food crops.

4.2.5. Water-Saving Irrigation Methods in Arid/ Semi-Arid Areas

Freshwater resources directly available for agricultural production are becoming less and less as the human economy grows and the population increases. This may result in the spread of arid regions, a worsening of the drought, soil erosion, and the decline of the environment. As a result, in arid and semi-arid regions, it has become more crucial than ever to implement the sustainable approach to water allocation in order to increase crop water productivity and lessen the pressure on soil water demand. In this context, several water-saving irrigation technologies have been researched as useful tactics in arid regions in order to better utilize water resources and balance the yield and economic benefits.

Researchers studying farmland water conservation are now interested in how to best utilize these water-saving irrigation technologies in accordance with local conditions.

4.2.5.1. Rainwater Harvesting Irrigation

About half of China's land area is made up of arid and semi-arid regions, making it one of the major water-poor nations. Precipitation has replaced surface and groundwater as the primary source of irrigation for agriculture in arid regions. However, in arid and semi-arid regions, the annual variation in rainfall is relatively high, and the efficiency of rainwater utilization is low. In addition to retaining and fully utilizing rainfall runoff as a source of irrigation water, rainwater harvesting irrigation reduces soil erosion and enhances the natural environment.

When crops have critical water demand periods, rainwater-harvesting irrigation techniques make use of scarce water resources for supplemental irrigation. These irrigation methods involve collecting, storing, moving, and using rainwater. In the past, people have collected rainwater using water cellars, dry wells, karez, and reservoirs; however, the development of rainwater-harvesting irrigation techniques has been sluggish due to technical limitations and topography.

Irrigation using rainwater harvesting has been growing quickly lately. Researchers have created a number of methods, including CIS remote sensing technology, to assess the viability of using rainwater in a given area, the imperviousness of surfaces that collect it, the composition and effectiveness of the catchment, and the combination of additional water-saving techniques (e.g., for drought resistance, seed preservation, and seedling preservation—drip irrigation and sprinkler irrigation. The "121 rainwater harvesting project" in Gansu province, China, has advanced significantly from single irrigation to water and soil conservation, water pressure reduction, and improved ecological benefits. Currently, rainwater harvesting irrigation has reached a certain scale in arid areas.

4.2.5.2. Plastic Film Mulching Technology

The current applications of plastic film mulching technology are primarily focused on furrow and border irrigation. When using border irrigation, the crop-growing border fields are covered with a membrane. Crops are planted on the groove slope or ridge back when using furrow irrigation, and the ditch bottom, groove slope, and portion of the ridge back are tiled with membrane. Water enters the soil through the manual watering holes or seedling holes in both techniques, passing through the upper layer of the membrane. This allows water to permeate the soil through the membrane gaps, resulting in loose soil and a low rate of ground roughness. These techniques can also maintain entropy and heat while enhancing the financial gains from crops. Furthermore, research has demonstrated that it is possible to fortify tomato seedlings by altering the mulch's color. Tomato seedlings treated with red light exhibit strong nutrient growth, a significant build-up of dry matter, and rapid expansion of their leaf area.

Despite its beneficial aspects, the usage of plastic film mulching can lead to environmental contamination. To address this environmental concern, researchers have been diligently exploring innovative eco-friendly materials in recent times, including photolysis membranes and biodegradable membranes, among others. The study of plastic film mulching technology also encompasses methods to enhance soil aeration, raise the soil's nutrient availability, and decrease greenhouse gas emissions associated with its application. Simultaneously, efforts have been made to advance the technology of mechanical recovery and reuse of the residual membrane. This not only supports sustainable growth but also boosts the potential for mechanization, further enhancing its capabilities.

4.2.5.3. Deficit Irrigation

Deficit irrigation is a method aimed at achieving a balance between crop yields and optimal water efficiency by leveraging the self-regulating and supplementary capabilities of plants. This technique does not adhere to a proportional water supply in relation to the overall water needs of the plant growth and development process. The core concept of controlled deficit irrigation involves subjecting crops to a specific level of water scarcity, which in turn restricts cell elongation growth due to water deficit and reduced cell turgor. Consequently, leaves become smaller, and the photosynthetic area shrinks. As the stress level intensifies, the water potential drops significantly, while the cellular concentration of abscisic acid (ABA) increases. This results in a decrease in the net photosynthesis rate and retention of dry matter in the crops. Post-rehydration, crops exhibit a positive response mechanism that boosts their growth rate and enhances their drought tolerance (Figure 4.6).

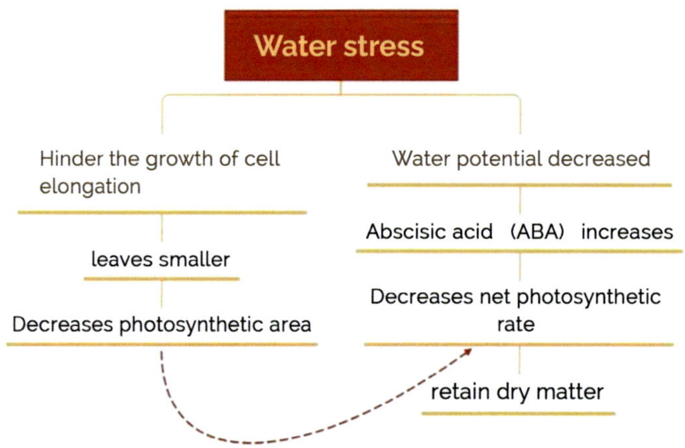

Figure 4.6. *The principle of regulated water stress.*

Source: https://www.e3s-conferences.org/articles/e3sconf/pdf/2020/51/e3sconf_reee2020_02001.pdf.

Each region should conduct research on various crops and soils to establish an optimized crop water deficit evaluation system. The gradual transition from extensive irrigation to precision irrigation has become the focus of research due to the varying water requirements of different crops and the need to prevent the development from a moderate water deficit to a serious water shortage in production. Current crop water deficit evaluation methods primarily consider ecological factors, but more researchers are hoping to influence crop genetic material to increase crop resistance to drought. In addition, controlled deficit irrigation, which affects crop yields without compromising crop quality, will influence crop yields of protein, sugar, fat, and other components and will become a future area of study.

4.2.5.4. Alternate Furrow Irrigation

Furrow irrigation remains the primary irrigation technique in the majority of developing nations, including China. Traditional furrow irrigation gave rise to alternate furrow irrigation, a suitable water-saving method. Water stress and alternate furrow irrigation work on similar principles. Since the mid-1960s, there has been extensive research conducted on alternate furrow irrigation in arid and semi-arid regions of certain developed nations, including the United States. However, there has been less theoretical investigation into how to achieve the mechanism of water-saving and crop yield increase, with the majority of these preliminary studies focusing on the relationship between irrigation water and yield.

With a wide range of field application possibilities, this water-saving irrigation technology—which is based on the advancement of the conventional furrow irrigation method—avoids the drawbacks of high startup costs and challenging operation associated with other water-saving techniques like micro- and sprinkler irrigation. Research has shown that regular irrigation with alternating water and fertilizer could preserve maize's high root activity as well as its typical physiological metabolism, which includes photosynthesis and nitrate reductase activity, and enhance leaves' water-use efficiency to conserve water and boost yield.

Crops with wide row spacing can be irrigated using alternate furrow irrigation. Consequently, researchers attempt to implement the narrow row close planting crop mechanism (e.g., wheat), increase crop disease prevention and water-saving effectiveness, allowing for the integration of alternate furrow irrigation and the development of ecological irrigation zones.

4.2.6. Comparison of Main Irrigation Strategies in Arid/Semi-Arid Areas

Table 4.2 illustrates our inadequate comparison of the benefits and drawbacks of four sustainable water-saving irrigation technologies based on their cffccts on crops.

Table 4.2. *Comparison of Advantages and Disadvantages of Main Irrigation Strategies in Arid/Semi-Arid Areas*

Irrigation strategies	Advantages	Disadvantages
Rainwater harvesting irrigation	• Impounding stormwater runoff as a source of irrigation • Make full and effective use of local stormwater runoff • Prevent soil erosion • Improve the ecological environment	• Lack of integration of the key techniques (i.e. harvesting, storage, transportation and utilization) and other water-saving irrigation measures • High cost of development and maintenance of concrete rainwater collecting surface • Low durability and potential environmental pollution of plastic rainwater collecting surface • Low rainwater utilization efficiency
Plastic film mulching technique	• Reduce surface evaporation and deep leakage • low ground roughness rate and the loose soil • Preserve heat and entropy • Improve the uniformity of irrigation	• Causes microplastic pollution to the terrestrial environment • The broken plastic film affects the absorption of soil nutrients • Reduce soil organic matter • Increase greenhouse gas emissions
Deficit irrigation	• Improve crop quality and drought tolerance • Balance economic benefits and output	• Excessive water shortage will lead to dead seedlings
Alternate furrow irrigation	• Keep the root soil loose and well ventilated • Prevent groundwater level from rising, soil nutrient loss and crop lodging • Serve as a drain during the rainy season	• Can not be widely used in field cultivation of narrow row crops(e.g.,wheat) • Difficult to design alternate irrigation time and its flow rate

Source: https://www.e3s-conferences.org/articles/e3sconf/pdf/2020/51/e3sconf_reee2020_02001.pdf.

Table 4.2 illustrates that the latter two methods rely on the physiological and genetic potential of the crops themselves, whereas the first two use technologies to accomplish the goal of saving water. In contrast, the latter two (biological water-saving irrigation techniques) are more likely to be developed in the future, be more affordable, and have stronger operability. Furthermore, in order to pursue accurate irrigation, future research would concentrate more on automatic intelligence. This is merely the author's partial comparison; there are still certain restrictions on the conclusion.

4.3. SIMPLE AND LOW-COST IRRIGATION SYSTEMS FOR (SMALL SCALES) IN ARID REGIONS

Roughly 62% of Sindh's land area is thought to be covered by desert, Thar Kohistan, and Nara, in addition to a sizable 350 km stretch of coastline and rural areas.

- Total Area = 14.09 Million Hectare
- Cultivated Area = 04.87 Million Hectare (35%)
- Un-cultivated Area = 06.77 Million Hectare (48%)
- Forest Area = 01.03 Million Hectare (7%)
- Culture able Waste = 01.42 Million Hectare (10%)

The advancement and superior development of the agriculture sector is a major factor in Sindh's economic development. According to the Agriculture Department of Sindh (2021), this province makes a major contribution to the nation's total agricultural output.

The frequency of prevailing droughts in Sindh's arid and semi-arid regions is rising daily, and this is negatively affecting both the province and the national economy. Sindh's production is threatening to decline annually as a result of the scarcity of irrigation water for small and cash crops. Only a few 1,000 cusecs will remain at the distal ends of water distributaries, enough only for drinking, after a massive amount of water from the Indus River has been used; most of this will perish en route to remote villages in Sindh (Talpur et al., 2001).

Productive crop production requires the use of innovative irrigation techniques. Subsurface drip irrigation and sprinkler systems, for example, can reduce irrigation water use by nearly half, but their adoption is hampered by sociocultural, technological, and financial issues. Scientists continue to face a significant challenge in adapting traditional, low-cost water-saving technologies for sustainable crop production in semi-arid and arid regions. Jug irrigation, deep pipe irrigation, porous capsule irrigation, tree canopy irrigation, wick irrigation, and other techniques are some of these techniques (Bhayo et al., 2018).

One of the best conventional irrigation techniques is jug irrigation. Water seeps into the root zone of the irrigated crop from the buried pitcher as a result of the pressure head gradient across the pitcher wall. The pressure gradient arises from the difference in height between the positive pressure inside the pot and the negative pressure outside the pot where it comes into contact with the soil. (Zreig and colleagues, 2006). Due to the pitcher's close relationship with its surroundings, which includes the soil, climate, and plants, it possesses high levels of self-regulation for the reasons mentioned above. In places where surface irrigation techniques are impractical or water is scarce, pitcher irrigation is utilized for small-scale field irrigation (Barthwal, 2005).

The conditions at the pitcher wall's exterior have a significant impact on seepage rates. According to some studies (Stein, 1998; Cliff-Hill, 1985; Chigura, 1994), pitchers may be able to regulate themselves in situations where seepage is regulated by the soil water pressure head, which is a function of the soil water content surrounding the pitcher. The hydraulic gradient and seepage rate across the pitcher wall increase as evapotranspiration causes the soil water to decrease and, in turn, the soil water pressure head to decrease. Thus, it is important to accurately assess how soil water pressure head caused by transpiration and evaporation affects pitcher seepage rates.

The most effective and environmentally friendly irrigation technology is wick irrigation, which is adaptable to all types of weather, terrain, soil composition, plant species, and plant life. Wick irrigation maintains moisture levels consistently, promoting crop growth (Bainbridge, 2007). Wick irrigation can significantly improve water use efficiency, which is especially important in regions of extreme water scarcity, and wicks help move water farther away from the clay pot to promote more root development (Bainbridge, 2012). Wick irrigation is a capillary flow-based subsurface irrigation system. Using this technique, water enters the root zone straight through the wick and from the water reservoir (Yeager et al., 2004).

The components of a wick irrigation system are a hose, a water tank, and a wick. The tube keeps the wick's moisture from evaporating. The wick is the primary component of a wick irrigation system; it provides water to the plant roots as needed. Water is drawn from the reservoirs into the root zone of the plant by the wick component. Cotton, nylon, blankets, or polyester textiles can be used to make the wick (Wesonga et al., 2014). Water is gradually used up in the soil, which lowers evaporation and ensures that the plant always has access to water. When the plant needs it, the wicks draw nutrients and water from the solution they are submerged in. Thus, this irrigation technique uses an automated irrigation system, which lowers the labor requirement (Joseph, 2016).

This study examined simple, inexpensive, and efficient irrigation techniques like wick and pitcher irrigation in light of Sindh, Pakistan's arid regions' lack of water for agriculture. They are therefore simple to construct, use, and maintain.

4.3.1. Pitcher Irrigation

Pitcher irrigation is an age-old method that has been used in many arid regions of the world, such as Africa, South America, Iran, and India (Mondal, 1974; Stein, 1997). A book written in China some 2000 years ago makes reference to pitcher irrigation (Sheng, 1974). It has been reported that watermelons in Pakistan and India have been irrigated using this technique (Mondal, 1974; Soomro, 2002).

In India's semi-arid Karnal state, pitcher irrigation is a low-cost, small-scale irrigation technique. The method is placing unglazed clay pots up to their necks in the ground. The natural pores in the pot's walls allow water to seep laterally into the soil when it is filled with water, establishing the moist conditions required for plant growth. Pitchers are refilled as needed to keep the plant root zone supplied with water continuously. Pitchers' ability to save water is one of the benefits of using them for irrigation. The fact that the scales are drastically different must be taken into account when comparing pitcher irrigation to flood or sprinkler irrigation.

Small-scale irrigation is done with pitchers; larger-scale irrigation is done with sprinkler and flood systems. Pitcher irrigation remains more effective in light of this. Since pitcher irrigation applies water directly to plant root zones rather than to larger areas of the field, it is a more efficient use of water than other methods. Since water is released from smaller areas when using pitcher irrigation, deep percolation losses are negligible. Additionally, the amount of water in each pitcher can be adjusted to control the rate of water loss at different sites. A field that is irrigated with pitchers may require even less water than one that is irrigated with drips because of the pitchers' extremely low hydraulic conductivity and decreased evaporation losses (on the same scale).

Advantage of Pitcher Irrigation

- Give moisture to nearby soil not the whole area.
- Evaporation minimum.
- Water seepage below the ground is also in minimum quantity.
- best for horticulture crops and vegetables.
- Once the pitchers are fixed, irrigation can be done for six years, which reduces expenditure.
- Don't need skilled labor.

4.3.1.1. Effect of Pitcher Irrigation on Crops Cultivation

High tomato yields of 27 t ha^{-1} were achieved in India using saline irrigation water, EC 10.2 mmhos cm^{-1}, while typical yields in this region with fresh water, EC 0.4 ranged from 15-25 t (Mondal, 1983). In Kenya, 61% of normal crop yield was achieved with irrigation water of EC 8 dS m^{-1} when typical irrigation failed at EC 4 dS m^{1} (Okalebo et al., 1995).

Balakumaran and associates. In a thorough investigation of cucumber production, (1982) found that irrigation with pitchers at a rate of 1.9 mm ha^{-1} produced yields that were equivalent to hand irrigation at a rate of 7.3 mm ha^{-1}. Pachpute (2010) also came to the conclusion that the pitcher irrigation method, among other water management techniques, increased overall yield by 203% and produced a water use efficiency of 12.06 kg m^{-3}. Even with saline water, pitcher irrigation is thought to be more effective than surface, drip, and sprinkler irrigation. Dubey and associates. (1990) studied the effects of increasing salinity on the germination of scalloped gourd (*Luffa acutangula*) and discovered that the highest yield (4.45 kg/pitcher) was obtained at 0 dS m^{-1} irrigation.

4.3.2. Wick Irrigation

The newest method created by scientist Kamalam Joseph at Kozhikode's Center for Water Resource Development and Management (CWDRM) is wick irrigation, which aims to make farming easier even in times of water scarcity.

"Wick irrigation reduces evaporation by supplying water to plant roots below the soil's surface through the use of ropes as wicks. A rope wick (typically housed inside a plastic tube to prevent evaporation) can be used to transfer water from a pot or raised water container into the soil. The idea is comparable to a wick-equipped kerosene lantern that, once lit, continuously draws fuel up the wick. Water first passes through the wick through capillary action. Afterwards, water is continuously drawn through the wick to replenish the water that the roots of the plants have absorbed. It is possible to use the system in one location and then move it to another. It functions well on terrain that slopes steeply or flatly. Planning, design, and highly skilled labor are not necessary for installation. Hopefully, plants absorb water when needed, so no computations or forward planning is necessary for this system. It's time to add more water if the bottle is empty (Pedro Ochoa Mena and Baracoa, 2010) Wick irrigation is a sustainable technology that is easily accessible.

This system can be classified in two ways:

- surface wick irrigation method: plastic bottles/buckets are placed on the soil's surface, and ropes are buried into the soil near plant roots to provide the necessary amount of moisture to plants.
- Subsurface wick irrigation method: plastic bottles are half-buried in soil and ropes are wrapped around the root zone to increase the depth of water supply to plants.

Advantages of Wick Irrigation System

- Water consumption is reduced since evaporation is all but eliminated
- Less time is needed for refilling containers
- there may be fewer problems with salt build-up, since the water will not evaporate at the surface and leave behind salt.

4.3.2.1. Effect of Wick Irrigation on Crops Cultivation

It is being considered whether to use a suction irrigation system, which uses water-specific suction devices to reduce pumping pressure. An acre-sized suction irrigation system is the subject of a theoretical model development, and the system's water and energy efficiency is compared to that of a typical drip irrigation system. A suction system can function at a significantly lower power and could even be a power source, while a wick irrigation system has a larger distribution of water supply from the wicks than a drip system. The wick system would provide even greater energy savings if a direct connection could be made between the plant roots and the wick, removing the need for water to travel through the soil. (Kuntz et al., 2013).

A CWIS was successfully implemented to simulate the wick discharge for various potted crops using the Solver program. The results of the study may help smallholder farmers produce crops in greenhouses by using CWIS, a low-cost, highly effective micr-oirrigation system (Roonjho et al., 2022). The wick irrigation method is found to be more effective than the pitcher irrigation method in terms of crop yield and crop water productivity, but it uses more water. Wick irrigation should only be used by farmers during the warm season, when high rates of evapotranspiration prevent pitchers from meeting crop water requirements. (Bhayo & Co. 2018). It is very effective and impressive how ductile the area beneath the wick and pitcher is.

REFERENCES

1. Abioye, E. A., Abidin, M. S. Z., Mahmud, M. S. A., Buyamin, S., Ishak, M. H. I., Rahman, M. K. I. A., Otuoze, A. O., Onotu, P., Ramli, M. S. A. (2020). A review on monitoring and advanced control strategies for precision irrigation. *Computers and Electronics in Agriculture, 173*, 105380.
2. Abu-Zreig, M. M., Abe, Y., & Isoda, H. (2006). The auto-regulative capability of pitcher irrigation system. *Agricultural Water Management, 85*(3), 272–278.
3. Adamala, S. (2016). Traditional to smart irrigation methods in India: Review. *American Journal of Agricultural Research*, *1*(1), 1–10.
4. Adamala, S. (2017). Development of smart automated irrigation system. *International Journal of Environment, Agriculture and Biotechnology*, *2*(1), 219–224.
5. Adamala, S., Raghuwanshi, N. S., & Mishra, A. (2014). Development of surface irrigation systems design and evaluation software (SIDES). *Computers and Electronics in Agriculture*, *100*, 100–109. https://doi.org/10.1016/j.compag.2013.11.004.
6. Agarwal, A., & Narain, S. (1997). *Dying Wisdom: Rise, Fall and Potential of India's Traditional Water Harvesting Systems.* State of India's Environment–A Citizens' Report, No. 4. Centre for Science & Environment.
7. Agarwal, M. (2018). What we can learn from traditional Indian rainwater harvesting systems? Retrieved from: https://tahaanefforts.org/blog/what-can-we-learn-from-traditional-indian-rwh-systems (accessed on 8 October 2024).
8. Al-Ghobari, H. M., & Dewidar, A. Z. (2017). Deficit irrigation and irrigation methods as on-farm strategies to maximize crop water productivity in dry areas. *Journal of Water and Climate Change*, *9*(2), 399–409.
9. Alemi, M. H. (1980). Distribution of water and salt in soil under trickle and pot irrigation regimes. *Agricultural Water Management*, *3*, 195–203.
10. Bainbridge, D. A. (2001). Buried clay pot irrigation: A little known, but very efficient traditional method of irrigation. *Agricultural Water Management*, *48*(2), 213–215.
11. Bainbridge, D. A. (2002). Alternative irrigation systems for arid restoration. *Ecological Restoration*, *20*(1).
12. Bainbridge, D. A. (2006). Beyond drip irrigation: Hyper efficient irrigation systems. *Proceedings of the American Society of Agricultural and*

Biological Engineers Annual International Meeting, Portland, Oregon, ASABE #062073. St. Joseph, Michigan.

13. Bainbridge, D. A. (2007). *A Guide for Desert and Dryland Restoration: New Hope for Arid Lands*. Island Press.
14. Bell, J. M., Schwartz, R., Howell, T., Cristine, L. S., & Morgan, C. L. S. (2018). Deficit irrigation effects on yield and yield components of grain sorghum. *Agricultural Water Management*, *203*, 289–296.
15. Bhalge, P., & Bhavsar, C. (2007). Water management in arid and semi-arid zones: Traditional wisdom. In *International History Seminar on Irrigation and Drainage* (pp. 423–428). Tehran, Iran.
16. Bhattacharya, S. (2015). Traditional water harvesting structures and sustainable water management in India: A socio-hydrological review. *International Letters of Natural Sciences*, *37*, 30–38.
17. Bhayo, W. A., et al. (2018). Water saving and crop yield under pitcher and wick irrigation methods. *Pakistan Journal of Agriculture, Agricultural Engineering and Veterinary Sciences*, *34*(1), 57–67.
18. Borthakur, S. (2008). Traditional rainwater harvesting techniques and its applicability. *Indian Journal of Traditional Knowledge*, *8*(4), 525–530.
19. Brebbia, C. A., Marinov, A. M., & Bjornlund, H. (2010). Sustainable irrigation management, technologies and policies III. *WIT Transactions on Ecology and the Environment*, *134*, 251.
20. Chaturvedi, A. K., Surendran, U., & Chandran, K. M. (2021). Exploring growth, physiological status, yield and water use efficiency of vegetables grown under wick method of irrigation. *Plant Physiology Reports, 26*(1), 64–73.
21. Chen, S. H., Chou, J. H., & Li, J. (2002). Optimal grey-fuzzy controller design for a constant turning force system. *International Journal of Machine Tools & Manufacture, 42*(3), 343–355.
22. Chhabra, B., & Gai, G. (1981). Inscription of the early Gupta kings in *Corpus Inscriptionum Indicarum*, Vol III. Archaeological Survey of India.
23. Cuello, J. P., Hwang, H. Y., Gutierrez, J., Kim, S. Y., & Kim, P. J. (2015). Impact of plastic film mulching on increasing greenhouse gas emissions in temperate upland soil during maize cultivation. *Applied Soil Ecology, 91*, 48–57.
24. Dhawan, B. D. (2002). *Technological Change in Indian Irrigated Agriculture: A Study of Water Saving Methods*. Commonwealth Publishers.
25. Dhiman, S. C., & Gupta, S. (2011). *Rainwater Harvesting and Artificial Recharge*. Central Ground Water Board, Ministry of Water Resources.

26. Domingues, J., Valerio, D., & Costa, J. S. D. (2011). Rule-based fractional control of an irrigation canal. *Journal of Computational and Nonlinear Dynamics, 6*(2), 24503.
27. Dubey, S. K., Gupta, S. K., & Mondal, R. C. (1990). Response of ridge gourd to saline irrigation water applied through earthen pitchers. *Transactions of the Indian Society of Desert Technology, 15*, 53–61.
28. Eapen, D., Barroso, M. L., Ponce, G., et al. (2005). Hydrotropism: Root growth responses to water. *Trends in Plant Science, 10*(1), 44–50.
29. El-Naggar, A. G., Hedley, C. B., Horne, D., et al. (2020). Soil sensing technology improves application of irrigation water. *Agricultural Water Management, 228*, 105918.
30. Eugenio, C., Maria, G., Nunzio, F., et al. (2020). Appraisal of biodegradable mulching films and vegetal-derived biostimulant application as eco-sustainable practices for enhancing lettuce crop performance and nutritive value. *Agronomy, 10*(3), 383.
31. Geerts, S., & Raes, D. (2009). Deficit irrigation as an on-farm strategy to maximize crop water productivity in dry areas. *Agricultural Water Management, 96*(9), 1275–1284.
32. Government of India (GOI). (2004). *Report of Task Force on Micro Irrigation* (Chairman: N. Chandrababu Naidu). Ministry of Agriculture.
33. Gutierrez-Gines, M. J., Mishra, M., McIntyre, C., et al. (2020). Risks and benefits of pasture irrigation using treated municipal effluent: A lysimeter case study, Canterbury, New Zealand. *Science of the Total Environment, 727*, 118530.
34. Heidari, H., Zarei, Z., & Mohammadi, K. (2022). Improving water use efficiency and biomass in maize, foxtail millet, and bitter vetch by wick irrigation. *Water SA, 48*(3), 264–270.
35. Iyengar, S. (2006). *Waternama: A Collection of Traditional Practices for Water Conservation and Management in Karnataka*. Communication for Development and Learning. http://www.indiawaterportal.org (accessed on 8 October 2024).
36. Jasrotia, A. S., Majh, A., & Singh, S. (2009). Water balance approach for rainwater harvesting using remote sensing and GIS techniques, Jammu Himalaya. *Water Resources Management, 23*, 3035–3055.
37. Kahinda, J. M., Taigbenu, A. E., & Sejamoholo, B. B. P. (2009). A GIS-based decision support system for rainwater harvesting (RHADESS). *Physics and Chemistry of the Earth, 34*, 767–775.
38. Kang, S., Liang, Z., Pan, Y., Shi, P., & Zhang, J. (2000). Alternate furrow

irrigation for maize production in an arid area. *Agricultural Water Management, 45*(3), 267–274.

39. Kemper, W., Ruffing, B., & Bondurant, J. (1982). Furrow intake rates and water management. *Transaction of the ASAE, 25*(2), 333–339, 343.
40. Kulkarni, S. A. (2005). Looking beyond eight sprinklers. In *Proceedings of the National Conference on Micro Irrigation* (pp. 1–10). G.B. Pant University of Agriculture and Technology, Pantnagar, India.
41. Kundu, D. K., Neue, H. U., & Singh, R. (1998). Comparative effects of flooding and sprinkler irrigation on growth and mineral composition of rice in an Alfisol. In *Proceedings of the National Seminar on Micro Irrigation Research in India: Status and Perspective for the 21st Century* (pp. 1–10). Bhubaneswar, India.
42. Li, C. M., & Gao, S. H. (2002). The evolution law and supply and demand of water resources in arid and semi-arid areas in North China. *Journal of Soil and Water Conservation, 17*(2), 68–71. (In Chinese)
43. Ministry of Rural Development. (2004). *Water Harvesting and Artificial Recharge*. Rajiv Gandhi National Drinking Water Commission, Government of India.
44. Mishra, A. (2011). *The Radiant Raindrops of Rajasthan*. Research Foundation for Science, Technology, and Ecology.
45. Mohan, L. (2012). Traditional rainwater harvesting method dying slow death. *The Tribune*. http://www.tribuneindia.com/2012/20121217/himachal.htm (accessed on 8 October 2024).
46. Mondal, R. C. (1974). Farming with pitcher: A technique of water conservation. *World Crops, 26*, 91–97.
47. Mondal, R. C. (1983). Salt tolerance of tomato grown around earthen pitchers. *Indian Journal of Agricultural Sciences, 53*(5), 380–382.
48. Muhammed, H. H. (2015). *Modeling of Capillary Wick Irrigation System for Potted Plant and Small Scale Plantation* (Master's thesis, Universiti Putra Malaysia).
49. Okalebo, J. A., Home, P. G., & Lenga, F. K. (1995). Pitcher irrigation: A new irrigation technique to curb the effects of salinization. In *Proceedings of the 7th Annual Conference Kenya Society of Agricultural Engineers* (pp. 15–21). Jomo Kenyatta University of Agriculture and Technology, Kenya.
50. Oweis, T., & Hachum, A. (2006). Water harvesting and supplemental irrigation for improved water productivity of dry farming systems in West Asia and North Africa. *Agricultural Water Management, 80*(1–3), 57–73.

51. Pachpute, J. S. (2010). A package of water management practices for sustainable growth and improved production of vegetable crops in labor and water-scarce Sub-Saharan Africa. *Agricultural Water Management, 97*(9), 1251–1258.
52. Pillai, G. M., & Ouseph, P. P. (2003). Anthropogenic impact on water quality: A case study from Kavaratti Island, Lakshadweep, Union Territory of India. In V. P. Singh & R. N. Yadava (Eds.), *Environmental Pollution* (pp. 69–82). Allied Publishers Pvt. Limited.
53. Pu, G. B., Liu, S. Q., Liu, L., & Ren, L. H. (2005). Effects of different light quality on the growth and physiological characteristics of tomato seedlings. *Acta Horticulturae Sinica, 32*(3), 420–425. (In Chinese)
54. Roonjho, S. J., Rowshon, M. K., & Rehman, A. R. (2022). Modeling capillary wick irrigation system for greenhouse crop production. *Agricultural Water Management, 274*, 107927.
55. Sengupta, N. (1985). Irrigation: Traditional vs modern. *Economic and Political Weekly, 20*(45/47), 1919–1938.
56. Sharma, A. (2006). Water harvesting context in the Indian subcontinent. In *Proceedings of UNESCO G-WADI Meeting on Water Harvesting* (pp. 63–70). Aleppo, Syria.
57. Soomro, A. A. (2002). Viability of pitcher irrigation. *Daily Dawn Karachi*. http://www.dawn.com/2002/05/06/ebr12.htm (accessed on 8 October 2024).
58. Soomro, S. A. (2013). *Effect of Saline Water on Soil Properties and Yield of Bottle Gourd Using Pitcher Irrigation Method* (M.E. thesis). Land and Water Management, Sindh Agriculture University, Tandojam.
59. Srinivasan, R. K., & Babu, S. S. V. (2000). *A Water Harvesting Manual for Urban Areas*. Centre for Science and Environment, New Delhi, India.
60. Subrahmaniyan, K., & Mathieu, N. (2012). Polyethylene and biodegradable mulches for agricultural applications: A review. *Agronomy for Sustainable Development, 32*(2), 501–529.
61. Tagar, A. A., Soomro, A., Bhatti, S. M., Memon, S., & Soothar, R. K. (2016). Comparative study of scientific and traditional water application practices for wheat crop. *Pakistan Journal of Agriculture, Agricultural Engineering and Veterinary Sciences, 32*(2), 202–211.
62. Talip, T., Ustun, S., Salih, E., Erdal, D., Erdal, G., & Rafet, A. (2019). The deficit irrigation productivity and economy in strawberry in the different drip irrigation practices in a high plain with semi-arid climate. *Journal of Science Horticulture, 245*.

63. Talpur, M. A. M. (2001). Water shortage in Sindh: Causes and consequences. *Sangat (SANA), 17*(1), 9–11.
64. Tan, J. L., Wang, L. Q., Wang, X. N., et al. (2010). Effect of alternate irrigation with water and fertilizer in different furrows on physiological indexes of summer maize. *Acta Botanica Boreali-Occidentalia Sinica, 30*(2), 344–349.
65. Wagan, F. H., & Khoso, S. (2013). Water shortage: Its causes, impacts, and remedial measures. In *6th International Civil Engineering Congress* (Vol. 28).
66. Wang, J., Lv, S., Zhang, M., Chen, G., Zhu, T., Zhang, S., & Luo, Y. (2016). Effects of plastic film residues on occurrence of phthalates and microbial activity in soils. *Chemosphere, 151*, 171–177.
67. Ward, F. A., & Pulido-Velazquez, M. (2008). Water conservation in irrigation can increase water use. *Proceedings of the National Academy of Sciences, 105*(47), 18215–18220.
68. Xie, J. H., Zhang, R. Z., Li, L. L., Chai, Q., Luo, Z. Z., Cai, L. Q., & Qi, P. (2018). Effects of plastic film mulching patterns on maize grain yield, water use efficiency, and soil water balance in the farming system with one film used for two years. *Acta Ecologica Sinica, 29*(6).
69. Yang, B., Yao, H., Zhang, J. X., Li, Y. Q., Ju, Y. L., Zhao, X. F., Sun, X. Y., & Fang, Y. L. (2020). Effect of regulated deficit irrigation on the content of soluble sugars, organic acids, and endogenous hormones in Cabernet Sauvignon in the Ningxia region of China. *Journal of Food Chemistry, 312*.
70. Yang, H., Zhang, X., & Zehnder, A. J. (2003). Water scarcity, pricing mechanism, and institutional reform in northern China irrigated agriculture. *Journal of Agricultural Water Management, 61*(2), 143–161.
71. Yi, H., Qin, L., Wei, Q. J., Chang, R. Y., & J, W. (2020). Agricultural plastic mulching as a source of microplastics in the terrestrial environment. *Journal of Environmental Pollution, 260*.
72. Zarei, Z., & Heidari, H. (2017). Investigating water use efficiency in sunflower under furrow and wick irrigation methods. *Environmental Stresses in Crop Sciences, 10*(4), 521–530.
73. Zhang, P., Wei, T., Li, Y., Zhang, Y., Cai, T., Ren, X., & Jia, Z. (2019). Effects of deficit irrigation combined with rainwater harvesting planting system on the water use efficiency and maize (*Zea mays L.*) yield in a semiarid area. *Irrigation Science, 37*, 611–625.

Chapter **5**

Drip Irrigation Systems

Contents

INTRODUCTION

For all life on Earth, water is a necessary and indispensable resource. It makes up about 70% of the surface of the Earth, the majority of which is made up of oceans and seas. It is essential to conserve water and provide it precisely to the root zones of crops because both the population growth and the food demand are reliant on it. Adopting water-efficient irrigation technologies, like drip irrigation, and promoting water-saving techniques, like crop rotation and mulching, are two ways to improve water management in agriculture. The average share of freshwater withdrawals from agriculture is 70%, but in certain developing nations, this percentage can reach 95%. Using a system of emitters and pipes, drip irrigation delivers tiny drops of water straight to the roots of plants. This system's goal is to minimize waste and use less water while giving the plants a steady supply of nutrients and water. Compared to conventional irrigation techniques, it offers a number of benefits, such as increased water efficiency, decreased soil erosion, and enhanced plant growth and yield. Water resources can be conserved by delivering water directly to the root zone, which lowers evaporation and runoff. Through regulated water flow, drip irrigation provides plants with a balanced uptake of water, making it the most beneficial technique for preserving food security and sustainable agriculture.

5.1. DRIP IRRIGATION: THE MOST EFFICIENT IRRIGATION SYSTEM

A growing issue in light of climate change and an increasing world population is the use of water in food production. The largest user of the limited water resources on our planet is agriculture, which is a resource that is becoming more and more scarce. 70% of the fresh water supply on Earth and 95% of all water withdrawals in certain developing nations are used for agriculture.

The most water-efficient irrigation method is drip irrigation, which can significantly lower a farm's water usage while improving crop quality and yields. However, drip irrigation isn't a panacea, just like any other irrigation system. Lower-value crops might not be a good fit for drip irrigation due to its labor-intensive and costly installation costs. Because of its many advantages, drip irrigation is a sustainable irrigation technique that is gaining popularity in agriculture. Drip irrigation delivers water directly to the plant's root in an organized manner, which can reduce water consumption and improve soil health. This is in contrast to traditional irrigation methods, which can lead to wasteful water usage, soil erosion, and loss of nutrients. Typically, a drip irrigation system is made up of a system of pipes connected to a reservoir or well for water. The water is then gradually released through drippers and emitters at the base of every plant, enabling the water to seep into the ground and get to the roots of the plants. By lowering the quantity of water lost to evaporation and runoff, this controlled water delivery significantly saves water and improves efficiency (Van der Kooji et al., 2013). Drip irrigation is the practice of directly applying water to a plant's root zone through a system of tiny tubes or pipes. It can be applied in many different contexts, ranging from modest home gardens to extensive agricultural endeavors. It can greatly reduce water usage and increase crop yields, making it especially helpful in places with scarce water resources or high-water costs. One benefit of drip irrigation is its automation, as timers and sensors can be used to regulate how much water is given to the plants. When compared to more conventional irrigation techniques, this can save time and labor. All things considered, it's a water-saving, effective, and sustainable way to irrigate plants that can also increase crop yields and quality. In comparison to other irrigation systems, drip irrigation provided the highest water use efficiency for tomato and cucumber crops, increasing water use efficiency by up to 70% and reducing energy costs by 50% when compared to conventional irrigation techniques (Niaz et al., 2009).

5.1.1. What Is Drip Irrigation?

A low-pressure method for precise water delivery is drip irrigation. It makes use of a network of emitters, sprinklers, pipes, and tubing. Water is applied in small amounts over an extended period of time using drip irrigation, which is also referred to as trickle irrigation or micro-irrigation. In contrast to conventional irrigation techniques like flood irrigation or center pivot irrigation, a drip or micro-irrigation system can lower energy costs by half and boost a farmer's water efficiency by up to 70% (Figure 5.1).

Figure 5.1. *Drip irrigation.*

Source: https://www.rainbird.com/sites/default/files/inline-images/stakedrippers544_650x434_0.jpg.

With drip irrigation, two major inefficiencies in other irrigation systems are eliminated: water runoff and evaporation loss are greatly reduced because the irrigation occurs at ground level. An end cap is used to seal off drip irrigation pipes or tubing, enabling water pressure to build up along the entire length of the pipe. As a result, pressure is created, forcing water to spray or drip from the emitter points. While PVC and galvanized iron pipes can also be used for drip irrigation, flexible polyethylene, sometimes known as "black roll pipe," is the most popular method of distributing water in drip irrigation systems. Drip irrigation lines can be buried or placed on top of soil, depending on the crop, kind of soil, and farming methods.

5.1.2. Is Drip Irrigation the Same as Micro Irrigation?

Micro-irrigation and drip irrigation are two low-pressure, precisely delivered irrigation techniques that differ slightly. Micro-spray is another term for micro-irrigation systems.

True drip irrigation systems actually cause water to "drip." Either from drip irrigation emitters positioned at the base of a plant or from the slits in the tubing itself. The same low-pressure delivery system and pipework used for drip irrigation are also used for micro-sprinkler irrigation. However, in a micro-irrigation system, the drip tubing is staked into the ground and small, low-pressure sprinkler devices are mounted on it.

The diameter of micro-irrigation sprinkler heads can be adjusted, with the most common setting being between 3 and 10 feet. It is possible to modify the spray pattern to suit varying pattern diameters, such as a partial or complete circle. The sprinkler heads can be changed, allowing operators to alter the spray pattern. Options include streams, misters, bubblers, and spray pattern sprinkler nozzle heads.

For crops with a wide distribution, like strawberries, or if farmers need to use their irrigation system to help cool their crops during hot weather, a micro-irrigation system can be beneficial. Sand soils are also better suited for micro-irrigation systems because they distribute water over a greater surface area. On sandy soils, however, water supplied through a true drip system tends to flow straight down instead of in a horizontal pattern, which may not provide enough soil moisture to support crop growth.

5.1.3. Drip Tape – How Does It Relate to Drip Irrigation?

One particular way that water is delivered in a drip irrigation system is through drip tape. Drip tape does away with the need for separate emitters by purposefully creating slits of cuts inside a drip tube pipe with thin walls.

The thickness and distance between slits of drip tape vary. Emitters should be placed at least 12 inches apart in sandier soil or among closely spaced crops. Drip tape is primarily used by farmers who produce cut flowers or annual vegetables. Installed, taken down, and disposed of in a single season, it is typically regarded as a single-use product.

5.1.4. Crops Most Commonly Irrigated with Drip Irrigation

Drip irrigation is a viable method of crop irrigation. However, the production of high-value specialty crops like vegetables or perennials like berries and fruit trees is where drip irrigation is most frequently utilized.

Drip irrigation kits and micro irrigation systems are frequently used by home gardeners for their vegetable gardens and landscape plantings.

Drip irrigation helps commercial vegetable row crop producers in a number of ways, including lower water and energy consumption, less weed pressure, and higher-quality produce that is less vulnerable to pest and disease pressure. This can help offset the higher cost of a drip irrigation system and reduce food waste during harvest time.

5.2. SUITABILITY OF DRIP IRRIGATION

Drip irrigation systems can be adopted under the following conditions:

5.2.1. Crops

With the exception of paddy crops, vegetables, fruits, sugarcane, and cereal crops are best suited for drip irrigation. Fruit crops and other high-value crops allow for an early return on the capital invested in installing a micro-irrigation system. Plantation crops like coconut, coffee, cardamom, cumin, citrus, grapes, and mango are also appropriate for these systems. Closely spaced crops will require more investment; otherwise, these systems can be installed with ease.

5.2.2. Slopes

It is possible to use drip irrigation on any cultivable slope. Typically, contour lines are used to plant crops and laterals. By using this technique, the impact of shifting land elevations on emitter discharge is reduced to a minimum.

5.2.3. Soils

For the majority of soil types, drip irrigation works well. For instance, irrigation water should be applied gradually on clay soils to prevent ponding and runoff.

Higher emitter discharge will be necessary on sandy soils to guarantee lateral water movement into the soil. It can be used to irrigate crops grown on slopes with limited soil depth and undulating land topography.

5.2.4. Irrigation Water

Emitter clogging is one of the main issues with drip irrigation. The diameter of the tiny holes in all emitters ranges from 0.2 to 2.0 mm, and using dirty water can clog them. Installing filters is therefore necessary to ensure that irrigation water is clear of sediments. Clogging can be partially resolved with micro sprinklers. Algae, fertilizer deposits, and dissolved chemicals that precipitate, like calcium and iron, can also cause clogging in water systems. Some of the materials may be eliminated by filtering. Saline water, which is of low quality, can also benefit from drip irrigation. Watering individual plants implies that the technique can be effective in increasing water use efficiency, making it most appropriate in situations where water resources are limited.

5.3. COMPONENTS OF A DRIP IRRIGATION SYSTEM

The same fundamental parts and design considerations are used in the setup of all drip irrigation systems, regardless of whether they use emitters, micro-sprinklers, or drip tape. The various parts of a drip irrigation system are shown in Figure 5.2.

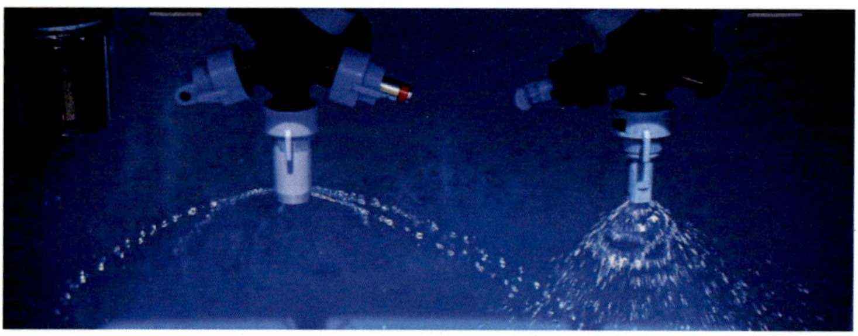

Figure 5.2. *Components of a drip irrigation system.*

Source: https://www.agrivi.com/blog/drip-irrigation-as-the-most-efficient-irrigation-system-type/.

5.3.1. Water Resources

Since drip irrigation reduces water loss from evaporation and runoff, it is a very efficient way to deliver water to crops. The water requirements of the crops being grown as well as the available water resources must always be taken into account when designing a drip irrigation system. When choosing a water source for a drip irrigation system, keep the following things in mind:

- **Water Source:** Determine where your irrigation system gets its water from. It might come from a surface water source, a well, rainwater collection, or a municipal water supply.
- **Water Quality:** Water quality is important for drip irrigation systems because blockages in the drip lines and emitters can interfere with the system's functionality. To find out the pH, salinity, and mineral content of the water, conduct tests.
- **Water Availability:** Checking the water flow rate and total volume available will help you determine how much water is available for irrigation. This will assist in figuring out the drip irrigation system's size and schedule.
- **Crop Water Requirements**: The type of crop, the weather, and the stage of growth will all affect how much water the crops require. See a crop specialist to find out how much water your crop needs to grow as healthily as possible.
- **Irrigation System Efficiency**: Take into account the drip irrigation system's efficiency as it will impact the quantity of water needed. By supplying water straight to the crop's root zone, drip irrigation systems can minimize water loss from evaporation and runoff.

In general, the implementation of a drip irrigation system requires careful planning and management of available water resources.

5.3.2. Pumping System

The water is transported from the source to the field by a pumping system. Pumping systems can be powered by electricity, gas, diesel, or even solar energy. Unlike center pivot or sprinkler irrigation systems, drip irrigation systems are low-pressure systems, meaning that even gravity can be used to pressurize drip lines. Farmers using a gravity-fed system can draw water from barrels or a mounted water tank.

5.3.3. Distribution System

The network of pipes, valves, and fittings that carries water from the pumping system to each drip emitter is referred to as the distribution system in a drip irrigation system. The responsibility for distributing water to the plants under irrigation in an even and effective manner rests with the distribution system (Figure 5.3). In a drip irrigation system, the distribution system's major parts are as follows.

- **Mainline**: This is the main pipe that distributes water to the different irrigation system zones from the pumping system.
- **Sub-main**: These are smaller pipes that emerge from the mainline and supply water to certain irrigation system locations.
- **Lateral Lines**: These are the system's tiniest pipes; they deliver water to each individual drip emitter.
- **Valves and Fittings**: These parts serve to control the water flow and guarantee that it is dispersed uniformly throughout the irrigation system.

The layout and dimensions of the area to be watered, the type of soil, and the water needs of the plants to be grown will all influence the design of the distribution system. Since this will affect plant growth and yield, it is crucial to make sure the distribution system is appropriately sized and designed to deliver water to the plants being irrigated efficiently and effectively.

5.3.4. Mainline Tubing

One essential part of a drip irrigation system is mainline tubing. Water is delivered to the plants via a large-diameter pipe that travels from the water source to the irrigation area. Mainline tubing is usually composed of polyethylene and comes in different diameters; the most popular sizes are 3/4 inch and 1 inch.

The drip irrigation system's mainline tubing serves as its structural support and is in charge of distributing water to each drip line or emitter. Burying it beneath the surface will shield it from harm and keep it hidden. Although the exact depth of burial will depend on the soil and climate where it is placed, it should generally be buried at least 6 inches deep.

Figure 5.3. *Pipelines in drip irrigation.*

Source: https://cdn.shopify.com/s/files/1/0663/9613/files/Comp2-16_480x480.jpg?v=1628836863.

A crucial factor to take into account during mainline tubing installation is the water source's pressure. Low pressure, usually between 20 and 30 PSI (pounds per square inch), is used by drip irrigation systems. Water waste and possible plant damage can result from emitters malfunctioning or even bursting due to excessive water pressure. A pressure regulator needs to be installed between the mainline tubing and the water source in order to control the water pressure. This will make it more likely that the system will function effectively and that the water pressure will stay within the intended range. All things considered, mainline tubing is a vital part of a drip irrigation system, and the system's success depends on its installation and upkeep.

5.3.5. Drip Tape or Drip Tube

In drip irrigation systems, drip tape is a kind of tubing that is used to deliver water directly and neatly to plant roots. The material used to make drip tape is a flexible plastic with tiny emitters or holes spaced evenly along its length. With the help of these emitters, water can be delivered to the plant root zones—where it is most needed—at a slow, steady rate.

Usually placed above ground, drip tape is arranged in rows alongside plants. It works well with a broad range of crops, such as fruits, vegetables, and flowers. An effective method of irrigating crops is with drip tape, which minimizes water waste and increases crop yields by applying water exactly where it is needed. Compared to other irrigation techniques, it uses less water and lowers the chance of soil erosion, making it an environmentally friendly choice.

5.3.6. Pressure Regulator

A drip irrigation system's pressure regulator is a crucial part. It is employed to control the system's water pressure and keep the emitters' flow rate steady (Figure 5.4). Generally operating at 20 to 30 PSI (pounds per square inch), drip irrigation systems are less pressurized than traditional irrigation systems. An excessively high-water pressure can force the emitters to release water too quickly, which can cause an uneven water distribution and possibly harm the plants.

Usually, the pressure regulator is placed in between the drip system and the water supply. It functions by bringing the incoming water pressure down to the appropriate level and keeping the system pressure constant. This guarantees that each plant gets the right amount of water and that the emitters release water at a constant rate. It is crucial to select the appropriate pressure regulator for your drip irrigation system as they differ in size and pressure range.

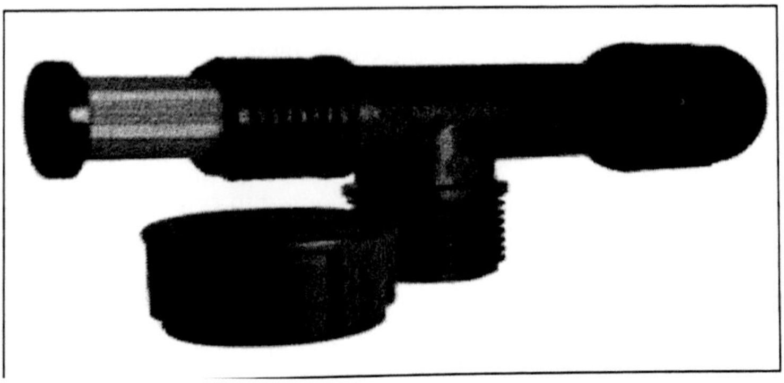

Figure 5.4. *Pressure regulator.*

Source: https://zenodo.org/records/7985739.

5.3.6.1. Drippers/Emitters

The emitters regulate the water flow that emerges from the laterals and are attached to them. The concept of energy losses due to abrupt changes in fluid flow velocity in the emitter path forms the basis of the design of single outlet emitters. The abrupt expansion and contraction of the disc element's intended flow path causes the subsequent change in velocity. The necessary configurations are made with a larger cross-sectional area in the form of a circle for flow expansion and a smaller cross-sectional area in the form of a rectangular channel for flow contraction. In order to decrease the hydraulic radius and flow velocity in the flow passage, the path's bottom is made flat to increase the wetted perimeter of the passage (Figure 5.5). When the flow channel abruptly widens to a circular path with a greater diameter, energy losses take place. Two mechanisms account for the energy loss: the first is the impulse momentum that occurs when water moves from a narrower rectangular passage to a wider circular passage, and the second is the movement of water from a wider circular section to a narrower channel section. Many eddies are created during this process, which results in a significant energy loss. A regulated flow of water emits in distinct drops nearly at atmospheric pressure as a result of a significant reduction in the high pressure from the lateral line that is transferred into the emitter. Poly-propylene or polyethylene are used to make the various types and designs of emitters that are sold on the market.

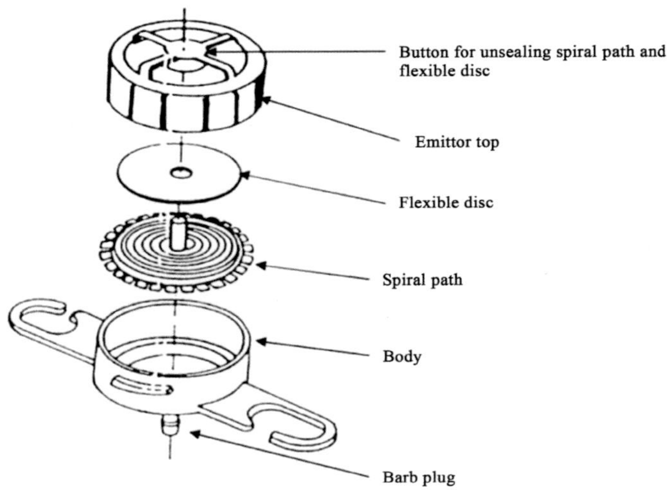

Figure 5.5. *Components of a point-source online pressure compensating dripper.*

Source: Ajai Singh. (2012). An Introduction to Drip Irrigation Systems. New Delhi Publishers. First Edition. ISBN: 978-93-81274 -18-7.

Emitters that are line-source or point-source can function above or below the surface of the earth. On-line or in-line emitters make up the majority of point source emitters. An in-line emitter passes through the entire required flow downstream of the emitter, which is the main distinction between on-line and in-line emitters. Due to obstruction caused by the barbs of on-line emitters, there is greater head loss along a lateral with on-line emitters than one with in-line emitters. Two crucial components of the quality and safety of drip irrigation systems are the percentage of the area that is wetted and the emitters' dependability in the face of obstructions and malfunctions. Durability, affordability, dependable performance, a relatively low rate of uniform discharge, and a sizable or self-flushing passageway to lessen or eliminate clogging are the characteristics of an ideal set of emitters.

5.3.6.2. Types of Drippers/Emitters

1. **In-Line Emitters:** The lateral line is fixed, and so are in-line emitters. After cutting the pipe, the dripper is fixed so that it forms a continuous row between the cut ends. Generally speaking, they have a straightforward thread or labyrinth flow path (Figure 5.6).

Figure 5.6. *In-line dripper.*

Source: Ajai Singh. An Introduction to Drip Irrigation Systems. New Delhi Publishers. First Edition 2012. ISBN: 978-93-81274 -18-7.

2. **On-line Emitters:** By drilling holes in the pipe of the right size, these are fastened to the lateral. They belong to the following categories:

- **Simple Type/Laminar Flow:** The discharge of this kind of dripper is precisely proportionate to the pressure. To dissipate energy, they have zigzag, labyrinth, vortex, and simple thread types as well as float type arrangements.
- **Turbo Key Drippers:** These come in 2, 4, and 8 lph discharge capacities and are composed of virgin and stabilized polymers. They are pressure-compensating and offer resistance to obstruction.
- **Pressure Compensating Drippers:** The pressure range covered by this kind of dripper is 0.3 atmospheres to 3.5 atmospheres, with a reasonably uniform discharge. Their superior rubber diaphragm regulates pressure, making them ideal for steep inclines and challenging terrain.
- **Built-in Dripper Tube:** In this system, when polyethylene pipes are extruded, drippers are welded inextricably to the inside of the tube. To avoid clogging, they feature an incredibly wide water passage and an independent pressure-compensating water discharge mechanism. Additional accessories consist of manifolds, joints, tees, rubber grommets, end plugs, and take-out/starter.

5.3.7. Filter

The size of your drip irrigation system and the local water quality will determine the kind of filter you select. In order to make sure your filters are operating correctly and not causing any blockages, it's critical to clean and maintain them on a regular basis. A drip irrigation system's filter is a necessary part (Prathysha and Suman, 2012). It aids in preventing drip emitter clogging, which can lower system efficiency and result in uneven water distribution (Figure 5.7).

Figure 5.7. *Filters.*

Source: https://www.researchgate.net/publication/371140308_Drip_Irrigation_System_An_Effective_Approach_for_Increasing_Water_Use_Efficiency.

5.3.7.1. Selection of Filters

The following criteria are taken into account when choosing filters for the drip irrigation system.

- The physical quality of water such as the concentration and pattern of the impurities, suspended solids and organic matter.
- The chemical nature of water such as pH level, and presence of sediments forming chemical elements and possible reaction with the injected fertilizers when fertigation is applied.
- Discharge and allowable head losses in the system.
- Reliability and durability of the filters.
- Cost of the filter.
- The total surface area of the filtration element is very important. The filtration area needed for moderately dirt water is in the range of 60–150 cm^2 for drip irrigation.

5.3.7.2. Types of Filters

A drip irrigation system can use a variety of filters, such as the following:

5.3.7.2.1. Settling Basins

In stream water used for irrigation, settling basins can remove suspended material ranging from sand (2,000 mm) to silt (200 mm). It gets rid of a lot of silt and sand. Basins are designed to minimize turbulence and allow water to retain at least 15 minutes as it travels from the basin inlet to the pumping system intake. Extended periods of retention are necessary to facilitate the smaller particle settling. A 57 lps stream needs a basin that is 1.2 m deep, 3.3 m wide, and 13.7 m long to provide a one and a quarter hour retention time. In order to reduce the effective retention time caused by short circuit current, the settling basin should be relatively long and narrow. Sitting basins should not be used if the water source is groundwater; however, if canal water is being used for irrigation through drip irrigation systems, they might be very beneficial.

5.3.7.2.2. Media Filter

When irrigating with water that has a high organic load, like water that has been reclaimed or pumped from open bodies of water, media filters are used. It is made up of fine gravel and pre-sorted sand that are arranged inside a pressurized tank. The layout for laying sand and gravel is displayed in Figure 5.8. Media filters can remove comparatively large amounts of suspended solids before cleaning is required because they are difficult to clog by algae. It can hold particles that are between 25 and 200 mm in size. In general, 10 to 18 lps of water should pass through each square meter of filtration surface area at the filters. After backwashing, a secondary screen filter ought to be placed after the media filter to stop contaminants from lingering. In most drip irrigation filters, the sand media filter denoted by the numbers is comprised of crushed granite (numbers 8 and 11) and silica sands (numbers 20, and 30). For numbers 8, 11, 16, 20, and 30, the mean granule size in microns is roughly 1,900, 1,000, 825, 550, and 340, respectively, for each media number.

The water exits the filter at the bottom after passing through a layered bed as it flows from the top. The organic contaminants gather in the filter tank after adhering to the media particles' surface. The pressure head loss across the filter is heightened when there is an excessive buildup of dirt in the layered bed; therefore, it is important to backwash the accumulated dirt promptly to prevent excessive head losses. The pressure drop across the filter indicates when backwashing is necessary. Backwashing involves changing the direction in which water passes through the filter and avoiding the effluent. The media filter's pressure drop should not be greater than 70 kPa, according to American Society of Agricultural Engineers recommendations. They don't get rid of bacteria or extremely tiny silt and clay particles.

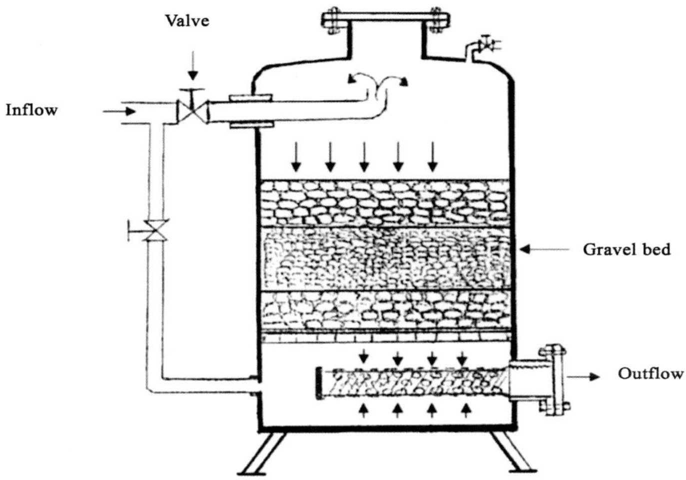

Figure 5.8. *Arrangements of gravels in media filter.*

Source: Ajai Singh (2012). An Introduction to Drip Irrigation Systems. New Delhi Publishers. First Edition. ISBN: 978-93-81274 -18-7.

5.3.7.2.3. Centrifugal Filter

Sand separator-assisted particle sedimentation is the most effective method of treating water containing soil particles. Hydro-cyclones, centrifugal filters, and sand separators are examples of synaminous devices that eliminate suspended particles with a specific gravity of more than 1.2. The centrifugal sand separator uses the centrifugal force of water entering a conic tank tangentially to separate sand and other heavy particles from the water (Figure 5.9). The centrifugal force propels the sand against the conic wall, where it settles and gathers in a collector at the bottom. When the collector is full, it is manually cleaned out. Through an outlet at the top of the separator, the clean water is released spirally. In a centrifugal filter, the top and bottom diameters of the conical shape are engineered to match the water flow. A significant amount of sand particles are successfully removed by the centrifugal filters. They are frequently positioned upstream of screen filters or media.

Figure 5.9. *Centrifugal filter.*

Source: https://captainpolyplast.com/images/userFiles/products/irrigation-filters2.jpg.

5.3.7.2.4. Screen Filter

If the water source is ground water, screen filters are installed right after the pumping unit and no additional filter is needed. Either plastic or metal is used to construct the casing. Water inlet, outlet, draining valve, and cover are its four apertures. It is composed of a synthetic cloth, metal, or plastic screen housed in a special house designed to restrict the largest possible particle size. The number of openings per inch of a screen determines its classification, with a standard wire size for every screen size (Table 5.1). For drip irrigation systems, most manufacturers advise using screens with a mesh count of 100 to 200. Less than 135 lps per square meter of screen openings is the typical discharge through the screens. The amount of open area in a standard 200 mesh stainless steel screen is just 58%, while the same size opening in an equivalent nylon mesh only has 24% of open area. The efficiency and operating limits of a screen filter are determined by the mesh size and the total open area.

Table 5.1. *Specifications of Filter Screens*

Mesh No.	Hole Size, Micron	Wire Thickness, Micron
40	420	250
50	300	188
80	177	119

100	149	102
120	125	86
155	100	66
200	74	53

Source: https://www.eng.uc.edu/~beaucag/DripIrrigation/Books/Singh,%20 Ajay%20-%20An%20introduction%20to%20drip%20irrigation%20systems-New%20Delhi%20Publishers%20(2012).pdf.

Therefore, when sizing a filter for a particular system discharge, it is crucial to take the percentage of open area into account. To get rid of solid contaminants like fine sand, dust, etc., it can be installed in series with a gravel filter. from the water. Water containing inorganic contaminants can be treated with screen filters. A high concentration of biological and organic contaminants may quickly clog the screen. The screen filters can be manually or automatically cleaned, just like media filters (Figure 5.10).

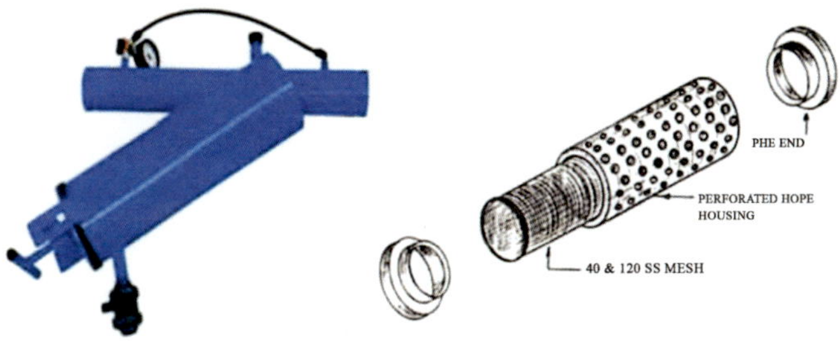

Figure 5.10. *Schematic views of components of screen filter (Jain irrigation systems).*

Source: Ajai Singh. An Introduction to Drip Irrigation Systems. New Delhi Publishers. First Edition 2012. ISBN: 978-93-81274 -18-7.

5.3.7.2.5. Disk Filters

This works better for water that has a combination of organic waste and inorganic solid particles as contaminants. Either plastic or metal make up the casing. The filtering element consists of a series of ring-shaped, grooved disks that are stacked and secured together with a threaded cap. As water passes through the grooves, it is filtered. Finer particles and organic debris stick to the inner grooves of the stack, while coarser particles are trapped on its exterior. When in back flushing mode, the flow direction is reversed and the discs are compressed together during filtration. When the discs begin to rotate, all retained impurities are eliminated. The discs are pressed together after being assembled in a cylinder. They can be installed in batteries in order to handle high flow rates. When comparing screen filters that meet the same specifications, the dirt retention capacity of disk filters is significantly higher. Before the water enters the system, these filters are frequently employed as the last stage of filtering.

5.3.7.3. Filter Characteristics and Evaluation

Determining the pressure differential across the filter and the filtration efficiency are two aspects of the field evaluation of a drip irrigation system. Assessing the degree of emitter plugging is another indirect way to gauge the efficacy of the filtration system.

The filtration efficiency (F_r) of a filtration system is estimated by the following formula:

$$F_r = \left(1 - \frac{S_{out}}{S_{in}}\right) \times 100$$

Where; S_{out} = the concentration of suspended solids (mg/l) in filter outlet; and S_{in} = the concentration of suspended solids (mg/l) in filter inlet.

After the system is turned on, the water sample needs to be taken at least 30 minutes later. The differential pressure that occurs before and after filter backwashing is the second factor to consider when evaluating filters in the field for a drip system. A loss of flow may transpire if the differential pressure is extremely high, thereby impacting the downstream pressure. The degree of emitter plugging is another method for assessing the system's efficacy. The uniformity of water application can be impacted by a minor percentage of emitter plugging. In order to lower the higher system pressure to the necessary system pressure regulators are typically utilized. It only uses one method to regulate the pressure, i.e., from high to low.

5.3.8. Emitter Tubing

With built-in emitters or drippers spaced regularly along its length, emitter tubing is a type of tubing used in drip irrigation systems. These emitters irrigate plants right at their roots by releasing water gradually and uniformly. Usually composed of polyethylene, emitter tubing is available in a range of sizes and emitter spacing configurations to accommodate different plant and soil types. It is simple to install and can be buried underground for a more permanent installation or placed on the ground. Emitter tubing has the benefit of precisely delivering water to each plant, which minimizes runoff and lowers the possibility of overwatering. Furthermore, the emitter tubing's gradual and even water release guarantees that the water seeps deeply into the soil, encouraging strong root development. To avoid clogging and guarantee ideal water low, it is crucial to make sure the emitter tubing is regularly flushed and maintained. For the drip irrigation system to operate properly, emitter tubing should also be periodically checked for damage and replaced as necessary.

5.3.9. Sprayers and Emitters

Two different kinds of parts are used in drip irrigation systems: emitters and sprayers. Although they both have the same objective of providing water to plants, they do it in different ways. Water can be sprayed in a fine mist or spray pattern using a sprayer. They are usually utilized in places like vegetable gardens or fruit trees where there are more plants or where more water is required. Sprayers can be placed above or below the soil's surface and connected to a drip irrigation system with hoses or tubing.

Conversely, emitters are tools that send water straight to the roots of plants; typically, they do this by creating tiny holes or charnels within the emitter. Emitters are usually placed at or close to the base of the plant and work best on smaller plants, such as flowers or shrubs. Two additional types of emitters are pressure-compensating and non-pressure-compensating emitters. While non-pressure-compensating emitters may deliver more or less water based on the system pressure, pressure-compensating emitters maintain a constant flow rate regardless of the water pressure.

Ultimately, there are benefits and drawbacks to both sprayers and emitters, and which one to use will depend on the drip irrigation system's design as well as the particular requirements of the plants that need to be watered.

5.3.10. Chemical Injector

Chemicals can be injected into a drip irrigation system using a chemical injector. It is commonly used to treat water with fertilizers, pesticides, or other chemicals to enhance plant growth, guard against illnesses or pests, or offer other advantages. The injector works by creating a pressure differential that draws the chemicals into the water stream. It is usually installed between the drip irrigation system and the water source. One way to regulate the amount of chemical injected is to alter the concentration of the chemical solution or the injector's flow rate. Drip irrigation systems can be used with a variety of chemical injectors, such as peristaltic pumps, diaphragm injectors, and venturi injectors. The kind and concentration of chemicals being used, the size of the drip irrigation system, and the intended flow rate are just a few examples of the variables that will determine which kind of injector is appropriate for a given scenario. When handling chemicals, it's critical to use chemical injectors correctly and adhere to all safety precautions. Additionally, to make sure the injectors are operating correctly and to avoid contaminating the irrigation system or the crops being grown, it is crucial to clean and maintain them on a regular basis.

5.3.11. System Controller

The system controller in a drip irrigation system is in charge of controlling the water flow to the plants. The controller modifies the water flow based on information it receives from a variety of sensors, including weather and soil moisture sensors (Figure 5.11). Drip irrigation systems use a variety of controller types, such as:

- **Manual Controllers**: These are simple controllers that need the water flow rate to be manually adjusted. They might work well in small-scale systems where watering requirements are regular and predictable.
- **Automatic Controllers**: Utilizing sensors, these controllers determine the soil's moisture content and modify the water flow rate accordingly. In order to modify watering schedules according to the current weather, they might also have weather sensors.
- **Smart Controllers**: In order to optimize watering schedules and reduce water consumption, these controllers analyze data from a variety of sources, including soil moisture sensors, weather sensors, and plant data, using sophisticated algorithms and machine learning.

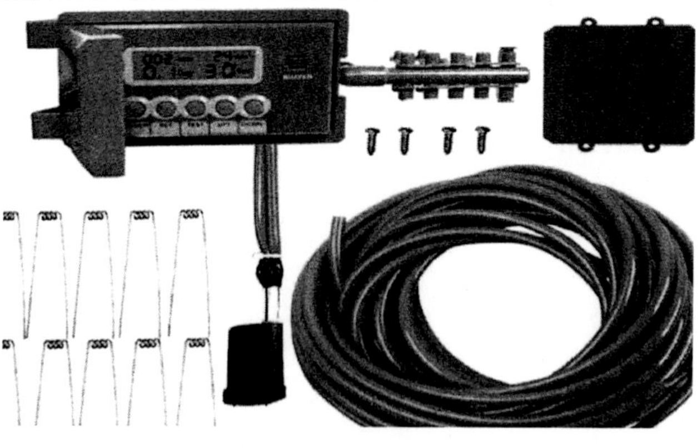

Figure 5.11. *System controller.*

Source: https://www.researchgate.net/publication/371140308_Drip_Irrigation_System_An_Effective_Approach_for_Increasing_Water_Use_Efficiency.

The main objective is to minimize waste and reduce water usage while making sure plants receive the right amount of water at the right time, regardless of the type of controller used.

5.3.12. Backflow Preventers

Drip irrigation systems require backflow preventers as a necessary component. By keeping water from returning to the main water supply system, these devices shield it from contamination. In drip irrigation systems, backflow preventers come in a variety of forms, including

- **Pressure Vacuum Breaker (PVB)** - An air inlet valve and a spring-loaded check valve are used by this backflow preventer to stop backflow. It is appropriate for small commercial and residential applications and is usually installed above ground.
- **Reduced Pressure Zone Device (RPZ)** -There are two check valves and a relief valve on this backflow preventer. It can be used for larger commercial and industrial applications and is installed underground.
- **Double-Check Valve Assembly (DCVA)** -This backflow preventer is appropriate for smaller commercial and residential applications and features two check valves that are spring-loaded. Usually, it is installed above ground.

Whatever kind of backflow preventer is used, it must always be installed, maintained, and tested on a regular basis to make sure it is operating as intended. This will keep your drip irrigation system operating smoothly and help prevent contamination of the water supply. The way that drip irrigation's parts come together allows for the effective irrigation of crops.

5.3.13. Goof Plugs, Valves, Flush Valves, and More

Depending on its complexity, a drip system can have several parts. Goof plugs fill in unnecessary holes in mainline tubing. Various sections are turned on and off using valves. In the event of a blockage, sediment or debris can be flushed out using drip lines or flush valves at the end of mainlines. Two lengths of drip tape or mainline pipe are joined together by connectors. A larger diameter tube is connected to a smaller one using adapters.

5.4. WETTING PATTERN IN DRIP IRRIGATION

The area that receives moisture from the water emitted by the drip emitters is known as the drip irrigation system's wetting pattern. Numerous variables, such as the kind of emitter, the water flow rate, the kind of soil, and the emitter spacing, will affect the wetting pattern. Drip irrigation systems are typically made to produce a pattern of wetting that corresponds to the root zone of the plants they are watering. This indicates that the emitters are arranged in a way that guarantees water gets to the plant roots directly. The shape of the wetting pattern is usually rectangular or circular, and the distance between the emitters determines how big the pattern gets.

Because water will spread out more readily in high permeability soil, the wetting pattern may be shallower and wider (Figure 5.12). Because water penetrates low permeability soils more slowly, the wetting pattern becomes narrower and deeper (Ahmed et al., 2016).

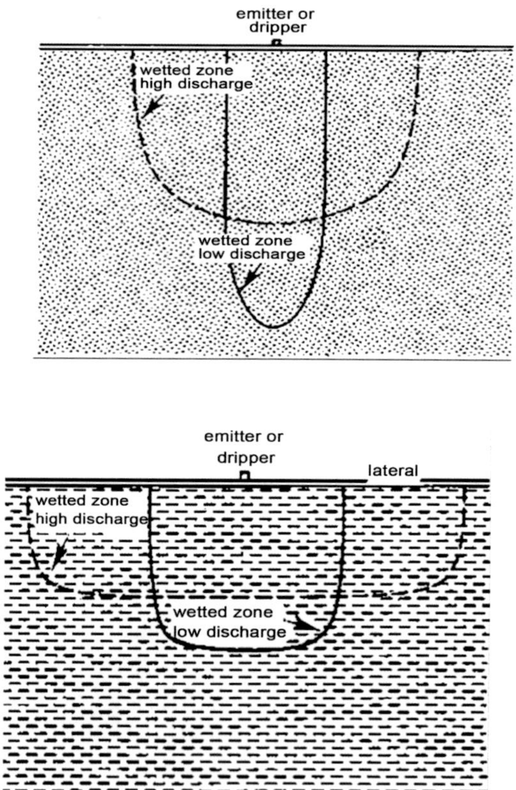

Figure 5.12. *Sand and clay soils with high and low discharge rates exhibit different wetting patterns.*

Source: https://vikaspedia.in/agriculture/agri-inputs/farm-machinary/drip-irrigation-system.

It is significant to remember that when the soil gets saturated and water starts to spread laterally, the wetting pattern may alter over time. In order to guarantee uniform water application, this may lead to uneven wetting patterns and necessitate system modifications. The wetting pattern can be guaranteed to stay constant over time with regular system maintenance and monitoring.

5.5. DRIP IRRIGATION IN INCREASING WATER USE EFFICIENCY

The effectiveness with which water is used in a specific system or process is measured by its water use efficiency. It is commonly defined as the volume of water required to generate a given unit of output or yield, like a tone of steel, a kilogram of crop, or a liter of milk (Martinez and Reca, 2014).

By providing water directly to the plant's root zone, drip irrigation systems aim to minimize water loss due to evaporation, runoff, and deep percolation (Wallace, 2000). These are a few ways that drip irrigation systems can increase the effectiveness of water use.

- **Precise Water Delivery**: By delivering water straight to the roots of plants, drip irrigation systems reduce water loss from evaporation, wind drift, and surface runoff. This maximizes the amount of water available for plant growth and enables precise water delivery, minimizing water waste.
- **Reducing Evaporation**: By applying water gradually, a drip irrigation system increases soil absorption efficiency and decreases evaporation loss. This is especially crucial in places where evaporation rates can be high due to high temperatures and low humidity.
- **Improved Plant Growth**: Water is delivered to the rootzones directly by drip irrigation systems, guaranteeing that the plants get the moisture they require to develop and flourish. Higher plant yield and better plant growth may arise from this (EI-Rahman, 2009).
- **Reduced Weed Growth**: By supplying water only to the roots of plants, drip irrigation systems reduce the amount of water that weeds can access. This may lessen the competition for water resources and the growth of weeds.
- **Reduced Nutrient Leaching**: A drip irrigation system lowers the chance of nutrient leaching by distributing water slowly and evenly. In other words, plants are able to absorb nutrients more efficiently, which lowers the need for fertilizer and the risk of water pollution.

5.6. ADVANTAGES OF DRIP IRRIGATION SYSTEM

Using drip irrigation, plants are irrigated by slowly and precisely applying water directly to their roots. The following are a few benefits of drip irrigation:

- **Water Use Efficiency**: The best way to irrigate plants with minimal water usage is through drip irrigation. It ensures that every drop of water is used efficiently by reducing evaporation and runoff and delivering water straight to the roots of plants (Sinobas and Rodriguez., 2012).
- **Improved Plant Growth**: Water is directly delivered to the roots of plants through drip irrigation, which encourages healthier and more robust plant growth and, as a result, increases yield by 20–60% (Sivanappan et al., 1974).
- **Reduced Weed Growth**: Drip irrigation inhibits the growth of weeds in the garden by irrigating the plants alone, excluding the surrounding soil.
- **Reduced Disease and Pest Problems**: Because drip irrigation doesn't wet plant leaves and stems, which can encourage the growth of harmful organisms, it lowers the risk of disease and pest issues.
- **Savings in Time and Labor**: More work and less time are needed for drip irrigation than for other types of irrigation.
- **Conserves Energy**: Because drip irrigation requires less pressure to deliver water than other irrigation techniques, it uses less energy overall.
- **Flexibility**: All types of soil can be used with drip irrigation, and it can be tailored to fit gardens of various sizes and shapes.

All things considered, drip irrigation is a very successful and efficient way to water plants that has several advantages for farmers and gardeners.

5.7. DRAWBACKS OF DRIP IRRIGATION SYSTEM

There are a few more possible downsides to think about:

- **Initial Installation Costs**: Installing drip irrigation systems can be more costly than installing other irrigation techniques. This is a result of the requirement for specific tools and supplies.
- **Clogging**: Debris or accumulation in the water can cause clogging in drip mineral irrigation systems. This may lessen the water flow and have an impact on how well the system functions.
- **Maintenance**: For drip irrigation systems to function properly and

avoid clogging, regular maintenance is necessary. This may entail inspecting for leaks and cleaning filters.

- **Uneven Water Distribution**: Drip irrigation systems can cause uneven water distribution if they are not installed correctly, which can cause some plants to be overwatered or underwatered (Lamm., 2002).
- **Vulnerability to Damage**: Animals such as rodents and pests can cause damage to drip irrigation systems by puncturing or chewing through the tubing.
- **Dependency on Water Pressure**: For drip irrigation systems to work effectively, there must be constant water pressure. Variations in water pressure can impact the system's functionality.

5.8. PLANNING AND DESIGN OF DRIP IRRIGATION SYSTEMS

To provide the necessary amount of irrigation water, a drip irrigation system's planning and design are crucial. The amount of water that a plant takes up from the soil and the amount that evaporation of the soil near the root zone in a given day determines the plant's daily water requirements. The amount of leaf area, growth stage, climate, soil conditions, etc., all affect how much water a plant takes in. But the soil-water budget method or methods based on plant indicators in the soil can be used to calculate the amount of water needed and the timing of irrigation. The desired hydraulic pressure in the pipe network must be maintained in order to apply the necessary amount of water uniformly to every plant in the field. The size of the pumping unit, emitter discharge, main pipeline, manifolds, laterals, and sub-main are all part of the drip irrigation system's hydraulic design. The following are important steps in drip irrigation system design.

- Collection of general information
- Layout of the drip irrigation system
- Crop water requirement
- Hydraulic design of the system

5.8.1. Collection of General Information

When designing a drip irrigation system, general information about the water source, crops to be grown, topography, soil type and texture, and climate data are crucial. The selection of the emitter and its rated discharge rate will be aided by the knowledge of the soil type. If pan evaporation data are not available, then climatic data are needed to estimate the potential evapotransiration.

5.8.2. Layout of the Drip Irrigation System

The drip irrigation system's layout is determined by the field's length, width, and water source location. The crop's characteristics determine the number and spacing of the laterals. One lateral can serve one row, two rows, or more than two rows. Usually, laterals are laid across the slope or along the field's contour lines, while the sub-main line follows the slope.

5.8.3. Crop Water Requirement

The daily crop water requirement is estimated by using the equation

$$V = E_{PAN} \times K_p \times K_C \times A \times W_p$$

Net volume of water to be applied,

$$V_n = V - R_e \times A \times W_p$$

Where;

V = volume of water required, liters

E_{PAN} = mean pan evaporation for the month, mm/day

K_c = crop coefficient

K_p = pan factor

A = area to be irrigated, m^2

R_e = effective rainfall, mm

W_p = Percentage wetting

The crop growth stage and season have an impact on the crop coefficient (K_c). Drip irrigation system design takes crop productivity into consideration.

5.8.4. Hydraulic Design of Drip Irrigation System

Hydraulically steady, spatially varied pipe flow with decreasing discharge characterizes drip irrigation lines. In relation to the line's length, the overall discharge in the main, sub-main, and lateral sections is declining. While the main line is designed based on input pressure, required pressure, and the slope of the energy gradient line that will give a total energy higher than that of required at any sub-main for irrigation, the lateral and sub-main lines share similar hydraulic characteristics. A drip irrigation system that has all of its emitters delivering the same amount of water in the allotted amount of time is ideal. Although this objective is challenging to accomplish in the field, the hydraulic design can regulate the emitter flow variation in the lateral line. The pipes used in drip irrigation systems are smooth and composed of plastic. The Hazen-Williams empirical equation or the Darcy-Weisbach equation can be used to calculate the pressure drop caused by friction or frictional head loss.

Flow carried by each lateral line (Q_l) = Discharge of one emitter × No. of emitters per lateral

Flow carried by each sub-main line (Q_s) = Q_l × No. of lateral lines per sub-main

Flow carried by main line (Q) = Q_s × No. of sub-main lines

The hydraulics of pipe flow is used to determine the diameter of the main, sub-mains, and laterals. The following process is used to estimate the sizes of mains, sub-mains, laterals, and pumps

Total head loss due to friction (H_f) = friction head loss in mains + friction head loss in sub-mains + friction head loss in laterals + friction head loss in accessories and fittings

Operating pressure head required at the dripper $=H_e$

Total static head = H_s

Total pumping head (H) = $H_f + H_e + H_s$

5.8.4.1. Development in Design of Lateral Pipe

A high degree of uniformity in the application of water is a requirement for a drip irrigation system's proper design. The pipes that the emitters are placed on are called lateral lines. They are often constructed of LLDPE with a diameter ranging from 12 to 20 mm, and they receive their water from the sub-main line. The lateral pressure determines the emitter discharge. The distribution network's low discharges and low pressure heads enable the use of smaller, lower-rated pipes, which lowers expenses. Water is applied slowly and disperses over time, reducing peak discharges and necessitating smaller pipes and pumps, which reduce wear and extend network life. The land slope, the emitter's characteristics, and the pressure variation in the laterals are the main causes of the irregularity in emitter discharge. Both partial or total emitter plugging and the temperature of the water have an impact on an emitter's discharge. It is challenging to alter the pipe network's design and system layout once it is installed. Therefore, it is crucial to ensure that pipe sizes and frictional losses are calculated accurately.

Numerous studies on the design of drip irrigation laterals have been published in peer-reviewed journals. We have made an effort to include the different approaches taken when designing laterals. If we go back, Christiansen (1942) employed poly plots, or graphical methods. Due to the widespread availability of computers, this approach became outdated, and Wu and Gitlin (1974) created a computer model based on the average discharge. In order to determine the pressure at any point along the lateral, Keller and Karmeli

(1974) developed a computational model that tested a wide range of emitter exponent values. Assuming that the emitter discharge is constant along the lateral simplifies computations significantly. The conservation of energy and the law of continuity have been used to create mathematical models. Perolt (1977) converged the solution through an iterative procedure based on the back step method. Solomon and Keller (1978) attempted the piezometric curve-based computation. Researchers started to become interested in the hydraulic analysis of drip irrigation laterals as a way to improve design efficiency. A methodical numerical process known as the finite element method (FEM) has been employed to examine the hydraulics of the lateral pipe network. Bralts and Segerlind (1985) created a finite element computer model to examine micro-irrigation sub-main units. Their method's benefits included low computer storage requirements and broad application to a vast micro-irrigation network. Bralts and Edwards (1986) evaluated micro-irrigation sub-main units in the field using a graphical method and compared the outcomes with computed data. A spatially variable discharge function was one of the alternative treatments that Yitayew and Warrick (1988) offered as part of the fundamental solution for drip irrigation lateral design. They presented two analyses: a numerical Runge-Kutta solution and an analytical solution for non-linear differential equations.

Bralts et al. (1993) used a microcomputer program to further analyze the design of drip irrigation systems. This program supplied the system's pressure head and flows at every emitter. In addition, the program offered a number of helpful statistics as well as an assessment of hydraulic design using basic economic and statistical criteria. A method for creating virtual nodes by merging several emitters and lateral lines was presented by Bralts et al. (1993) in response to the high number of laterals in such a system. Following the development of these nodal equations, nodal pressure heads at all emitters were numerically solved using the FEM technique. This node number simplification simplified the equations and made it simple to compute on a personal computer. The majority of numerical techniques for analyzing drip irrigation systems use the back step procedure, which is an iterative method for calculating pressure heads and flow rates in a lateral line using an assumed pressure at the line's end. Because drip irrigation network programs involve a large number of matrix equations, they require a large amount of computer memory and computation time. Hills and Povoa (1993) also developed a mathematical model for a microcomputer to analyze the hydraulic characteristics of flow in a drip irrigation system. For the numerical solution of non-linear second order differential equations, Bralts et al. (1993) employed the finite element method. A thorough explanation of alternative techniques is given in their articles. The pressure head and discharge distribution along lateral and sub-main pipes were also analyzed using the finite element method by Kang and Nishiyama (1994). The operational pressure heads of the lateral that corresponded to the necessary uniformity of water application were located using the golden section search method. A straightforward formula for lateral hydraulics calculation was presented by Valiantzas (1998). The computations are predicated on the supposition that the emitter outflow profile is not uniform. A computer model based on the control volume method and the back step procedure was presented by Lakhdar and Dalila (2006) as a way to solve non-linear algebraic equations simultaneously. A different iteration procedure that reduced the complexity of the drip irrigation system design model was created.

5.8.4.2. Design of Sub-Main

The concepts used in the design of lateral hydraulics are comparable to those of sub-main line hydraulics. Assuming the laterals are analogous to emitters on lateral line, the sub-main hydraulics characteristics can be calculated. Since PVC and HDPE pipe are typically used, the hydraulic characteristics of sub-main and mainline pipes are typically taken to be smoothly flowing. The roughness coefficients (C) for Hazen-Williams range from 130 to 150. The sub-main energy loss can be calculated using techniques similar to lateral computations.

5.8.4.3. Design of Main

The submain inlet is typically where the pressure controls or adjustments are located. Therefore, system uniformity shouldn't be impacted by energy losses in the mainline. In the event of a main line, there is no outlet, hence the reduction factor is not multiplied. The formulas for Hazen-Williams, Darcy-Weibach, or Watters and Keller are used in the same manner to determine the frictional head loss in the main pipeline.

5.8.5. Design Tips for Drip Irrigation

Even though the design process has already been discussed, there are a few things to keep in mind when organizing a drip irrigation system for crop production. The uniformity of fertilizer and chemical application will be impacted by the drip irrigation systems' water application efficiency, since drip systems are also utilized for fertigation and chemical injection. These are a few things.

- Lateral lines shouldn't be longer than what is advised. Water application will be uneven and of poor uniformity if laterals are too long. Verify the amount of water supplied by the emitters in the final segment of the lateral as an added precaution.
- It is important to carefully choose the main and sub-main line sizes to prevent excessive pressure losses and velocities. Because of the

too-small diameter, excessive water velocity in the lines can cause a water hammer, which can harm the main and submain lines.

- Use pressure compensating emitters where necessary.
- Choose the irrigation area zone in conformity with the existing water pump.
- The uniformity of water application in drip irrigation systems is reduced due to emitter clogging. The kind of emitters, the water filtration system, and the water quality all affect how clogged the system is. An irrigation system ought to have an injection port so that chlorine can be injected to clean the systems on a regular basis.
- A suitable backflow mechanism must be installed in any irrigation system that injects chemicals or fertilizers in order to prevent contaminating the water source.
- Monitoring pressure variations in the irrigation line is necessary. Emitter blockage or any other obstruction, like a broken valve or pressure regulator, can result in a rise in pressure. A drop in pressure could be the result of a burst pipe, leaking valves, or improper flush valve closure.

5.9. INSTALLATION, MANAGEMENT AND MAINTENANCE OF DRIP IRRIGATION SYSTEMS

The malfunction of well-designed systems can be brought on by improper drip irrigation system maintenance. The systems are more complex and require careful maintenance due to the emitter's sensitivity to clogging, the network of lateral pipes, the filtration units, and the fertigation systems. If a system is installed correctly, it is maintainable. An easy installation can make maintenance easier. Let's talk about drip irrigation system installation first.

5.9.1. Installation

We took it for granted that there was already a water supply. There are three steps involved in installing a drip irrigation system: (i) fitting filter units; (ii) laying mains and sub-mains; and (iii) laying laterals with emitters.

- **Installation of Filter Unit:** When choosing the location of the filter unit, the following factors should be taken into account.
  - There should be minimum use of fittings such as elbows and bends.
  - The filter unit should be located on the delivery side of the pump.

  - The filter size should be in accordance with the capacity of the system.

- **Installation of Mains and Sub-Mains:** Both mains and sub mains are buried, with the exception of fully portable systems, at a minimum depth of 50 to 60 cm, so as to be unaffected by heavy agricultural machinery or cultivation. Sand padding can be used in trenches prior to PVC pipe installation in order to prevent compacted soil expansion and contraction. It is customary to install permanent underground sub-mains, even for systems with portable laterals removed at the end of each season. Sub-mains typically run in the opposite direction of the rows.
- **Laying of Laterals:** In a field that slopes, laterals are typically positioned along contours and laid on the ground's surface. To allow the twisting that has formed in the reel to be released, laterals placed from the reels should be left on the ground for five to six hours. The lateral's downstream end can be sealed by merely folding the pipe back and using an end plug to seal it. This is easily slippable in order to flush. The lateral should be placed straight into the sub-main for the lowest-pressure system connection. It is preferable to use a metal twist drill or a wood auger bit to create a slightly undersized hole in the sub-main. The tapered tool is used to enlarge the hole, and once the taper is removed, the lateral is swiftly inserted. At the end, the lateral is cut at an angle of roughly 450. Remember that the lateral shrinks by 1 meter for every 100 meters when determining the lateral length for installation. Punch is used to create holes on the lateral in the desired spacing. After the laterals are spread out on the ground, the on-line point source emitters should be adjusted so that the outlets face upward.

5.9.2. Testing of the System

Regular inspections will guarantee that the drip systems are operating correctly. Depending on the characteristics of the system's components and the water quality, this can be done on a weekly, monthly, or seasonal basis. At the very least, every three years, pump efficiency must be checked. More than 80% pump efficiency is preferred because it will reduce the steadily rising cost of energy. Prior to testing, all of the flush and gate valves need to be open. Backwashing the filter is recommended until clean water exits through the flush valve. After flushing the sub-main, the flush valve needs to be closed. At the sub-main, the air release valve's functionality is examined. By controlling the bye-pass valve, the filter's inlet and outlet pressure should be achieved. Emitter discharges are measured at standard operating pressure in order to ascertain emission uniformities and discharge variation.

5.9.3. Maintenance of Drip Irrigation System

Taking a sample of the water to be used from the source is the first step in maintaining a drip irrigation system. This will assist in figuring out how often the drip system needs to be flushed. When getting the maintenance program ready, there are three different kinds of clogging issues that need to be addressed. Physical, chemical, and biological are these. The system may be physically hazardous due to sand and silt. To remove particles larger than the maximum permitted particle size for the emitters being used, the proper filtration technique is needed. One of the risks associated with chemical clogging is the accumulation of an excessive amount of salt. More than 0.1 parts per million of sulfur dioxide in irrigation water can promote the growth of sulfur bacteria in the irrigation system. The development of bacterial slimes and algae inside drip lines and emitters causes biological clogging hazards. They block the emitters when combined with clay particles. A drip irrigation system needs to have regular preventive maintenance in order to function properly. The following procedures comprise general maintenance.

- The emitter functioning, wetting pattern and leakage of pipes, valves, and fittings should be checked regularly.
- The placement of emitters should be ascertained. If the placement is disturbed, put the emitters in proper location.
- Leakages through filter gaskets in the lids, flushing valves and fittings, etc., should be monitored regularly

5.9.3.1. Flushing the Mainline, Sub-Main and Laterals

Flush the main and sub-main lines before attaching the drip laterals during the installation process. It is advised to flush the dirt from lines by releasing the ends of the main and sub-main lines and flushing them for at least 10 minutes in the event of a failure or burst in the main or sub-main pipe. Generally speaking, filters don't stop water-soluble materials and suspended particles from entering. A layer of sediment or organic matter gradually accumulates in the drip laterals and main line over time. After two to four weeks of initial operation, it is advised to flush the lines because the end of the lateral line accumulates the most sediment. It is generally advised to flush at least once every fourth operation when using dirty water (ponds or canals) for irrigation, and three to four times during the season when using high-quality water (shallow or deep tube well). In order to achieve the intended performance during the crop-growing season, daily checks on the media/screen filter's performance as well as the system's pressure and flow rate should be made. It is possible to make weekly observations regarding screen and strainer cleaning and emitter clogging. Every month, attention needs to be paid to the media/screen filter's maintenance requirements, the comparison of the lateral pressure to the designed pressure, and the emitter discharge rate measurement at specific locations.

5.9.3.2. Filter Cleaning

A drip irrigation system's filter is its central component, and if it fails, the entire system will become clogged. The irrigation system's lifespan can be extended by keeping it clean and in working order. The accurate indicator of when to clean the filter is the pressure differential across the filter. Here are some instructions for maintaining the filters.

- Examine screens and seams to ensure that they are undamaged.
- To stop lateral water leaks, make sure the filter's body, cover, and screen are all properly tightened. Make sure the disc assembly is securely tightened if you're using a disc-style filter.
- It is bad for the filter and the filtration quality to operate with a partially clogged filter. The system needs to be flushed and cleaned when the pressure differential reaches about five meters.
- Avoid entry of dirt into the pipe when the filter is opened for cleaning.
- Prior to the commencement of irrigation, flush the filter. Clean the filter more frequently if it is extremely dirty. Perhaps automatic back-flushing is necessary.

The following lists the cleaning methods for different types of filters.

- **Media Filter:** Every day, the filter needs to be backwashed for five minutes to get rid of the silt and other debris that accumulated from the irrigation the day before. Every seven days, when backwashing, the backwash water ought to flow through the lid rather than the backwash valves. Without endangering the filter candles, the sand in the filter bed is raised to them. Any dirt that has gathered far down in the sand bed will break free and escape through the lid with the water. Add enough chlorine to the container to remove the aggregates and algae, then let it soak for a full day. Pour off the water after rinsing, then store the filter somewhere dry and airy. Regular maintenance involves repairing and lubricating to prevent rusting.
- **Screen Filter:** It is necessary to inspect the screens and sieves to make sure they are in good condition. Make sure the filter's screens,

cover, and body are properly tightened to stop lateral water leaks. To clear the dirt and silt, the flushing valve on the filter link is opened prior to the drip irrigation system being turned on. After being removed from the filter, the filter element is cleaned under running water. The rubber seals are removed from both sides. Care must be used when replacing the rubber seals to prevent damage.

5.9.3.3. Chemical Treatment

Chemical treatment is advised when standard flushing is insufficient to thoroughly clean the system and provide consistent emission. Emitters and bi-wall orifice clogging or plugging is caused by precipitation and build-up of specific dissolved salts, such as bi-carbonates, carbonates, iron, calcium, and manganese salts. The clogging is also caused by microorganisms and the associated sulphur and iron slimes produced by bacteria and algae. Chemical treatment is typically used to clear clogged emitters. Common chemical treatments for drip irrigation systems involve adding acid or chlorine to the water source. The level of clogging and water quality determine how often chemicals are treated. Generally, a 10-day acid treatment and a 15-day chlorine treatment are carried out.

- **Acid Treatment:** The precipitation brought on by dissolved salts can be removed by flushing with an acidic solution. The drip system is injected with 30% hydrochloric acid at the rate recommended by the water quality analysis report. Prior to the acid treatment, the system's discharge is checked and all submains and laterals are flushed. A tenth (1/10) of the injector pump's maximum discharge is equal to a solution, which is prepared and treated until the farthest lateral shows a pH of 4. Irrigation is then maintained for a further 30 to 60 minutes.
- **Chlorine Treatment:** To break down organic matter and stop bacteria and algae from growing, oxidation treatment is used. As an oxidizing agent, chlorine destroys bacteria and other organic materials and stops accumulation. The two most prevalent forms of chlorine are calcium hypochlorite (a solid) and sodium hypochlorite (a liquid). There are three primary ways to apply chlorine:
  – **Continuous Treatment:** This is the most straightforward way to clear blockages in a drip irrigation system. To keep organic material out of the system, the water is continuously treated. The necessary concentration changes based on the quality of the water. Typically, the injection rate is set so that, at the end of the lateral that is furthest from the pump, 1 ppm of chlorine

is detectable. Typically, the starting concentration ranges from 5 to 10 ppm (mg/l).

- **Super chlorination:** To get rid of the organic slime that builds up inside the pipeline after an irrigation system has been running for a while, super chlorination is required. 500 parts per million is the typical injection concentration of chlorine. After turning off, the system is left for a full day. Subsequently, the system is flushed in stages: main line, submain, and lastly, through the end of the laterals.
- **Intermittent Treatment:** Microorganisms like fungi, bacteria, and algae may get past the filter system's operational barriers and enter the irrigation system. Periodic sterilization with chlorination is required to avoid an accumulation of this organic matter in the laterals. A chlorine concentration of 10–20 ppm is added to the irrigation water for 30 minutes during the final hour of irrigation. Until the following irrigation, the chlorine is kept in the pipe.

REFERENCES

1. Ah Koon, P. D., Gregory, P. J., & Bell, J. P. (1990). Influence of drip irrigation emission rate on distribution and drainage of water beneath a sugar cane and a fallow plot. *Agricultural Water Management, 17*, 267–282.
2. Ahmed, A. M., Ogaidi, A., Wayayok, A., Rowshon, M. K., & Abdullah, A. F. (2016). Wetting patterns estimation under drip irrigation systems using an enhanced empirical model. *Agricultural Water Management*, 203–213.
3. Anwar, A. A. (1999). Factor G for pipelines with equally spaced multiple outlets and outflow. *Journal of Irrigation and Drainage Engineering, ASCE, 125*(1), 34–38.
4. Assouline, S., Cohen, S., Meerbach, D., Harodi, T., & Rosner, M. (2002). Microdrip irrigation of field crops: Effect on yield, water uptake, and drainage in sweet corn. *Soil Science Society of America Journal, 66*, 228–235.
5. Badr, A. E., Ebabi, F. G., & Eltomy, E. O. (2006). Fertigation methods effects on water and fertilizer uniformity in drip irrigation. *Misr Journal of Agricultural Engineering, 23*(1), 122–136.
6. Bernstein, L., & Francois, L. E. (1973). Leaching requirement studies: Sensitivity of alfalfa to salinity of irrigation and drainage water. *Soil Science Society of America Proceedings, 37*, 931–943.
7. Blaney, H. F., & Criddle, W. D. (1950). Determining water requirements in irrigated areas from climatological irrigation data (Technical Paper No. 96). US Department of Agriculture, Soil Conservation Service, Washington, D.C.
8. Bowen, I. S. (1926). The ratio of heat losses by conduction and by evaporation from any water surface. *Physical Review, 27*, 779–787.
9. Bralts, V. F., & Segerlind, L. J. (1985). Finite element analysis of drip irrigation submains units. *ASAE, 28*(3), 809–814.
10. Bralts, V. F., Kelly, S. F., Shayya, W. H., & Segerlind, L. J. (1993). Finite element analysis of micro-irrigation hydraulics using virtual emitter systems. *ASAE, 36*(3), 717–725.
11. Cardon, E. G., & Letey, J. (1992). Soil-based irrigation and salinity management models: II, Water and solute movement calculations. *Soil Science Society of America Journal, 56*, 1887–1892.
12. Carter, A., & Howell, J. (2000). An overview of drip irrigation. Department of Plant and Soil Sciences, University of Massachusetts.
13. Christiansen, J. E. (1942). Hydraulics of sprinkling systems for irrigation. *ASCE, 107*, 221–239.

14. Clapp, R. B., & Hornberger, G. M. (1978). Empirical equations for soil hydraulic properties. *Water Resources Research, 14*, 601–604.
15. Coelho, E. F., & Or, D. (1996). A parametric model for two-dimensional water uptake by corn roots under drip irrigation. *Soil Science Society of America Journal, 60*, 1039–1049.
16. Coelho, E. F., & Or, D. (1999). Root distribution and water uptake patterns of corn under surface and subsurface drip irrigation. *Plant and Soil, 206*, 123–136.
17. Cote, C. M., Bristow, K. L., Charlesworth, P. B., Cook, F. J., & Thorburn, P. J. (2003). Analysis of soil wetting and solute transport in subsurface trickle irrigation. *Irrigation Science, 22*, 143–156.
18. Dalton, J. (1802). Experimental essays on the constitution of mixed gases: On the force of steam or vapour from water or other liquids in different temperatures, both in a Torricelli vacuum and in air; on evaporation; and on expansion of gases by heat. *Manchester Literary and Philosophical Society Memoirs and Proceedings, 5*, 536–602.
19. Drnevich, V. P., Ashmawy, A. K., Yu, X., & Sallam, A. M. (2005). Time domain reflectometry for water content and density of soils: Study of soil-dependent calibration constants. *Canadian Geotechnical Journal, 42*, 1053–1065.
20. El-Rahman, A. (2009). Water use efficiency of wheat under drip irrigation systems at Al-Maghara area, North Sinai, Egypt. *Journal of Soil Sciences and Agricultural Engineering, 34*(3), 2537–2546.
21. Evett, S. R., Howell, T. A., & Tolk, J. A. (2005). Time domain reflectometry laboratory calibration in travel time, bulk electrical conductivity, and effective frequency. *Vadose Zone Journal, 4*, 1020–1029.
22. Hahn, R. H., & Rosentreter, E. E. (1995). *American Society of Agricultural Engineers (ASAE) Standards*. St. Joseph, Michigan.
23. Hargraves, G. H., & Samani, Z. A. (1982). Estimating PET. *Journal of Irrigation and Drainage Engineering, 108*(3), 225–230.
24. Hargraves, G. H., & Samani, Z. A. (1985). Reference crop evapotranspiration from temperature. *Applied Engineering in Agriculture, 1*(2), 96–99.
25. Hartz, T. K., & Hochmuth, G. J. (1996). Fertility management of drip-irrigated vegetables. *HortTechnology*, 168–172.
26. Hartz, T. K., Smith, R. F., LeStrange, M., & Schulbach, K. F. (1993). On-farm monitoring of soil and crop nitrogen status by nitrate-selective electrode. *Communications in Soil Science and Plant Analysis, 24*, 2607–2615.

27. Hathoot, H. M., Al-Amoud, A. I., & Mohammad, F. S. (1993). Analysis and design of trickle irrigation laterals. *Journal of Irrigation and Drainage Division, American Society of Civil Engineers, 119*(5), 756–767.
28. Hawkins, A. J. (1985). Electrical sensor for sensing moisture in soil. *US Patent 5179347*. Date issued: 12th January.
29. Helmi, M. H., Ahmed, I. A. A., & Fawzi, S. M. (1993). Analysis and design of trickle irrigation laterals. *ASCE, 119*(5), 756–767.
30. Hochmuth, G. J. (1994). Sufficiency ranges for nitrate-nitrogen and potassium for vegetable petiole sap quick tests. *HortTechnology, 4*, 218–222.
31. Howell, T. A., & Barinas, F. A. (1980). Pressure losses across trickle irrigation fittings and emitters. *Transactions of the ASAE, 23*(4), 928–933.
32. Howell, T. A., & Hiler, E. A. (1974). Designing trickle irrigation laterals for uniformity. *Journal of Irrigation and Drainage Division, ASCE, 100*(IR4), 443–445.
33. Howell, T. A., & Hiler, E. A. (1974). Trickle irrigation lateral design. *ASAE, 17*(5), 902–908.
34. James, G. L. (1988). *Principles of Farm Irrigation System Design*. John Wiley & Sons, Inc., USA.
35. Jensen, M. C., & Fratini, A. M. (1957). Adjusted 'F' factors for sprinkler lateral design. *Agricultural Engineering, 38*(4), 247.
36. Jensen, M. E. (1980). *Design and Operation of Farm Irrigation Systems* (Monograph 3). American Society of Agricultural Engineers, St. Joseph, Michigan.
37. Jeppson, R. W. (1977). *Analysis of Flow in Pipe Networks*. Ann Arbor Science, Michigan.
38. Kang, Y., & Nishiyama, S. (1994). Finite element method analysis of micro-irrigation system pressure distribution. *Transactions of the Japanese Society of Irrigation, Drainage, and Reclamation Engineering, 169*, 19–26.
39. Kang, Y., & Nishiyama, S. (1996a). Analysis and design of microirrigation laterals. *Journal of Irrigation and Drainage Division, American Society of Civil Engineers, 122*(2), 75–81.
40. Kang, Y., & Nishiyama, S. (1996b). Design of microirrigation submain units. *Journal of Irrigation and Drainage Division, American Society of Civil Engineers, 122*(2), 83–89.
41. Keller, J., & Bliesner, R. D. (1990). *Sprinkler and Trickle Irrigation*. Van Nostrand Reinhold, New York.

42. Keller, J., & Karmeli, D. (1974). Trickle irrigation design parameters. *ASAE, 17*(4), 678–684.
43. Koenig, E. (1997). Methods of micro-irrigation with very small discharges and particularly low application rates. (In Hebrew). *Water Irrigation, 365*, 32–38.
44. Kushwah, M. (2021). Micro irrigation: A key to increase water use in agriculture. *Just Agriculture*. e-ISSN: 2582–8223.
45. Lakhdar, Z., & Dalila, S. (2006). Analysis and design of a microirrigation lateral. *Journal Central European Agriculture, 7*(1), 57–62.
46. Lamm, F. R. (2002). Advantages and disadvantages of subsurface drip irrigation. In *International Meeting on Advances in Drip/Micro Irrigation*, Puerto de La Cruz, Tenerife, Canary Islands, 1–13.
47. Martinez, J., & Reca, I. (2014). Water use efficiency of surface drip irrigation versus an alternative subsurface drip irrigation method. *Journal of Irrigation and Drainage Engineering, 140*(10), 04014030.
48. Meyer, A. F. (1915). Computing runoff from rainfall and other physical data. *Transactions of the American Society of Civil Engineers, 79*, 1055–1155.
49. Mohammed, A. I. (2010). TIWFC–Trickle Irrigation Wetting Front Pattern Calculator. *Fourteenth International Water Technology Conference, IWTC 14 2010*, Cairo, Egypt.
50. Morton, F. I. (1983). Operational estimates of areal evapotranspiration and their significance to the science and practice of hydrology. *Journal of Hydrology, 66*, 1–76.
51. Mwale, S. S., Azam-Ali, S. N., & Sparkes, D. L. (2005). Can the PR1 capacitance probe replace the neutron probe for routine soil-water measurement? *Soil Use and Management, 21*, 340–347.
52. Niaz, S., Ali, M. A., Ali, A., & Awan, S. I. (2009). Comparative water use efficiency of drip and furrow irrigation systems for off-season vegetables under plastic tunnel in rainfed areas. *Life Sciences International Journal, 2*, 952–955.
53. Penman, H. L. (1948). Natural evaporation from open water, bare soil, and grass. *Proceedings of the Royal Society of London, 193*, 120–145.
54. Perolt, P. R. (1977). Design of irrigation pipe laterals with multiple outlets. *ASCE, 103*, 179–193.
55. Phene, C., Hoffman, G. J., & Rawlins, S. L. (1971). Measuring soil matric potential *in situ* by sensing heat dissipation within a porous body I, Theory and sensor construction. *Soil Science Society of America Proceedings, 35*, 27–33.

56. Phocaides, A. (2000). *Technical Handbook on Pressurized Irrigation Techniques*. Food and Agriculture Organization of the United Nations.
57. Prathyusha, K., & Suman, M. C. (2012). Design of embedded systems for the automation of drip irrigation. *International Journal of Application or Innovation in Engineering and Management, 2*, 254–258.
58. Priestley, C. H. B., & Taylor, R. J. (1972). On the assessment of surface heat flux and evaporation using large-scale parameters. *Monthly Weather Review, 100*(2), 81–92.
59. Prueger, J. H., Hatfield, J. L., Aase, J. K., & Pikul, J. L. Jr. (1997). Bowen-ratio comparisons with lysimeter evapotranspiration. *Agronomy Journal, 89*(5), 730–736.
60. Ragab, R. (2002). A holistic generic integrated approach for irrigation, crop and field management: The SALTMED model. *Environmental Modelling and Software, 17*, 345–361.
61. Ram Nivas, Singh, D., & Rao, V. U. M. (2002). *Practical Manual on Evapotranspiration Estimation*. CCS Haryana Agricultural University, Hisar, India.
62. Rauschkolb, R. S., Rolston, D. E., Miller, R. J., Carlton, A. B., & Burau, R. G. (1976). Phosphorus fertilization with drip irrigation. *Soil Science Society of America Journal, 40*, 68–72.
63. Reddy, T. Y., & Reddy, G. H. S. (1992). Irrigation. In *Principles of Agronomy* (pp. 318–388). Kalyani Publishers.
64. Reeve, R. C., & Fireman, M. (1967). Salt problems in relation to irrigation. In R. M. Hagan, H. R. Haise, & T. W. Edminster (Eds.), *Irrigation of Agricultural Lands* (pp. 998–999). Madison, WI: American Society of Agricultural Publishers.
65. Scaloppi, E. J. (1988). Adjusted F factor for multiple outlet pipes. *American Society of Civil Engineers Journal of Irrigation and Drainage Division, 114*(1), 169–174.
66. Schartzman, M., & Zur, B. (1986). Emitter spacing and geometry of the wetted soil volume. *Journal of Irrigation and Drainage Engineering, 112*, 242–253.
67. Shareef, T. M. E., Ma, Z. M., & Zhao, B. W. (2019). Essentials of drip irrigation system for saving water and nutrients to plant roots: As a guide for growers. *Journal of Water Resource and Protection, 11*, 1129–1145.
68. Shedeed, S. I., Zaghloul, S. M., & Yassen, A. A. (2009). Effect of method and rate of fertilizer application under drip irrigation on yield and nutrient uptake by tomato. *Ozean Journal of Applied Sciences, 2*(2).

69. Shuttleworth, W. J., & Calder, I. R. (1979). Has the Priestley-Taylor equation any relevance to forest evaporation? *Journal of Applied Meteorology, 18*, 639–646.
70. Sinobas, L. R., Rodriguez, M. G., & Lee, T. S. (2012). A review of subsurface drip irrigation and its management. In *Water Quality, Soil and Managing Irrigation of Crops* (pp. 171–194).
71. Sivanappan, R. K., Rajgopal, A., & Paliniswami, D. (1974). Response of vegetables to drip irrigation. *Madras Agricultural Journal, 65*, 576–579.
72. Solomon, K., & Keller, J. (1978). Trickle irrigation uniformity and efficiency. *Journal of the American Society of Civil Engineers, 104*(3), 293–306.
73. Stagnitti, F., Parlange, J. Y., & Rose, C. W. (1989). Hydrology of a small wet catchment. *Hydrological Processes, 3*, 137–150.
74. Thornthwaite, C. W. (1948). An approach toward a rational classification of climate. *Geographical Review, 38*(1), 55–94.
75. Thornthwaite, C. W., & Holzman, B. (1939). The determination of land and water surfaces. *Monthly Weather Review, 67*, 4–11.
76. Thornthwaite, C. W., & Mather, J. R. (1955). The water balance. *Publications in Climatology, 8*(1). Drexel Institute of Technology, Centerton, New Jersey.
77. Valiantzas, J. D. (1998). Analytical approach for direct drip lateral hydraulic calculation. *Journal of Irrigation and Drainage Engineering, 124*(6), 300–305.
78. Van der Kooji, S., Zwarteveen, M., Boesveld, H., & Kuper, M. (2013). The efficiency of drip irrigation unpacked. *Agricultural Water Management, 123*, 103–110.
79. Verburg, K., Bridge, B. J., Bristow, K. L., & Keating, B. A. (2001). Properties of selected soils in the Gooburrum–Moore Park area of Bundaberg. *Technical Report 09/01, CSIRO Land and Water*, Canberra, Australia.
80. Vitosh, M. L., & Silva, G. H. (1994). A rapid petiole sap nitrate-nitrogen test for potatoes. *Communications in Soil Science and Plant Analysis, 25*, 183–190.
81. Wallace, J. S. (2000). Increasing agricultural water use efficiency to meet future food production. *Agriculture, Ecosystems & Environment, 82*, 105–119.
82. Watters, G. Z., & Keller, J. (1978). Trickle irrigation tubing hydraulics. ASAE Paper No. 78-2015. St. Joseph, MI: ASAE.

83. Wu, I. P., & Gitlin, H. M. (1974). Drip irrigation based on uniformity. *ASAE, 3*, 429–432.

84. Yitayew, M., & Warrick, A. W. (1988). Trickle lateral hydraulics II: Design and examples. *ASCE Journal of Irrigation and Drainage Engineering, 114*(2), 289–300.

85. Zella, L., & Kettab, A. (2002). Numerical methods of microirrigation lateral design. *Biotechnology, Agronomy, Society and Environment, 6*(4), 231–235.

86. Zella, L., Kettab, A., & Chasseriaux, G. (2006). Design of a micro-irrigation system based on the control volume method. *Biotechnology, Agronomy, Society and Environment, 10*(3), 163–171

Chapter 6

Sprinkler Irrigation Systems

Contents

INTRODUCTION

A sophisticated kind of irrigation known as sprinkler irrigation lets water spray into the air and fall to the ground like rain. Water is sprayed via a nozzle attached to a system of pipes that are pressurized with water. A sprinkler's application rate and coverage area are controlled by the system's pressure and nozzle size selection. These systems use sprinklers to spray water into the air, where it falls and falls on the plants below, distributing it through a network of pipes. This technique works well for a variety of crops and terrain types because it distributes water evenly over wide areas. Sprinkler systems are adaptable to a variety of agricultural requirements because they can be tailored to operate at different pressures and cover different ranges. By lowering evaporation and runoff, they also contribute to water conservation, especially when weighed against more conventional flood irrigation techniques. Sprinkler irrigation can also be automated, which lowers labor costs associated with watering crops and guaranteeing that plants get regular moisture levels. However, wind and evaporation may have an impact on the system's efficiency, and the initial installation costs can be significant. Sprinkler irrigation is still a common and useful technique for contemporary agriculture in spite of these difficulties, particularly in places where water conservation is a top concern.

6.1. SPRINKLER IRRIGATION

Water is applied through sprinklers, which are tiny holes or nozzles that allow water to flow under pressure and form a spray. The method's applicability in most topographic conditions is limited only by land use capability and economics due to the flexibility of modern sprinkler equipment and its effective control of water application. Applying water evenly to the soil's surface to replenish water that plants have removed is the main goal of sprinkler irrigation. The pressure is typically achieved through pumping, but if the water source is sufficiently high above the irrigated area, gravity may also be used to achieve the pressure.

There are two broad types of sprinkler irrigation systems. Sprinklers in fixed and periodic-move systems stay in place during irrigation, while sprinklers in continuous move systems travel in a straight or circular path. Hand-move and wheel-line laterals, hose-fed sprinkler grids, perforated pipes, orchard sprinklers, and gun sprinklers are examples of periodic-move systems. Sprinklers with linear or lateral movement and center pivot are the most widely used continuous-move systems. Water can be applied uniformly at a rate based on the soil's infiltration rate using carefully planned periodic-move and fixed systems, preventing runoff and the resulting damage to crops and land. Continuous-move systems can apply even more uniformly than fixed and periodic move systems, and they can apply light watering at a speed that minimizes or completely eliminates runoff.

6.1.1. Adaptability

Because sprinklers come in a variety of discharge capacities, sprinkler irrigation works well for most crops and can be applied to almost any type of soil that can be irrigated. Water can be applied at any chosen rate above 0.15 inches per hour (in/h) for periodic-move systems with the appropriate spacing between them. To apply water uniformly at low rates on extremely fine-textured soils, special consideration must be given to the selection of the right nozzle size, operating pressure, and sprinkler spacing. Periodic-move irrigation systems work well in regions where crop-soil-climate conditions don't necessitate more frequent irrigations than once every five to seven days. For crops with shallow roots and soils with limited water retention capacity, light, frequent irrigations are necessary. Fixed or continuously moving systems are more flexible for these kinds of applications; however, in low-permeability soil, some continuously moving systems—like the traveling gun and center pivot—may result in runoff issues. Apart from being adjustable to any kind of irrigation frequency, fixed systems can also be planned and run for crop cooling, bloom delay, and protection against frost and freeze.

6.1.2. Special Uses

Sprinkler systems come in a range of varieties and can be used for more than just regular irrigation to regulate soil moisture. The most adaptable multipurpose systems are center-pivot, solid-set, and automatic permanent systems. Because multipurpose systems allow for timely chemical applications and reduce the number of times machinery must cross a field, they save labor, material, and energy. The most significant multipurpose tasks, aside from standard irrigation, are the application of pesticides and herbicides as well as the application of fertilizers and soil amendments using irrigation water. The most significant special use systems manage the microclimate, remove waste waters, and guard against frost damage. In addition to preventing farm fires, sprinkler systems also provide moisture for the construction of earth fill, cooling and dust control for feedlots and poultry buildings, and log pile curing.

6.1.2.1. Federal, State, and Local Regulations

Government regulations are changing quickly, and the use of chemicals is tightly regulated. For information on which chemicals are approved for use in sprinkler irrigation water and on which crops, speak with a reliable chemical dealer, county agricultural agent, state agricultural extension specialist, State Department of Agriculture, or U.S Environmental Protection Agency (EPA).

6.1.2.2. Applying Fertilizers, Soil Amendments, and Pesticides

Soluble fertilizers can be dissolved in water and then applied with a sprinkler system; this method is inexpensive, simple, and efficient. The crop being irrigated can be fertilized with less effort than is necessary for mechanical application once the apparatus for adding the fertilizer to the irrigation water is set up. Minimal equipment is needed. The timing of fertilizer application in relation to overall irrigation can control how much of the fertilizer penetrates the soil. You can use liquid fertilizer or dissolve fertilizer in water in a barrel or other closed container at a rate of about one pound per gallon (120 gal/l).

The use of sprinkler irrigation systems to apply fertilizer has a number of benefits. Fertilization and irrigation can be completed with only a little bit more work than irrigation alone. This is especially crucial in arid and semi-arid regions where fertilizer and irrigation water applications can typically be planned to coincide. Both the lateral distribution and the depth at which fertilizer is applied are typically under strict control. Fertilizer distribution can only be as uniform as water distribution, but as long as the sprinkler system is functioning properly and has been designed correctly, fertilizer distribution should be acceptable.

6.1.2.3. Injection Techniques

Adding fertilizer to the system at the centrifugal pump's suction side is the most straightforward method of doing so with a sprinkler system. A hose or pipe is connected to the pump's suction pipe from a location close to the bottom of the fertilizer solution container. This line has a shutoff valve installed to control flow. An additional hose or pipe connects the pump's discharge side to the fertilizer container, making it simple to fill it, dissolve the fertilizer, and then rinse. The line from the discharge side of the pump can be left open and the solution's entry into the water controlled by the valve on the suction side of the line if a closed pressure type container—such as any of the many commercial fertilizer applicators—is being used. To prevent contamination of the water supply, a backflow prevention valve must be installed upstream of the injection point.

Sprinkler systems can also benefit from the addition of fertilizer using a small, high-pressure pump, like a gear or paddle pump. The small pump on the spray rig can be used to pump the fertilizer solution if one is available for orchards. When multiple sprinkler lines are in operation at once, this method can also be used to apply fertilizer to individual lines; however, it might be more difficult to move than other types of injectors. The empty fertilizer barrel or container should be filled with water and the water run through the pump in order to prevent corrosion after the fertilizer solution is pumped into the line. Repeating this process multiple times will ensure that the pump and barrel are completely cleaned. Using an aspirator unit is a popular technique for applying fertilizer through sprinkler systems. Bypassing the aspirator with a portion of the water discharged from the pump, suction is created, drawing the fertilizer solution into the line. The goal is to lower the pressure between the pressure-type container's intake and outlet so that water can pass through it and enter the sprinkler mainline or lateral. There are several of these commercial fertilizer applicators available. One of these makes use of the pressure gradient created by inserting a venturi section into the pipeline.

A second kind uses the pressure drop those results from enlarging a pipe to generate a significant pressure gradient without obstructing flow. In order to control the flow through the aspirator and the main line, valves are necessary. This kind of fertilizer applicator has the benefit of simplicity and being free of moving parts, and it costs roughly the same as a small positive displacement pump unit.

6.1.2.4. Fertilizer Materials

Sprinkler systems can be used to apply a wide variety of fertilizer materials, including liquid, dry, and liquid suspension. Convenience and expense of the required nutrients are the primary factors considered when choosing a fertilizer material. It is easy to handle clear liquid fertilizers from bulk storage tanks using pumps and gravity flow. These could have combinations of potassium (K), phosphorus (P), and nitrogen (N), or they could contain just one nutrient. There are numerous options for dissolving nitrogen, phosphorus, and potassium-containing soluble dry fertilizers into the sprinkler irrigation stream. The dry fertilizer products can be placed in a pressurized container that a section of the sprinkler stream passes through, or they can be dissolved by mixing with water in a separate, open tank and then pumped into an irrigation stream. In the latter case, the solid fertilizer is continuously dissolved by the water flow until the entire amount is applied. The use of sprinklers and better application equipment have led to an increase in the application of dry fertilizer materials and agricultural minerals.

The ability of suspension-type fertilizers to yield higher analysis and potassium-rich grades has sparked increased interest in these fertilizers in recent years. Plant nutrients in the suspension mixtures range from 11 to 133% higher than in the corresponding clear liquids. Suspensions are typically much less expensive to produce, handle, and apply than clear liquids due to their higher nutrient content. Suspensions also have the benefit of holding comparatively large amounts of micronutrients. In clear liquids, many micronutrient materials can only dissolve in trace amounts. To guarantee correct dispersion and even distribution, the irrigation water volume and velocity must be high enough to keep the fertilizers in suspension or solution. Urea-ammonium nitrate solutions, ammonium nitrate, ammonium sulfate, urea (potential N loss), calcium nitrate, potassium nitrate, liquid ammonium phosphates, some dry ammonium phosphates, potassium chloride (may be difficult to dissolve), and potassium sulfate (may be difficult to dissolve) are materials that are frequently applied using sprinkler systems. Sprinkler systems can be utilized to apply secondary and micronutrients such as molybdenum, magnesium sulfate, zinc sulfate and chelates, manganese sulfate and chelates, copper sulfate and chelates, iron sulfate and chelates, and soluble (boron).

Materials that should not be applied through sprinkler systems include:

- aqua ammonia (excessive N loss and calcium precipitation with hard water)
- anhydrous ammonium (excessive N loss and will precipitate with hard water)
- single super phosphate, concentrated or treble super-phosphate, and some dry ammonium phosphates (materials that will not dissolve)
- potassium sulfate and magnesium sulfate (hard to dissolve)
- almost all N-P-K dry fertilizers, liming materials, and elemental Sulphur (materials that will not dissolve)
- ammonium polyphosphate (precipitates with hard water)
- phosphoric acid or any acid (causes corrosion and precipitation)

6.1.2.5. Fertilizer Applications

Fertilizer should be applied in batches for sprinkler systems that are fixed or move periodically. The fertilizer needed for a specific area is placed in a tank or metered into the system using the batch method, and the solution is then injected into the irrigation water. The precise ratios don't matter, but high concentrations of solution should be avoided due to corrosion issues. A certain area is covered by the sprinklers that are running at any given time, and the amount that is metered into the system or put into the tank at that moment is the amount that should be applied to that area. It goes without saying that the entire batch needs to be used in one set.

The common procedure followed in applying fertilizer by the batch method consists of three timed intervals. The system functions normally during the first interval, soaking the soil and foliage. The fertilizer is injected into the system during the second interval. Rarely should this application last less than 30 minutes; an hour or more is preferable. This removes the chance that inefficient distribution could result from sprinklers rotating slowly or unevenly. Furthermore, a regular fertilizer application rate will result in a more diluted solution that flows through the system. This reduces the risk of corrosion damage to the system and foliage burn. The final time interval ought to be sufficient to fully rinse the system with clear water, remove all fertilizer from plant foliage, and move the fertilizer down into the crop root zone. In case of fast rates, the interval should last for 30 minutes, and for slower rates, for 90 minutes, depending on the system's operating rate.

Fertilizer application for continuously moving systems, like center pivots, requires the proportionate method. The system should be run through a full rinse cycle with clear water after fertilizer application. The rate of fertilizer injection plays a crucial role in determining the total amount of fertilizer applied when using the proportional method. The application precautions stem from the fact that a lot of commercial fertilizers and soil amendments tend to be toxic to plant leaves and corrosive to metals. With injection on the suction side of the pump, the approximate descending order of metal susceptibility to corrosion is:

- Galvanized steel
- Phosphor-bronze
- Yellow brass
- Aluminum
- Stainless steel

Stainless steel comes in a variety of grades, the best of which resist corrosion more or less. Protection is achieved by rinsing the area thoroughly and quickly after applying chemicals, as well as by diluting the fertilizer and reducing the amount of time in contact. Steer clear of using anything that contains heavy metals utilizing a sprinkler system to apply fertilizer. The table provides examples of applying urea-ammonium nitrate with 32% N and weighing 11.0 pounds per gallon (lb/gal) through a quarter-mile center-pivot lateral and a quarter-mile side-roll sprinkle system with 60-foot movements.

6.1.2.6. Soil Amendments

Sprinkler systems can be used to apply a variety of soluble soil amendments, including gypsum, sulfuric acid, lime, and soluble resins. In the San Joaquin Valley of California, gypsum needs to be applied to a lot of soils in order to lower the amount of soluble sodium, which can lead to poor soil particle dispersion and poor infiltration. Sprinkler irrigation systems are a common way to introduce gypsum in this area. Generally speaking, the procedures are the same as those for adding soluble fertilizers.

6.1.2.7. Applying Pesticides

Among the substances classified as pesticides are fumigants, herbicides, insecticides, fungicides, and rodenticides. State and federal laws govern these materials' use and sales. Many of the materials have been used successfully by researchers and growers; however, chemicals should not be used in sprinkler systems unless approved for use in the application method and conditions specified. This approach to pest control has a lot of promise, particularly when combined with fixed and center-pivot systems, but more study is required. During routine irrigation operations, the entire irrigated area can be treated with diluted solutions of basic fertilizers and herbicides. The program for most pesticides and trace element foliar applications is comparable to that of foliar cooling using fixed systems. To create diluted solutions, the chemicals are precisely added to the irrigation water. The system is then turned on and off for a sufficient amount of time to allow each application to dry and for the purpose of thoroughly wetting the foliage. The chemical is applied to the leaves in thin layers by repeating this process until the required amount is applied.

6.1.2.8. Disposing of Wastewater

Sprinkling wastewater onto land via sprinkler irrigation can be a more affordable option than traditional wastewater treatment. Municipal, industrial, and agricultural wastewaters are separated out. Most cities' wastewaters need to undergo a lengthy treatment process before being released. Industrial wastewaters may need a lot of pretreatments, from basic screening to primary and secondary treatment for metal removal, oil and grease removal, pH correction, and chlorination, among other things. Effluents from food processing facilities and animal production systems are included in agricultural wastewaters. For the purpose of applying sprinkle irrigation to land, the majority of animal wastes need to be treated, which may involve eliminating large fibrous solids. Wastewater from food processing facilities typically needs more thorough pretreatment, including solids, grease, and oil removal as well as pH adjustments.

6.1.2.9. Design Considerations

When applying wastewater to land using sprinkler systems, the main concerns are that the wastewater should be of high quality and applied in a way that doesn't damage or render ineffective the disposal site or contaminate nearby ground and surface waters. The soil and the vegetation cover can be harmed by oils, greases, and heavy metals. Moreover, too much solid matter can cause a mat to form on the surface, which will kill the vegetative cover. The type of soil, frequency of application, and effluent vegetative cover should all be taken into account when designing a sprinkler system. Deep sandy or loamy soils with good drainage are frequently ideal for applying wastes to the land. For some soils, subsurface drainage may be necessary. The majority of guidelines suggest a maximum application rate of 0.25–2/5 inches per hour (6.4 mm/h), which should not be greater than the soil's infiltration rate. A weekly total application of 1 to 4 inches (25 to 100 mm) is possible, with a higher application in the summer. However, there ought to be a gap of time between applications. Woodlands is a suitable location for wastewater disposal. The soil surface in woodlands is stable, and the surface cover aids in the breakdown of organic matter. Choose a grass that is appropriate for the location if grassland is to be used. On a disposal site, corn can also be grown, but only during specific seasons of the year can effluent be applied. Nitrogen application rates can vary, with corn requiring about 200 pounds per acre (224 kg/ha) per year and coastal Bermuda grass requiring 700 pounds per acre (785 kg/ha) per year. When plants receive more nitrogen than they need, nitrate can leach out and contaminate groundwater.

6.1.2.10. Hardware

Single nozzle sprinklers are used in the majority of land disposal sprinkler systems, which lowers application rate and minimizes nozzle clogging issues. Choose sprinklers that will function in subfreezing temperatures if the system is intended to run in that environment. For the main and lateral lines of periodic-move or fixed systems, either portable aluminum or buried pipe may be utilized; noncorrosive buried pipe is advised. Automating fixed systems that are meant to run constantly is advised. Air, water, or electricity can all be used to operate automatic valves. On the other hand, air or water valves that use clean water are the most ideal. The electric solenoid valves in wastewater frequently become clogged by solids. Fixed system valves ought to be numbered, color-coded, and kept in a valve box. Installing drain valves to empty the pipe system is a good idea if the location is exposed to freezing temperatures.

An air purge system that can be used to drain the water from the pipe is the most effective freeze protection system. The system needs to be flushed with fresh water after every use if it is only partially operated and the wastewater is corrosive or contains a high solids content. If wastewater is not removed from the pipes, it will turn septic and cause problems. Additionally, suspended solids may seriously clog the lines at low points as they settle and harden. For the land application of wastewaters, center-pivot and traveling sprinklers are occasionally utilized in addition to fixed irrigation systems and portable aluminum pipe. These two systems each have a comparatively high application rate. High suspended solids in the wastewater can clog the turbine or piston of water-drive traveling gun sprinklers.

Moreover, big impact sprinklers can cause serious drifting issues when they are used in windy conditions. Because of this, a lot of center pivot effluent disposal systems currently have spray nozzles pointing downhill. Because of the heavy water application, center pivot systems can also have traction issues. A sprinkle irrigation system intended for the land application of wastewater shares similarities with other types of sprinkle irrigation systems in terms of design. The effluent is a mixture of water and solids, so the designer must adhere to good design principles. Additionally, the system's lifespan will be shortened by abrasive or corrosive wastewaters. As a result, specialized tools might be needed.

6.1.2.11. Frost Protection

Frost protection can be achieved with sprinkler irrigation. The area that an ordinary system can cover at any given lateral line setting, however, limits it. Therefore, capacity must be added in order to water the entire field at once and provide adequate protection for the majority of areas. We previously presented the application rate and system capacity requirements for various protection levels. Small single-nozzle sprinklers work well since an application rate of roughly 0.1 inch per hour is usually adequate; double-nozzle sprinklers can be used by plugging one nozzle. Depending on the spacing, nozzle sizes ranging from three 30-seconds to a quarter inch have been effectively employed to prevent overhead frost. In order to prevent short-duration, light-radiant frosts (28 to 29 °F, or –2.2 to –1.7 °C), one can mist under trees or cycle an overhead system every 4 to 8 minutes with 2- to 4-minute applications, keeping half of the system running at all times. These systems use between 26 and 30 gallons per minute per acre (4 to 5 l/s/ha), which is about half of what full-coverage systems that run continuously need. Generally, when frost protection is possible, wind speeds are low. For the purpose of creating sprinkler spacing criteria, wetted diameters obtained from manufacturer catalogues can be combined with the standard reduction. Single-nozzle sprinklers that are commonly used in frost protection systems generate D profiles and can be positioned at a distance of 75% of the wetted diameter while still providing sufficient coverage. For optimal effects, impact sprinkler rotation speeds should be one revolution per minute (rpm) or higher, and sprinkler pressures should be kept on the higher end of the advised operating range.

6.1.2.12. Frost Control Operation

Water needs to be available continuously to achieve total frost control. In other words, water should always be left on the plants because the capacity of the water supply must be greater than the atmospheric potential for the water to freeze. The fact that water freezes at a higher temperature than plant fluids is essential to the mechanics of frost control. As a result, the temperature will remain above the freezing point of plant fluids, at roughly 32°F (0°C), as long as there is liquid water available to freeze. A wet surface will have a temperature equal to that of the dew point, also known as the wet bulb, which is lower than the ambient temperature. For this reason, when the outside air temperature gets close to 33°F, frost control systems ought to be activated. As long as water is applied, the field turns into a mass of ice that doesn't freeze because its temperature is still higher than that of the plant liquid. Damage may also result from turning off the water too soon after it reaches a temperature higher than 32°F (0°C). Therefore, keep applying water until the air temperature rises above 32°F (0°C) and all of the ice has melted off the plants in order to ensure adequate protection.

To help farmers know when to wake up at night to turn on the system, some kind of electric alarm system ought to be installed. This can be accomplished with a thermo-switch placed at plant level in the field, wired to the house, and equipped with a loud bell alarm. The switch ought to be adjusted such that the bell sounds when the temperature at the plant level approaches 34 °C. The system ought to be designed and tested well in advance of any potential use cases. Sprinkler-based frost protection has proven effective for low-growing vegetable crops like beans, cranberries, tomatoes, cucumbers, and peppers as well as trees, bushes, and vines. The ice that builds up on trees during frosts at low temperatures can be so heavy that it can snap branches. A similar buildup of ice could crush tall flowers, pole beans, sweet corn, and celery. Tall, thin plants are therefore typically not suited to ice encasement for frost protection.

6.1.2.13. Bloom Delay

Dormancy is the state that deciduous trees, vines, and bushes go into when they lose their leaves in the fall. Fruit buds do not form until the plants break dormancy, which typically occurs between midwinter and early spring. During this time, plants are typically unable to grow. The air temperature surrounding the buds determines how quickly the buds devclop. Trees that experience cool spring temperatures experience a delay in blossoming; on the other hand, when springtime temperatures surpass average levels, bud development quickens and trees blossom sooner.

The risk of freeze damage increases if an abrupt cold snap follows early bud development. For instance, during nine of the years between 1959 and 1973, freezes occurred after warm early spring temperatures caused the buds to develop to a vulnerable stage, resulting in losses for Utah's fruit growers. In the past, it was customary to use sprinklers to heat the orchard in order to prevent freezing, which happens after the buds have reached a vulnerable stage. A new practice is to sprinkle the trees to cool them down before the buds formation, keeping them dormant until after the main risk of freeze damage has passed. When the temperature rises above 40°F (4.4°C) after dormancy, the buds will begin to open. As the temperature rises, the rate of development accelerates until the outside air temperature reaches 77°F (25°C).

After that, the rate of development is not significantly affected by rising temperatures. Growing-degree hours are the units of measurement for the energy accumulation related to bud development over a period of days. The buds in the spring are more vulnerable to damage from cold temperatures as they continue to develop. According to tests, different fruit species need different amounts of chill units to finish dormancy and different amounts of growing degree hours to reach different phenological development stages. The section on capacity requirements for fixed systems contains information on the system capacity needed for bloom delay. The temperature of the tree buds, the difference in vapor pressure between the bud surface and the air, and the speed at which evaporated water is removed by convective mass transfer (caused by air movement) all affect how much evaporative cooling occurs on bare limbs. Therefore, it's essential to occasionally completely wet the buds and to let the majority of the water evaporate before rewetting in order to achieve maximum cooling with the least amount of water application.

6.1.2.14. Microclimate Control

Sprinkler irrigation is one way to provide crop or soil cooling. Applications can typically be made once or twice every one to three days to achieve soil cooling. Thus, center pivot systems with high-speed drives and regular fixed systems, with or without automated controls, are appropriate for cooling soil. Because foliar cooling necessitates two to four brief applications per hour, automated fixed systems are the only systems suitable for this use. The intermittent application of small amounts of water raises humidity, cools the air and plant, and theoretically enhances the quality and yield of the product. Water helps plants that would wilt in a hot afternoon by cooling them down and reducing transpiration. This allows the plants to continue growing and functioning as usual. To fully understand the use of sprinkler irrigation for crop cooling, more research is required.

When the humidity is between 20 and 40% and the air temperature is above 95°F (35°C), a 3-minute application at 0.1 inch per hour (2.4 mm/h) every 15 minutes has typically been sufficient to lower the temperature by 10 to 20°F (–12 to –7°C) on low crops and vines. It has been demonstrated that a 6-minute application every 30 to 36 minutes works well on larger trees. Foliar cooling is only possible with very good water. The section on capacity requirements for fixed systems describes the system design procedures and capacity requirements of fixed systems intended for foliar cooling.

6.1.3. Advantages of Sprinkle Irrigation

Some of the most important advantages of the sprinkle method are:

- Small, continuous streams of water can be used effectively.
- Runoff and erosion can be eliminated.
- Problem soils with intermixed textures and profiles can be properly irrigated.
- Shallow soils that cannot be graded without detrimental results can be irrigated without grading.
- Steep and rolling topography can be irrigated.
- Light, frequent waterings can be efficiently applied.
- Crops germinated with sprinkler irrigation may later be surface irrigated with deeper applications.
- Labor is used for only a short period daily in each field.
- Mechanization and automation are practical to reduce labor.
- Fixed systems can eliminate field labor during the irrigation season.
- Unskilled labor can be used because decisions are made by the manager, rather than by the irrigator.
- Weather extremes can be modified by increasing humidity, cooling crops, and alleviating freezing by use of special designs.
- Plans for intermittent irrigation to supplement erratic or deficient rainfall or to start early grain or pasture can be made with assurance of adequate water.
- Salts can be effectively leached from the soil.
- High application efficiency can be achieved by a properly designed and operated system.
- Tall, dense crops can be adequately watered with sprinklers.
- Chemigation and fertigation (application of agricultural chemicals) is convenient.

- The effective land application of treated animal, municipal, and industrial effluent is possible.

6.1.4. Disadvantages of Sprinkle Irrigation

Important disadvantages of sprinkle irrigation are:

Depreciation is required for relatively high initial costs when compared to surface irrigation methods. Based on 2008 prices, these costs range roughly from \$300 to \$500 per acre for simple systems (like hand-move sprinklers); from \$750 to \$1,200 per acre for mechanized and self-propelled systems; and from \$800 to \$1,000 or more per acre for semi-automated and fully automated fixed systems.

Pressure development (pumping) costs, when applicable, are approximately \$0.20 per acre-foot of water for every pound per square inch of pressure; this is based on the electricity cost of \$0.06 per kilowatt hour and assumes a 70% pump efficiency; the equation is:

$$\text{cost per acre-ft per lb/in}^2 = \frac{2.36(\$/\text{kWh})}{E_p}$$

Where;

The required pumping energy, in per kilowatt hour, for 1 acre-ft of water and 1 pound per square inch is $2.36/E_p$, or,

$$\text{energy} = \frac{(2.31\ \text{ft/lb/in}^2)(0.746\ \text{kW/HP})(43{,}560\ \text{ft}^3/\text{acre-ft})(7.481\ \text{gal/ft}^3)}{(3{,}960\ \text{gal/min-ft/HP})(60\ \text{min/h})E_p}$$

Where:

\$0.20 per acre-feet of water per pound per square inch is equivalent to \$0.024 per 1000 cubic meters per kPa

- Without a reservoir, large flows that are delivered sporadically are not cost-effective, and even a slight variation in flow rate can cause problems.
- When the soil has an intake rate of less than 0.15 inches per hour, sprinklers are not a good fit.
- Windy and excessively dry locations appreciably lower sprinkler irrigation efficiency.
- Mechanized sprinkler systems find it difficult to irrigate irregular field shapes; hand move laterals are much more flexible because each pipe section is moved independently, making it simple to add or remove pipes to fit the shape of the field.

- Cultural operations must be coordinated with the irrigation cycle.
- Higher potential irrigation efficiency may be achieved with surface irrigation techniques on appropriate slopes and soils.
- When the soil moisture deficiency is satisfied, the water supply must be able to be turned off at random times.
- To realize the method's high potential efficiency, careful management and maintenance are required.
- A qualified specialist must plan the systems, taking into account labor convenience, pipe size and operation economy, and effective irrigation.
- Water quality can introduce problems.
  - When used in overhead sprinklers on fruit crops, irrigation water that has high concentrations of bicarbonates may affect the production quality.
  - Saline water may cause problems because salt may be absorbed by the leaves of some crops.
  - Careful filtration of surface water supplies is necessary to prevent clogging.
  - Corrosion of metal components and resultant clogging can be problematic.
- Certain varieties of sprinkler irrigation systems are vulnerable to wind movement when not in use.
- In significantly undulating topographies, pressure regulation and/or flow control nozzles are necessary to achieve adequate water application uniformity.
- More frequent irrigation may result in higher evaporation losses from wet soil and plant surfaces.

6.1.5. Installation and Operation of Sprinkler Systems

6.1.5.1. Installation and Operation

Even the most well-thought-out plan won't help much with conservation irrigation goals or maximizing crop yields of superior quality unless the farmer invests heavily in the equipment the plan calls for, installs it correctly, and uses it as intended. Depending on the financial and practical arrangements made by the farmer, the engineer, dealer, farmer, or any combination of the three may be in charge of installing sprinkler irrigation systems.

The farmer should receive a system plan that includes a map of the design area or areas that indicates the location of the pumping plant and water supply, the supply lines, mainlines, and submains, the location and direction of movement of the lateral lines, the sprinkler spacing, and the pipe sizes and lengths needed for each size. A comprehensive list of materials need not be provided, but at least the minimum requirements for the equipment must be. These consist of the sprinklers' discharge, operating pressure, and wetted diameter; the pump's capacity at the design dynamic head; and the power unit's horsepower needs. Where appropriate, fittings for continuous operation should be specified. Sprinkler system designs created by NRCS engineers may be given to farmers, who may then buy equipment completely unrelated to what is called for in the plans. Even though NRCS employees have no influence or accountability over the farmer's sprinkler equipment purchase, it is crucial to stress how important it is to get a good system. A sprinkler system should be energy-efficient, provide the necessary uniformity, and be able to meet crop watering needs all season long. Instructions on mainline and lateral layout, sprinkler spacing, lateral line movement, lateral operation time, and maintaining design operating pressures should be provided to the farmer.

6.1.5.2. Irrigation Scheduling

In order to assess when and how much watering is required, the farmer should also be taught how to estimate the soil's moisture content. The goal of irrigation scheduling management should be to maximize agricultural yield while using the least amount of money and water possible. Center pivot and fixed systems should be able to achieve nearly perfect irrigation. The depth of irrigation and time between irrigations should be determined by taking into account the soil moisture, crop growth stage, and climatic demand. Every crop, soil, and climate combination has a perfect irrigation management plan. Estimates of climatic evaporative demand or devices that show the soil-plant water status should be used to guide irrigation scheduling. For instance, to record and/or monitor field conditions in real-time, soil-water sensors can be placed at representative field sites and linked to a data logger or telemetry system. For sprinkler system management, computerized and internet-based scheduling services that take climate demand into account work perfectly.

6.1.5.3. Screens and Sand Traps

The intakes need to have self-cleaning screens if water is pumped straight from lakes, rivers, or canals. The diameter of the stainless-steel screens should be roughly twice that of the pipeline that is attached, with a mesh opening of no more than 0.25 inches (6 mm). Revolving within the intake screen, internal pressurized water jets move debris away from the mesh openings as they clean. A stainless steel or galvanized screen with a mesh size of 0.1 inch (3 mm) or less should also be installed at the lateral inlet point of center-pivot and linear-move systems to prevent debris, algae, weed seeds, and other objects from clogging nozzles. A mechanism for hydraulically isolating the screen from the remainder of the system is required. These screens can be manually or automatically cleaned. In order to manually wash the screen in both scenarios, there should be a pressurized water supply for a hose.

Sand traps should also be positioned at the distal ends of center-pivot and linear move systems since sand and small gravel have a tendency to gather at the distal end of pipelines. These usually consist of a short piece of 4 inch (100 mm) pipe on a pipeline that points downward from a tee near the mainlines end. Longer pipes on a center-pivot or linear-move lateral are suspended from a tee close to the lateral's end (near the end gun), with a valve easily accessible from the ground. These pipes can swiftly remove the accumulated sand from the system using a 4-inch (100 mm) spring-loaded valve or another mechanism. Occasionally, while the system is running, a large diameter (0.25 in or 6 mm) nozzle spray plate arrangement and special hose are used to continuously flush the sand.

6.2. TYPES OF SPRINKLER SYSTEMS

The three main categories of periodic move systems are boom and gun sprinklers; side-move laterals with or without trail lines; and hand-move, end-tow, and side-roll laterals. Sprinklers that are fixed usually use small or gun sprinklers. Center pivots, traveling gun or boom sprinklers, and linear move (also known as lateral move) are the three main categories of continuous-move systems. In technical terms, sprinkler systems also include micro-spray systems. Nonetheless, drip or trickle systems' design, operation, and hydraulics are comparable to those of micro-spray systems.

6.2.1. Periodic Move Sprinkler Systems

6.2.1.1. Hand-Move Lateral

Hand-move laterals consist of individual aluminum tubing lengths with quick couplers attached to either end- or center-mounted sprinkler head riser pipes. Hand lines are another name for hand-move laterals. They receive water through subterranean or portable mainline pipe with valve outlets spaced uniformly apart. Systems consist of one or more lateral components. This kind of system, which is currently utilized on practically all crops and topographical types, was once used to irrigate a greater area than any other system. The system's high labor and high-pressure requirements are two of its main drawbacks. Every mechanized system was created using this system as their model.

A hand-move lateral's typical length is 1,280 feet (390 meters), and its typical lateral spacing, or the distance it is moved each time, is between 50 and 60 feet (15 and 18 meters). Sprinkler spacing along laterals is typically between 30 and 40 feet (9 to 12 meters), which matches the individual pipe lengths. Aluminum is typically used in lateral construction to facilitate easy mobility. Polyethylene (PE) pipe is used in certain lateral systems. Mainline systems are frequently buried, with laterals connected to risers that reach the ground's surface every 50 to 200 feet (15 to 60 meters). Operating pressures for sprinklers typically range from 40 to 60 pounds per square inch (270 to 410 kPa). Hand-move irrigation systems may require crops to have deep enough roots or a high enough soil water-holding capacity so that the laterals do not need to be moved back to a designated spot in less than five days. However, this is dependent on labor availability, sprinkler pipe availability, and water supply flow rate.

Moving a hand-move system is not nearly as simple as moving an end-tow, side-roll, or side-move system, and it takes more than twice as long per irrigated acre. A hand-move system requires even more labor to move each section if the wet soil is sticky or soft. Nevertheless, when the laterals come to an end of an irrigation cycle, a significant annoyance of mechanical move systems arises. The laterals at the field boundaries can be disassembled, loaded onto a trailer, and hauled to the starting position at the opposite boundary when this occurs with a hand-move system. Because a mechanical-move lateral is more difficult to disassemble, an irrigator may choose to deadhead it back to its initial position. This procedure takes a long time, particularly when trail tubes are involved. In actuality, though, side-roll laterals are frequently disassembled following each irrigation, particularly in cases where labor costs are not prohibitive.

6.2.1.2. End-Tow Lateral

Similar to a hand move lateral system, an end-tow lateral system is made up of rigidly coupled lateral pipe that is connected to a mainline. For ease of use, the mainline should be buried and placed in the middle of the field. From one side to the other, laterals are towed lengthwise across the mainline (Figure 6.1). A quarter-mile, 4-inch (100 mm) diameter lateral can be easily pulled by a tractor with 20 to 30 horsepower after the pipe has been drained using automatic quick drain valves.

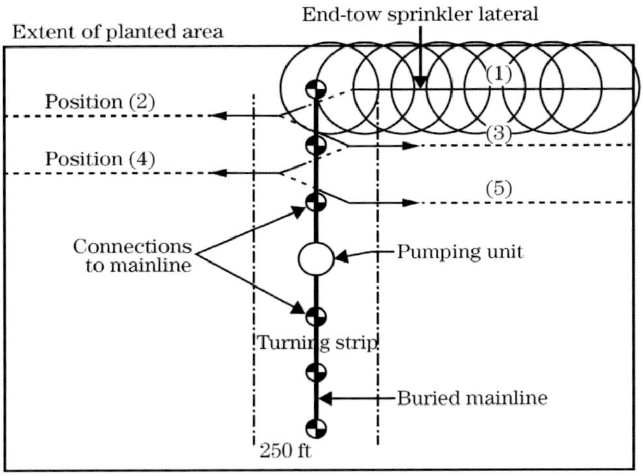

Figure 6.1. *Schematic of the move sequence between lateral positions for end-tow sprinklers with a buried mainline, and a pumping unit in the center of the field.*

Source: https://www.wcc.nrcs.usda.gov/ftpref/wntsc/waterMgt/irrigation/NEH15/ch11.pdf.

There are two kinds of carriages available for end-tow systems. One is a skid plate that is fastened to every coupler to protect the quick drain valve, raise the pipe off the ground a little, and offer a wear surface for the pipe while it is being towed. To keep the sprinklers upright on a quarter-mile lateral, two or three outriggers are needed. The other kind makes it easier to tow on sandy soils by adding tiny metal wheels at or in between each coupler. The least expensive mechanical move systems are end-tow laterals; however, they are not suitable for fields with physical obstacles, steep or uneven terrain, small or irregular areas, or row crops planted on the contour. They perform fairly well in row crops as well as in grasses, legumes, and other close-growing crops. However, negligent handling, such as moving the laterals before they have drained, making an excessively sharp "S" turn, or moving them too quickly, can easily harm them. As a result, they are not advised for tasks involving unpredictable labor quality.

A 200-250 foot wide turning strip must run the entire length of the mainline when used for row crops (Figure 6.1). Grass or alfalfa can be planted in the turning strip. To reduce crop damage in the turning areas, each time the lateral is towed across the mainline, make an offset equal to half the distance between lateral positions rather than a full offset every other time. A specific crop planting arrangement is needed for irrigation of tall crops like corn, such as 16 rows of corn followed by four rows of a low-growing crop that can be driven over by a tractor without causing much damage. A more recent kind of end-tow sprinkler system employs skid plates to support the upright placement of the sprinklers in pods that are uniformly spaced along a line. A tractor or other vehicle pulls the line to each irrigation location; it is not moved while the irrigation is taking place. Pods are usually used for pasture irrigation, and they can be made to not get damaged if horses or cattle step on the sprinklers.

6.2.1.3. Side-Roll Lateral

A system with hand-move laterals is similar to a side-roll lateral system, also called wheel lines. Each pipe section is supported by a large wheel, and the lateral pipes are coupled together rigidly (Figure 6.2). When a torque is applied, the lateral line, which serves as the axle for the wheels, rolls sideways.

An engine positioned in the middle of the line or an external power source at one end of the line are used to mechanically move this unit. In crops that grow slowly, side-roll laterals are effective. Although they are frequently utilized on irregular field shapes, they are most suited for rectangular fields with relatively uniform topography and no physical obstacles. It is important to choose the wheel diameters such that the lateral pipe passes over the mature crop and that the designated lateral move distance equals a complete rotation of the line. For instance, use three rotations of a 76.4-inch (1.9 m) diameter wheel for a 60 foot (18 m) move, and three rotations of a 64-inch (1.6 m) diameter wheel for a 50 foot (15.3 m) move. Occasionally, a 6-foot wheel diameter is employed to ensure adequate crop clearance. This means that one wheel rotation equals roughly 20 feet (6 meters), and five wheel revolutions equal approximately 100 feet. The sets then alternate between two and three revolutions in one irrigation and between three and two revolutions in the next irrigation, giving each pair of irrigations an effective lateral spacing of roughly 50 feet (15.3 meters).

Figure 6.2. *Side-roll sprinkler lateral (with drag lines, or trail tubes) in operation.*

Source: https://www.irriline.com/hs-fs/hubfs/equip_img.jpg?width=940&height=466&name=equip_img.jpg.

Smooth topography and closely planted crops can benefit greatly from the use of side-roll laterals up to 2,000 feet or longer. Laterals up to a quarter of a mile (400 m) long are advised for row crops with deep furrows, like potatoes, and for rough or steep terrain; however, the maximum length varies depending on a number of factors, including crop density, field topography, and pipe strength. Sprinklers are typically spaced 30 to 40 feet (9–12 m) apart along laterals. For laterals, 4- or 5-inch-diameter aluminum tubing is typically utilized. At least three lengths of 0.072-inch heavy walled aluminum tubing to either side of a center power unit for a standard quarter-mile lateral on a closely spaced crop should be made to avoid the pipe twisting and collapsing during set changes. Heavy walled tubing should be used for longer lines, in deep-furrowed row crops, or on steep terrain to allow the laterals to roll more evenly and smoothly with less chance of breaking. Each coupler on a well-designed side-roll lateral should have a fast drain so that the unit can be turned off, drained, and relocated in less than 30 minutes. Every sprinkler ought to have a self-leveling device installed so that every sprinkler will remain upright no matter where the lateral pipe is stopped. The lateral should also have a flexible or telescoping section to connect it to the mainline hydrant valves, as well as a minimum of two wind braces, one on each side of the power mover.

Trail tubes, also known as drag lines, are occasionally (though not often) added to 5-inch side-roll laterals with heavy walls. Compared to traditional side-roll lateral irrigation systems, the system can irrigate a larger area thanks to sprinklers positioned along the trail tubes. To roll the lateral forward, special couplers with a rotating section are required. Each trail tube must also have quick couplers at the end so that they can be disconnected when a lateral reaches its final operating position. To start a new irrigation cycle, the lateral needs to be rolled back to the starting point where the trail tubes are, and then reattached.

6.2.1.4. Side-Move Lateral

Side-move laterals, relatively rare, are moved periodically across the field in a manner similar to side-roll laterals. One notable distinction is that the pipeline functions as the axle instead of being carried above the wheels on tiny A-frames. In most cases, the wheel carriages are placed 50 feet (15 meters) apart, and the pipe is carried five feet (1.05 meters) above the ground. Each wheel carriage has a trail tube pulled behind it that is equipped with 11 sprinklers spaced 30 feet (9 meters) apart. As a result, a quarter-mile-long line can irrigate about 11 acres at a time using the system, which can wet a strip 320 feet (100 meters) wide. Low application rates and great consistency are the results of this system. For most field and vegetable crops, side-move lateral systems work well. But for field corn, the A-frames need to be extended to give a minimum of 7 feet of ground clearance, and the trail tubes cannot be utilized. A quarter-mile-long lateral can irrigate 4.5 acres (1.8 ha) per setting, and small gun sprinklers (60 to 100 gpm, or 4 to 6 lps) positioned at every other carriage will irrigate a strip that is 150 feet (45 m) wide. On the other hand, application rates—roughly 0.5 in/h or 1.3 cm/h—are rather high.

6.2.1.5. Fixed Sprinklers

A fixed-sprinkler system has enough sprinkler heads and lateral pipe so that, once installed in the field, none of the laterals needs to be moved for irrigation. Therefore, all that is required to irrigate the field is for the sprinklers to cycle on and off. Sequencing valve laterals, buried or permanent laterals, and solid-set portable hand-move laterals are the three primary types of fixed systems. While some systems use small gun sprinklers spaced 100 to 160 feet (30–50 m) apart, the majority of fixed-sprinkler systems use small sprinklers spaced 30 to 80 feet (9–24 m) apart.

- **Solid-Set Portable:** When a farm's crop rotation or irrigation plan changes, solid-set portable systems (Figure 6.3) are frequently used for potatoes and other high-value crops. These systems allow the system to be moved from field to field. Sometimes these systems are moved from field-to-field to germinate crops like lettuce, which need irrigation up to three times a day, but are irrigated in furrows after germination and root establishment. Although using specific trailers that allow the portable lateral pipe to be manually stacked can lessen the amount of labor needed to move the laterals into and out of a field. Following correct loading, the pipe is banded multiple times to create a bundle that is removed from the trailer at the farm storage yard using a mechanical lifter. When it comes to getting the laterals back on the field for the following season, the process is reversed. With furrow irrigation, laterals can be floated out of the germinated field in certain situations (such as in the Imperial Valley of California) where fields are flat and furrows are large. Section by section, the laterals are dismantled at the field edge. This lowers labor requirements and the amount of foot traffic that is necessary in wet fields.
- **Buried Laterals:** Permanent buried laterals are buried 18 to 30 inches (45–75 meters) below the surface, with the sprinkler head and riser pipe remaining above ground. Numerous citrus groves, orchards, vineyards, and other crops with comparatively high value use similar systems. The lateral of the sequencing valve can be hung from cables above the crop, buried, or placed on the soil's surface.

Every sprinkler riser has a valve at its core that responds to a control signal to either turn on or off the sprinkler. Certain systems rely on a shift in water supply pressure to trigger the valves. Electric or pneumatic valves that are activated by controllers can be used to automate the sequencing valve lateral systems, buried (or permanent) lateral systems, and portable lateral systems. Temperature and soil moisture sensors can trigger these automatic controllers, which can be set up for irrigation, crop cooling, and frost control.

- **Gun and Boom Sprinklers:** The nozzles of gun sprinklers, also known as giant or big guns, are 5/8 inch (1.6 cm) or larger and are connected to long discharge tubes (12 inches or more). The majority of gun sprinklers have a rocker-arm drive that allows them to rotate and can be adjusted to irrigate a partial circle. With boom sprinklers, a revolving 110–250 foot (35–75 m) boom is supported in the center by a tower that is trailer-mounted. Water jets released from nozzles rotate the boom, which is pivotal to the tower, once every one to five minutes. The nozzles are positioned and sized to spray water in a fairly uniform manner over a circular area that has a diameter of more than 300 feet (90 meters). Tower movement is manual and recurring. Gun or boom sprinkler systems are ideal for supplemental irrigation in a variety of comparable scenarios, as well as for use on fields with obstacles and irregular shapes. Every one of them has relative benefits and drawbacks. There are far more gun sprinklers in use than boom sprinklers because they are easier to use and significantly less expensive. Nevertheless, since guns need high pressure (80–150 lb/in^2, 500–1,000 kPa), their energy expenses are higher. Gun and boom sprinklers are typically operated independently rather than as sprinkler-laterals, and they can discharge more than 100 gallons per minute (6.3 lps). A standard sprinkler needs 80 to 100 pounds per square inch (550–700 kPa) of operating pressure or more to discharge 500 gallons per minute (30 lps).

The majority of crops can be irrigated with gun and boom sprinklers, although their large water drops and high application rates tend to compact the soil and seal the surface, causing issues with runoff. As a result, coarse-textured soils with high infiltration rates and reasonably mature crops that only require supplemental irrigation are the ideal candidates for these sprinklers. It is generally not advised to use boom or gun sprinklers in very windy conditions as their distribution patterns may become too warped. Similar to boom sprinklers, large gun sprinklers are typically trailer or skid mounted and are pulled by a tractor from one location to another. Boom sprinklers have the drawback of being unsteady and prone to toppling over uneven or inclining terrain when being towed.

Figure 6.3. *Hand move portable sprinkler system.*

Source: https://5.imimg.com/data5/SELLER/Default/2023/8/335050181/YI/ED/TV/772884/hand-move-portable-sprinkler-system.jpg.

6.2.2. Continuous-Move Sprinkler Systems

6.2.2.1. Center-Pivots

Water is sprayed by center pivot systems using a moving lateral pipeline. A large circular area (or part of it) can be irrigated by rotating the center pivot lateral, which is fixed at one end. The pivot point, which is the fixed end of the lateral, is attached to the water supply. The lateral is made up of a sequence of spans with steel trusses that are carried roughly 10 feet (3 meters) above the ground by drive units, also known as towers. The trusses range in length from 90 to 250 feet (25 to 75 meters). An A-frame supported by motor-driven wheels makes up a drive unit. All other drive units move sporadically to keep the lateral pipe roughly in a straight alignment, while the final downstream drive unit moves continuously or intermittently to set the rotation speed. The most popular center-pivot lateral is approximately 1,320 feet (1,320 meters) in length, uses 6 or 6–5/8-inch pipe, and irrigates 126 acres of a quarter section (160 acres) in a circular pattern. The additional 2 to 10 acres may be irrigated based on the end sprinklers' range. However, laterals are available that are up to 800 meters (half a mile) long and as short as 220 feet. Typically, one or more initial pipe spans of 8 or even 10 inches of pipe are found in laterals longer than a quarter mile.

The end-drive unit is configured to regulate the rotational speed, and a guidance system made up of components placed at each drive unit maintains the lateral in a line between the pivot and end-drive unit. The buried mainlines that supply center pivots can be single pipelines that connect the water source (a well, pond, or canal) to a single center-pivot system, or they can be networks that supply multiple center pivot systems. Sprinklers are installed on the moving lateral pipeline to distribute water throughout the circular field uniformly. As one moves toward the moving end, the lateral speed increases and the area to be irrigated by a nozzle grows progressively larger. Therefore, the sprinklers must be designed with increasingly larger discharges, closer spacing, or both, toward the moving end in order to provide uniform application. Near the moving end, the application rate is typically approximately 2.5 inches per hour (64 mm/h). With the exception of the initial few minutes of each wetting event, this is higher than the intake rate of many soils. The laterals are typically timed to rotate once every 14 to 84 hours, depending on the capacity of the system, the maximum desired soil moisture deficit, and the infiltration characteristics of the soil, in order to minimize runoff. Water pistons, air pistons, hydraulic oil motors, water spinners and turbines, and electric motors are the five power unit types that are frequently used to drive the wheels on center pivots. Although water pistons drove the first pivots, electric motors are now far more common due to their speed, dependability, and capacity for both forward and backward motion. Oil motors, both electric and hydraulic, enable the system to run dry (i.e., without irrigation).

Self-propelled, center-pivot sprinkler systems work well with nearly all field crops; however, they need fields free of overhead obstacles like electric power poles, buildings, trees, and phone lines. Their optimal application occurs in soils with elevated intake rates and consistent topography. Any runoff can result in erosion and puddles on soils with low intake rates and irregular topography, which can hinder the wheels' traction and uniform movement. If these systems are used on square fields, the unirrigated area must be put to other uses or a method for irrigating the four corners must be supplied. A center pivot system will not irrigate about 30 acres of a 160-acre quarter-section field unless the pivot is equipped with a specialized corner irrigating device. About 8 acres remain unirrigated when using certain corner systems. The majority of pivot systems are set up permanently in a particular field. Moving a typical quarter-mile center-pivot lateral back and forth between two 130-acre fields, however, might be feasible in areas with supplemental irrigation or for dual cropping. Remotely operated controls that use radio transmission and frequently internet delivery are a feature of many contemporary center pivot systems. When a system or water supply fails, some systems sound and light alarms and notify operators by radio or cell phone. In order to oversee numerous center pivots and manage operations and maintenance tasks, certain locations also feature control centers.

6.2.2.2. Traveling Sprinkler

The traveling sprinkler, also known as the traveler, is a high-capacity sprinkler that is placed on an autonomous chassis, travels in a straight line, and is supplied with water via a flexible hose. A gun sprinkler mounted atop a moving vehicle with a capacity of 150 to 500 gallons per minute (10 to 30 lps) that can wet an area more than 400 feet (120 m) in diameter is the most popular kind of traveler used in agriculture in the United States. The device frequently has a winch that reels in the cable, either powered by a turbine or a water piston. The unit is guided along a path by the cable while it tows a flexible, round or lay-flat high-pressure hose that is connected to a pressurized water supply system. The unit can travel up to 1,320 feet (400 meters) unattended thanks to the standard hose, which has a diameter of 4 inches (100 mm) and a length of about 660 feet (200 m). In certain systems, the hose can be flattened, drained, and wound onto a reel after use. In other situations, the gun-sprinkler carriage is pulled in a straight line by a tractor or truck from one end of the field to the other when the plastic hose flattens when empty. After that, the water is turned on, and the gun sprinkler is drawn in by the reel's slow rotation. A mechanism cuts the system off when the carriage that the gun is mounted on gets to the reel assembly. Once the entire field area has been irrigated, the unit is moved to the next row and the procedure is repeated. The water entering the unit powers a hydraulic motor that rotates the reel assembly, significantly lowering the pressure. There is also a large pressure loss due to the lengthy hose. Therefore, in order to run such a system, adequate supply-line pressure is required, and it is most practical to have the water supply at one end of the field rather than the center. The reel assembly's rotation speed and the irrigation water's desired depth can be changed by the operator via controls.

Certain mobile sprinklers are equipped with an independent pumping plant that moves with the vehicle and draws water straight from an open ditch. The hose is replaced by the supply ditches. Instead of carrying guns, travelers can also be outfitted with boom sprinklers. Boom sprinklers can be fixed or have revolving arms that are between 60 and 120 feet (18 and 36 meters) long, and they release water through the designated nozzles. Instead of wetting a circular area as a stationary sprinkler would, the traveler moves along its path, and the sprinkler drenches a strip of land about 400 feet (120 meters) wide. The unit is moved and positioned to water a nearby strip of land when it reaches the end of a travel path. The diameter of the area the sprinkler wets and the separation between travel paths determine how adjacent strips overlap. To ensure that the dry portion of the pattern is over the tow path, part-circle sprinklers are commonly used. This guarantees that the apparatus moves on dry land.

An 80-acre (32-hectare) field's typical traveling sprinkler layout is depicted in plain view in Figure 6.4. Eight tow paths, each measuring 1,320 feet (400 meters) in length and 330 feet (100 meters) apart, are used to irrigate the entire field. The highest pressures needed in any system are needed for traveling sprinklers. Hose friction losses increase the required system pressure by an additional 20 to 40 pounds per square inch, on top of the 65 to 100 pounds per square inch needed at the sprinkler nozzles. Travelers are therefore best suited for supplemental irrigation in areas with low seasonal irrigation requirements, which helps to reduce the high power expenses linked to high operating pressures. Traveling sprinklers have been applied to orchards as well as tall field crops like corn and sugarcane. Traveling sprinklers have a higher uniformity and a lower application rate than guns and stationary booms because they are mobile. They share many of the same benefits and drawbacks as guns and boom sprinklers. The field's center has only fair application uniformity, and the 100–200-foot-wide strips that run along its ends and sides are typically not well irrigated.

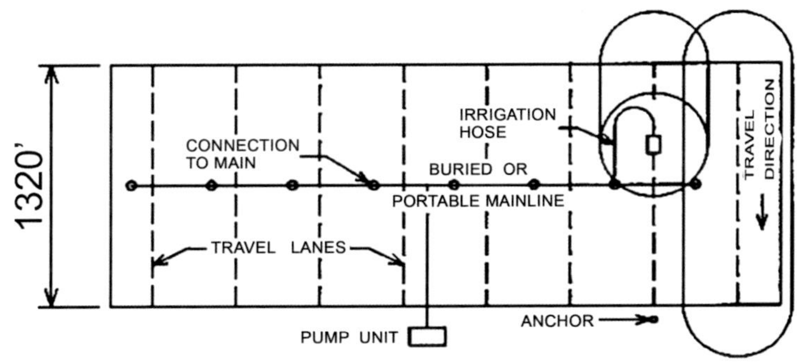

Figure 6.4. *Selection and management of efficient self-propelled gun traveler irrigation system.*

Source: https://content.ces.ncsu.edu/media/images/Screen%20Shot%202016-12-01%20at%202.png.

6.2.2.3. Linear-Move Lateral

Self-propelled linear-move laterals have a traveling water feed system akin to a traveling sprinkler combined with the structure and guidance system of a center-pivot lateral. Obstacle-free rectangular fields are necessary for linear-move laterals. The systems' measured water distribution has demonstrated that, in both calm and windy circumstances, single irrigations have the highest application uniformity coefficients. While some linear-move systems draw water from open ditches, others use flexible hoses to allow for some variation in ground elevation. Other systems use automated coupling mechanisms, where water is supplied via buried mainline pipe with hydrant valves spaced evenly. The automated coupling systems are appropriate for use in undulating topographies because they use swing arms with swivel joints to connect to the hydrants as the system moves down the field. However, they are significantly more expensive (and mechanically complex) than linear-move systems that draw water from a ditch or a flexible hose. Either way, to prevent shutdowns, the guidance system needs to be strong. Typical guidance systems run a raised cable or trench down one side of the field. To direct lateral movement, other systems make use of global positioning satellite (GPS) technology. Pumping from ditches requires linear-move systems to have a self-contained pump and motor, which are typically driven by an auxiliary electrical cable or an on-board diesel engine. The water source typically provides the required water pressure for hose-pull systems. Drive tower power is frequently supplied by an auxiliary electrical cable.

When compared to center-pivot systems, a significant drawback of linear-move systems is the difficulty of returning the lateral to its initial position. Since the linear-move lateral moves from one end of the field to the other, its ending position is the furthest distance from its starting position; if the machine is returned dry (not applying water), downtime occurs. The center-pivot lateral operates in a circle, so each irrigation cycle automatically ends at the start of the next. Systems that use linear motion can occasionally be run in both directions, applying half of the maximum soil water deficit on each pass. On the other hand, center-pivot systems can irrigate only a roughly circular area, while linear-move systems can irrigate a rectangular field in its entirety. When approaching the field's end, a relatively new kind of linear-move system has the ability to pivot at one end, making the lateral half as wide as the irrigated area. Half of the field is irrigated by the lateral on one pass, and the other half is irrigated on the return pass. The lateral pivots once more to go back to the beginning position when the irrigation is finished. However, just like with conventional linear-move systems, the lateral moves linearly when irrigating. Some characteristics of center pivots and linear moves are combined in this kind of system.

6.2.3. Other Sprinkler Systems

Micro sprinklers, hose-fed sprinklers that operate in a grid pattern, and perforated pipe are becoming more popular due to worries about the cost and availability of energy. These systems provide a low-pressure irrigation method (5–20 lb/in^2, or 30–140 kPa). Gravity (elevation differential) pressure is frequently enough to run these systems without the need for pumps. Additionally, the water can be distributed using low-cost, thin-wall plastic and unreinforced concrete pipes. The drawback of these systems is that they require a lot of labor if they are moved frequently as opposed to being fixed in one spot.

- **Perforated Pipe:** Despite being less expensive than other types of sprinkler systems, this kind of irrigation is not frequently utilized for agricultural purposes. Water is sprayed from holes with a diameter of 1/16th of an inch or less, drilled at consistent intervals along the top and sides of a lateral pipe or hose, in perforated pipe systems. In order to apply water evenly between neighboring lines of perforated pipeline, the holes are sized and spaced accordingly. Because of their high application rate, these systems are only suitable for coarse-textured soils like loamy sands with a high infiltration capacity. They can function well at pressures ranging from 5 to 30 pounds per square inch. Compared to most sprinkler systems, the water application uniformity is relatively low because of the rapid changes in application rate that occur along the lateral.
- **Hose-Fed Sprinkler Grid:** The mentioned systems employ hoses to distribute tiny sprinklers, which function at pressures ranging from 5 to 10 pounds per square inch. These sprinklers can operate effectively if arranged in a systematic grid pattern with adequate overlap. However, their usage is not widespread beyond home gardens and turf irrigation, despite their potential. In fact, they show great potential for application on small farms in underdeveloped nations, where financial and energy resources are scarce and labor is abundant.
- **Orchard Sprinklers:** Little spinners or impact sprinklers called orchard sprinklers are made to cover the space between trees with minimal to no overlap between the areas that are sprayed by nearby sprinklers. The pressure range at which orchard sprinklers operate is 10 to 30 pounds per square inch (70 to 200 kPa), with a typical coverage diameter of 5 to 30 feet (2 to 10 m). The range of discharge rates is 20 to 375 lph (20 to 100 gallons per hour). Sprinklers for orchards are positioned beneath the tree canopies to supply roughly equal amounts of water to every tree without soaking the foliage. While some of the soil surrounding each tree may receive little to no irrigation, water should be applied fairly evenly to areas that have been wetted. An orchard sprinkler can be installed for each position on lateral lines that resemble those used for drip irrigation,

or individual sprinklers can be supplied by aluminum laterals or hoses and periodically moved to cover multiple positions. The main reasons why orchard sprinklers are fixed are to lower labor costs and offer a certain level of frost protection.

6.3. PLANNING CONCEPTS OF SPRINKLER SYSTEMS

Complete farm sprinkler system planning takes into account the kinds of soil present in the designated design area, water quality, and crops and crop rotations. Sprinklers and associated hardware, such as laterals, submains, mainlines, pumping plants and boosters, filtration equipment, operation control equipment, and other accessories needed for the efficient application of water, are all part of a farm sprinkler irrigation system. Other changes that farm systems may go through include cropping pattern changes or the expansion of irrigated areas. Because of this, a lot of seasoned system designers install supply and mainlines with larger capacity than necessary to accommodate potential future expansions or higher flow rate requirements. In the long term, this method can save costs by avoiding the need to install additional pipe capacity. A sprinkler lateral system that rotates around buried mainlines in a periodic-move configuration. Multiple field systems typically comprise large farm systems. When comparing field systems to full farm systems, factors like labor utilization, power economy, and distribution efficiency may change significantly. Field systems can be stationary, semi-portable, or totally portable. Both types of systems have been poorly planned as a result of the planner's and the owner's (or both) incapacity to understand the essential distinction between field and farm systems.

Many piecemeal systems with low distribution efficiencies, high initial costs, and high annual water application charges have resulted from failure to estimate the necessary capacity of the final layout. Although the system planner may not always be aware of whether future expansion is intended, it is still the planner's responsibility to notify the owner of any potential future development considerations when creating a field system plan.

6.3.1. Sprinkle Irrigation System Design

- A sprinkler irrigation system's design must take into account a number of important factors to guarantee effective water distribution and ideal crop growth.
- **Site Analysis**: Assessing the field's topography, type of soil, crop needs, and accessibility to water sources should come first. Determining the system's design parameters, such as water pressure

and flow rate, requires an understanding of these factors.

- **Water Source and Supply**: Locate a dependable water source, such as a reservoir, river, or well. Make sure there is enough water available to meet the needs of the entire irrigation system. It is important to take into account the water quality as well, since contaminants can clog sprinklers and lower system efficiency.
- **System Layout**: By segmenting the field into zones and allocating a distinct set of sprinklers to each, you can plan out the irrigation system. All zones should have uniform water distribution according to the design. Sprinkler type, water pressure, and intended coverage area should all be taken into consideration when determining sprinkler spacing and arrangement.
- **Sprinkler Selection**: Based on the kind of crop, the size of the field, and the requirements for water distribution, select the right kind of sprinkler heads. Sprinklers come in rotary, impact, and stationary varieties, each with varying rates of water application and coverage patterns.
- **Pipe Network Design**: Create a system of pipes that will move water from the sprinklers to the source. In order to minimize frictional losses, the pipe sizes should be selected to maintain sufficient pressure and flow throughout the system. Make use of materials that are both suited to the local environment and long-lasting.
- **Pressure Regulation and Control**: Install controllers, valves, and pressure regulators to control water flow and pressure in various zones. This makes it possible to precisely control the irrigation schedule and guarantees that the right amount of water is applied to each zone.
- **System Automation**: To maximize water usage and minimize manual labor, think about implementing automation features like timers, sensors, and smart controllers. Watering schedules can be modified by automated systems in response to crop requirements, soil moisture content, and weather.
- **Installation and Testing**: Installing the system components should be done after the design is complete. Test the system thoroughly after installation to make sure everything works as it should, with even water distribution and no leaks or pressure problems.
- **Maintenance Plan**: Create a maintenance schedule to maintain the system's effectiveness. Check sprinklers, pipes, and valves frequently for wear and tear. Replace or clean any damaged parts.

Maintaining the system properly will guarantee consistent irrigation performance and increase its lifespan.

6.3.2. Application Depth Requirements

Water requirements for crops and the necessary depths for irrigation application. To be thorough, it is appropriate to include some of the fundamental computations here. Conventionally, the required net application depth per irrigation for irrigation system design has been considered as a function of the crop root depth (Z), the soil water holding capacity (W_a), and the management allowed deficit (MAD), as follows:

$$AD = \frac{MAD}{100} W_a Z \tag{1}$$

Where;

AD = the maximum net depth of water to apply per irrigation, in or mm, also referred to as allowable depletion

MAD = management allowed deficit (usually 25 to 60%)

W_a = the water-holding capacity, in/ft or mm/m, a function of soil texture and structure, equal to field capacity minus wilting point

Z = the effective crop root depth, ft or mm

Field capacity (FC) for the majority of agricultural soils is reached 1–4 days following a full irrigation that leaves no water deficit in the root zone. If irrigation frequency is higher than required during the peak-use period, the actual water depth applied might be less than AD; however, this is not important for system design. Due to the imprecise nature of many values (such as crop, weather, soil, and other data), MAD can also act as a safety factor. It can be presumed that when actual soil water falls below MAD (for longer than approximately one day), crop yield and crop ET start to decline below maximum potential levels.

Agricultural soils typically have a water-holding capacity of 10–20% by volume. W_a is also known as available water-holding capacity (AWHC), water-holding capacity (WHC), and total available water (TAW). Keep in mind that, depending on the crop type, it might be more appropriate to base calculations of net irrigation depth on soil-water tension rather than soil-water content. This is a standard criterion for using tensiometers to schedule irrigations.

In the case of layered soils having different values for Wa, AD can be calculated by summing the W_a values according to thickness of each layer within the effective rootzone and applying the same MAD to each layer:

$$AD = \frac{MAD}{100}\left[W_{a1}Z_1 + W_{a2}\left(Z - Z_1\right)\right] \quad (2)$$

Where;

W_{a1} and W_{a2} = the water-holding capacities of layers 1 and 2

Z_1 = the depth of the top soil layer

Z = the effective root zone depth, assuming that $Z \leq Z_1 + Z_2$

In the case of three or more layers, equation 2 can be further expanded.

When designing periodic-move systems, the intervals between irrigations are typically maximized during the period of peak water use. This lowers the number of laterals needed, which lowers capital costs, and it also reduces labor requirements. Therefore, the maximum net application depth, d_n, used for the initial iteration of hardware design, is generally set equal to the allowable depletion, AD:

$$d_n = AD \quad (3)$$

Where;

d_n = allowable net water depletion from soil between irrigations, in (mm)

When designing fixed systems like solid set or continuous-move systems like center pivots, the amount of d_n applied during each irrigation cycle may be less than AD in order to minimize runoff potential and reserve a buffer in the soil reservoir to catch any substantial rainfall that may fall during the irrigation period.

6.3.3. Root Depth

Many references provide information on the typical plant feeder root and total root depth; however, soil conditions can affect the actual depths of rooting of different crops, so it is important to check these values on the spot. It can be used as a guide to estimate the effective root depths of different crops in situations where local data are unavailable and there are no anticipated restrictions on root penetration. The maximum rooting depths listed are expected, average values that annual crops reach close to the peak water use (and peak crop coefficient) period. When grown in a deep, well-drained, and adequately irrigated soil, the values indicate the depth at which crops will receive the majority of the water required. When there is less root depth, irrigation should occur more frequently with correspondingly smaller application depths at each irrigation.

6.3.4. Management-Allowed Deficit (MAD)

Irrigation should occur as infrequently as possible with periodic-move and low frequency continuous move systems, like linear moves and traveling sprinklers, in order to minimize labor expenses. For crops grown in arid and semiarid regions, it is generally recommended that the allowable depletion (AD) or soil moisture deficit in the root zone not be less than 50% of the total amount of water that can be held in the soil. The management allowed deficit (MAD) is the name given to this percentage. In general, especially with periodic-move systems, it is preferable to return the moisture level to field capacity with each irrigation. Every instance of wetting lasts the same amount of time. The planner or designer is encouraged to determine values locally based on local usage and production practices, as these are MAD percentages.

Although values of 20% or less are occasionally used for high-value, water-sensitive crops, a MAD of 40% is typically used for high-value crops or crops with shallow rooting depths. Certain low-value crops having deep roots may be treated with a 60% MAD. The MAD of 50% is commonly applied. The quantity and rate of water applied for irrigation are determined by local soil conditions, soil management, water management, and economic factors. The conventional design method has been to calculate the total amount of water required to fill the root zone to field capacity, then apply a larger amount in one application to account for leaching, evaporation, and application efficiency. The conventional method for determining the frequency of application is to set a value for MAD, compute the equivalent depth of water in the root zone reservoir that can be extracted without causing moisture stress, and estimate the duration of this supply based on the plant's daily consumptive use rate. As a guide for irrigation requirements and system operation, this kind of approach is helpful. For a system to be designed and operated as best it can, a number of other factors may influence the quantity of irrigation water and the timing of applications. In humid regions, it may be necessary to manage soil moisture to capture rainfall during the irrigation period; however, the MAD limitation on soil moisture depletion should be followed for design purposes.

6.3.5. Irrigation Interval

The maximum irrigation frequency is:

$$f_x = \frac{d_n}{U_d} \tag{4}$$

Where;

f_x = the maximum interval (frequency) in days

U_d = the average daily crop water requirement during the peak use period

The range of f_x values for agricultural crops is usually:

$$0.25 < f_x < 8 \text{ d} \quad (5)$$

The value of f_x is then rounded down to the closest whole number of days to determine the nominal irrigation frequency, or f′. If the values are conservative and f_x is close to the next highest integer value, then it is sometimes acceptable to round up. Also, f′ can be fractional if the sprinkler system is automated, and it can be further reduced to account for non-irrigation days, whereby $f \leq f'$.

The most demanding conditions are typically taken into account when designing an irrigation system, and the crop type and weather will determine how much water the crop needs during the peak-use period. For the same crop type, crop water requirements can therefore vary from year to year. Due to land preparation requirements, some crops may have their highest water requirements at the start of the growing season; however, these crops are typically irrigated by surface systems. When determining the system capacity for irrigating different crops, whether in the same or different seasons, the crop with the highest peak evapotranspiration (ET) should be considered. Because weather variations cause peak ET for a given crop and location to vary from year to year, it is important to take into account design probabilities for ET during the peak use period. Think about deficit irrigation as well; in situations where water supplies are scarce or costly (in relation to crop value), this may be a viable option.

6.3.6. Leaching Requirement and Gross Depth

Leaching might be required if the root zone is not sufficiently flushed by yearly rainfall and/or snowmelt, or if irrigation's deep percolation is minimal (i.e., high application efficiency and/or uniformity). If the irrigation water's electrical conductivity (EC_w) is relatively high, leaching might also be required. It might not be required to take leaching into account when designing a system if the irrigation water has a low electrical conductivity (system capacity). The design equation for leaching is:

$$LR = \frac{EC_w}{5\,EC_e - EC_w} \quad (6)$$

Where;

LR = leaching requirement

EC_w = EC of the irrigation water, dS/m or mmho/ cm

EC_e = estimated saturation extract EC of the soil root zone for a given yield reduction value.

When LR > 0.1, the leaching ratio increases the depth to apply by 1/(1–LR); otherwise, LR does not need to be considered in calculating the gross depth to apply per irrigation, nor in calculating system capacity. Gross application depth, d_g, is calculated when LR ≤ 0.1:

$$d_g = 100\left(\frac{d_n}{E_{pa}}\right) \quad (7)$$

And, when LR > 0.1,

$$d_g = \frac{90\,d_n}{(1-LR)E_{pa}} \quad (8)$$

For a full irrigation, the inevitable deep percolation resulting from nonuniformity is taken into account by the 0.9 value, which is multiplied by 100 to get the coefficient, 90, in equation 8. The irrigation water is too salty when the computed value of LR is negative, and the crop will either perish or suffer greatly.

6.3.7. Irrigations Per Season

The number of irrigations per season is estimated as:

$$N_{is} = \frac{U - P_e}{d_n} \quad (9)$$

Where;

N_{is} = the approximate number of irrigations per season

U = seasonal crop water requirement, in or mm

P_e = effective precipitation during the irrigation season, in or mm

The net depth is expressed in millimeters or inches per irrigation. Nis is the approximate number of irrigations that occur in a given season. In addition to performing other economic analyses when comparing alternative designs, the value of Nis can be used to estimate labor requirements for non-automated sprinkler systems.

6.3.8. System Capacity Requirements

The size of the area to be watered (design area), the quantity of water to be applied, and the amount of time allotted to apply that quantity of water determine the necessary capacity of a sprinkler system. Put another way, the minimum capacity is determined by the total area that is irrigated, the gross depth of water applied during the irrigation design, and the net operating time that is permitted for applying this depth. The required minimum capacity of a sprinkler system can be computed by equation 10, which is derived from a volume balance in which the product of average flow rate and duration is equal to the product of area and application depth:

$$Q = K\frac{Ad_g}{fT} \qquad (10)$$

Where;

Q = required minimum system discharge capacity, gpm or lps

A = design area, a or ha

d_g = gross application depth, in or mm

f = irrigation interval or frequency, days

T = actual operating time per day, h/d

K = 453 for English units, or 2.78 for metric units

The number of laterals, or sprinklers, that are operational simultaneously will be determined in later stages of sprinkler system design. A more precise estimate of the necessary system capacity can be obtained by multiplying the nominal sprinkler flow rate by the sprinkler count. This number will either equal or exceed the capacity determined by equation 10. The computed number of laterals in a periodic-move sprinkler system may be rounded up to a whole number, which would increase the necessary system capacity. Additionally, the number of sprinklers needed to cover the entire field area multiplied by the nominal sprinkler flow rate will determine the required capacity for a fixed (solid set) sprinkler system. This will almost always result in a larger value of Q than equation 10. In less than f days, irrigations can be finished when the actual system capacity needed exceeds that determined by equation 10.

Parameter f is less than or equal to f′, the maximum interval between irrigation events, determined by allowable net water depletion from soil between irrigations (d_n) and the design rate of daily water use (u_d):

$$f' = \frac{d_n}{u_d} \qquad (11)$$

Where;

d_n = net application depth, in (mm)

u_d = design rate of daily water use, in/d (mm/d)

After determining f′, the duration required to finish a single irrigation cycle is computed by deducting any days of inactivity from the operational time, which encompasses periods of system downtime due to malfunctions, instances where a sufficient water supply is unavailable, or occasions when the system is not utilized:

$$f = f' - \text{days off} \tag{12}$$

The type of irrigation system should be compatible with parameter f. Partial days, for continuous-move systems (e.g., f must be rounded down to the closest set interval of the system for periodic move systems, but 4.7 days) are practical. If hand lines are to be moved only once a day, for instance, f needs to be rounded to the nearest half-day (e.g., 6 days) or to a full day (an integer) for lines that will be moved twice daily, 6.5 days).

If downtime is necessary for lateral moves, filter cleaning, or other maintenance and/or repairs, Parameter T must be less than 24 hours per day. Because maintenance and sporadic emergency repairs must be made, even during the peak use period, the value of T in designs is nearly always less than 24 hours per day. Additionally, power outages can occasionally occur during irrigation, especially if the source is electrical lines. In certain instances, the computation of T can be significantly reduced to account for the billing rates for electrical consumption based on the time of day. This can be achieved by comparing the annualized fixed costs with the uniform annual operating costs, which include the power required for pumping, over the system's expected useful life.

6.3.9. Fixed Systems

The ability to utilize fixed systems for crop cooling, frost protection, high-frequency irrigation, and regular irrigation provides additional benefits. For the application of other agricultural chemicals and water-soluble fertilizers, all fixed systems are perfect. However, when estimating the system capacity needed for each of these uses, extra care must be taken.

- **Ordinary Irrigation**: A few permanent crops have fixed systems installed, and because of the significant root development of the crop and long-term crop health concerns, relatively long irrigation intervals are frequently utilized. Because there is no downtime during lateral moves, the capacity of such systems can be 5 to 10% less than that of traditional periodic-move systems in the same area. When the system is run 24 hours a day, 7 days a week, the capacity should be enough to apply the peak net crop water requirements for low frequency irrigations (every one or two weeks). These systems, which can be automated or operated by hand valves, can be used to apply fertilizer and other chemicals.

- **High Frequency**: A larger system capacity is needed if the system is intended to apply irrigations once or twice a day, for example, to regulate soil temperatures during germination and to maintain the soil-moisture content within a specific range. Because the crop may be consuming water at the peak daily potential ET rate, the net system capacity should be increased by 10 to 20% over a conventional periodic-move system. In contrast, because the average number of days with a somewhat lower potential ET occurs during lower frequency irrigation, the average consumptive use rate over a longer period of time is lower than the peak daily potential rate. Maintaining the crop at its peak ET rate to boost crop quality and yield could be one of the main goals of such a system. It is obvious that crops that are not particularly good candidates for solid-set (fixed) sprinkler systems are those that do not respond well to continuously high soil moisture conditions (due to disease, molds, and other issues). Although hand valve operation is possible, high-frequency systems are typically automated. Applications of water-soluble fertilizers and other agricultural chemicals are usually performed by automatic valve systems.
- **Crop Cooling**: Foliar cooling high-frequency systems need to have up to twice the capacity of standard high-frequency systems, automatic valving, and high-quality water. The arrangement of foliar cooling systems keeps the leaves wet. The leaf surfaces are wetted with water, allowed to dry slightly, and then wetted again. Depending on the system capacity, crop size, and weather, this usually entails running one-sixth to one-quarter of the system at a time and cycling the system once every 15 to 40 minutes. A tree cooling system, for instance, could run six times every 30 minutes, sprinkling a fifth of the area at a time. It is imperative that foliar cooling systems possess adequate capacity to meet the minute-by-minute evaporation demand during peak use days and hours. In order to achieve this, the system capacity needs to be 1.5–2.5 times larger than what is needed for a traditional periodic-move system. Except from offering complete frost protection, these systems can be used for all the purposes previously mentioned.
- **Frost Protection**: The lowest expected temperature, the type of frost (radiant or advective), relative humidity, wind direction, crop height, and sprinkler cycle time (or turning speed) all affect the

system capacity needed for frost protection. For overhead freeze control to function, there must always be a steady supply of water available. The main source of the protective effect of sprinkling is the latent heat released during the actual freezing of the water, which is equivalent to 144 British thermal units (BTU) per pound (334 kJ/kg). Furthermore, as the water cools to the freezing point, it releases a tiny quantity of heat (1 BTU per pound of water per degree Fahrenheit temperature drop, or 4.2 kJ/kg/°C). Humidity and temperature effects can be combined with dew point temperature. Generally speaking, an application rate of 0.15 inches per hour (4 mm/h), or 65 gallons per minute per acre (10 l/sec/ha), should provide overhead freeze protection down to a dew point temperature of 20°F (–6.7°C), provided that sprinklers are turning faster than 1.0 RPM and winds up to 1 miles per hour (0.45 m/sec). The application rate can be adjusted by 0.01 inches per hour (0.25 mm/h) for each degree above or below a dew point temperature of 20°F (–6.7°C). This is equal to 4.3 gallons per minute per acre (0.67 lps/ha).

Frost protection systems must be activated prior to the dew point temperature falling below freezing and must continue to run until the next morning, when all of the ice has melted. In areas where dew point temperatures are likely to drop significantly over several days, trees may be so severely damaged by the ice load that overhead freeze control is not feasible. Use over-crop sprinkler systems with limited capacity where they are rapidly sequenced and under-tree sprinkler systems with every other sprinkler turned on to protect against minor frosts with dew point temperatures of 28 or 29°F (–2°C). Only 25 to 30 gallons per minute per acre (4 to 5 lps/ha) may be needed for such systems. Even so, these rates are four or five times higher than those of systems intended solely for irrigation. Pump, mainline, and lateral sizes are therefore noticeably larger.

- **Bloom Delay:** By cooling woody plants with a light sprinkle during the dormant season, bloom delayers postpone budding until there is a reduced chance of a damaging frost. These systems resemble crop cooling systems, but they usually cycle so that only half of the system is using energy at once. Considerations pertaining to equipment and distribution uniformity control the system's ability to accomplish this. An application rate of 0.10 to 0.12 inch per hour (2.5 to 3 mm/h) is about as low as can be practically achieved with ordinary impact sprinklers. Operating half of such a system simultaneously over a field may require 22 to 26 gallons per minute per acre (3.4 to 4 lps/ha).

6.3.10. Continuous-Move Systems

The fully automatic nature of center-pivot systems makes it reasonably simple to regulate soil moisture levels.

- **Ordinary Irrigation**: For regular irrigation, center-pivot systems possess the same characteristics as fixed systems. But the likelihood of a mechanical failure is higher. When designing center-pivot designs, it is customary to assume irrigation 90% of the time (21.6 h/d), with 10% of the time set aside for any necessary repairs and maintenance during the peak-use period. The system capacity is reduced by assuming that a center pivot will run continuously during peak use, but there is a chance that it won't be able to apply the necessary water if the system is unable to run continuously for any reason—mechanical, electrical, or otherwise.
- **High Frequency**: Fixed and center-pivot irrigation systems should have comparable capacities when utilized for the same purposes. These remarks also apply to situations in which a low water-holding capacity soil-crop system is utilized in arid areas to minimize runoff through high-frequency irrigation. The system's maximum speed and coverage area will set a limit on how often irrigation is permitted.
- **Limited Irrigation**: Irrigation rates approximately 10% lower than the peak-use rate can be used during the peak-use period on crop-soil systems with 5 inches (125 mm) or more of water storage capacity without significantly affecting the yields of many crops. When the system capacity is insufficient to meet crop moisture withdrawal rates during peak-use periods, it is feasible to gradually reduce deep soil moisture by using light, frequent irrigation. However, light, frequent irrigation can lead to significant water loss due to evaporation from wet soil and vegetation and may be an inefficient use of a limited water supply when subsoil moisture storage is insufficient. In certain circumstances, regular irrigation may exacerbate salinity issues as well. In these circumstances, fewer, deeper irrigations could result in higher yields.

6.3.11. Intake and Optimum Application Rates

The rate at which water should be applied depends on:

- The amount of time needed for the soil to absorb the applied depth without runoff under the specified soil, slope, and cover conditions. The maximum application rate is calculated by dividing the depth of application by the necessary time.

- The lowest possible application rate that, in the majority of climate scenarios, will produce even distribution and acceptable efficiency—or that is feasible given the system chosen. The minimum rate of application for the majority of irrigated crops is roughly 0.15 inches per hour (4 mm/h) in order to achieve high efficiency and a reasonably good distribution under ideal climatic conditions. In order to lessen issues related to wind drift, the minimum application rate needs to be higher in areas where high temperatures and wind speeds are frequent. It takes judgment and experience to determine the minimum application rates for the circumstances in a given area.
- How long does irrigation take to accomplish effective use of labor available in harmony with other farm operations?
- The application rate was modified based on how many sprinklers were running in the most practical system configuration.

Generally speaking, the longer the soil is wet, the lower its rate of infiltration. Plan your application rate so that by the end of the irrigation, it doesn't exceed the soil's ability to hold onto water. Applying water at the expected sprinkling intensity of the system selected on crops, soils, and slopes similar to the expected site conditions is the ideal way to develop intake versus time of application information. Although it can be challenging to set up an experiment to observe it, this information is frequently obtained by looking at an existing system. Due to the effects of surface sealing by sprinkler drops and the fact that the sprinkle application rate is typically lower than a ponded infiltration rate, intake curves for sprinkler irrigation may differ from those obtained from ponding studies. Drop impact seals the surface and lessens infiltration on bare soils. A falling drop's kinetic energy is calculated by multiplying its mass by half and its velocity by square. The terminal falling velocities of drops vary from approximately 6 to 30 feet per second (2 to 10 m/sec), with sizes ranging from 0.5 to 5.0 mm. Most impacts from impact sprinklers are nearly at their terminal velocities, with a typical fall distance of roughly 10 to 20 feet (3 to 6 meters).

The amount and kind of crop cover, the application rate, the texture and structure of the soil, and the drop impact all affect how much the surface seals and how much infiltration is reduced. Three distinct bare soils were used to test the general relationship between drop size and reduction in infiltration rate at an application rate of roughly 0.5 inches per hour (13 mm/hr). About 20 minutes after the application started, the reduction in the infiltration rate on the recently tilled, heavy textured soil approached its maximum level. Because of the increased velocity and turbulence inside the nozzle orifice and in the jet, drop size decreases as pressure rises or as nozzle size decreases. Jet breakup can also be produced by reducing drop sizes without the use of high pressure.

For instance, some nozzle designs use sharp orifices or triangular, rectangular, or oval orifices instead of tapered, circular nozzles; others use impinging jets of water. Some even have pins that stick into the jet close to the nozzle orifice. Fixed or rotating plates are frequently used by low pressure nozzles on continuous move systems to split and deflect drops. Due to the high cost of energy, smaller (i.e., low pressure drop size (medium) has grown in popularity.

With impact sprinklers, a circular wetted area is created. However, only a tiny portion of the entire wetted area receives all of the water in the jet at any given time. The concentrated area's application rate is higher than the soil's rate of instantaneous infiltration. The extra water ponds briefly and forms a film on the soil's surface, lubricating the particles and removing any surface tension forces that could otherwise hold the grains of surface soil in place. Soil particles are dislodged by droplets hitting a wet surface, become suspended, and eventually settle out on the soil's surface. Compaction and surface sealing result from some of these particles being absorbed into the soil by seeping water. Impact sprinklers concentrate drop energy more than spray and rotating plate nozzles used on center pivots because they typically distribute water in all directions. Nonetheless, certain nozzle systems generate large drop sizes in an attempt to extend the drop travel's distance, which can result in comparatively high energy levels for each individual water drop.

REFERENCES

1. Al-Baaj, A. A. A., & Lewis, A. (2019). Variable pulsed irrigation algorithm (VPIA) to reduce runoff losses under a low-pressure lateral move irrigation machine. *Horticulturae, 5*(1), 10.
2. Benami, A., & Ofen, A. (1984). *Irrigation Engineering*. Irrigation Engineering Scientific Publications (IESP).
3. Beyaert, R. P., Roy, R. C., & Coelho, B. R. B. (2007). Irrigation and fertilizer management effects on processing cucumber productivity and water use efficiency. *Canadian Journal of Plant Science, 87*(2), 355–362.
4. Hallmark. (2004). *Hallmark Aquaequipment Pvt. Ltd.* 208, Rash Behari Avenue, 2nd Floor, Kolkata.
5. Howser, H. (2010). Rainbird–California's gold (12002). *California's Gold.* Chapman University Huell Howser Archive.
6. Liang, Z., Liu, X., Zou, T., & Xiao, J. (2021). Adaptive prediction of water droplet infiltration effectiveness of sprinkler irrigation using regularized sparse autoencoder–adaptive network-based fuzzy inference system (RSAE–ANFIS). *Water, 13*(6), 791.
7. Michael, A. M. (1978). *Irrigation Principles and Practices*. Vikas Publishing House Pvt. Ltd.
8. Million, J. B., & Yeager, T. H. (2015). Capture of sprinkler irrigation water by container-grown ornamental plants. *HortScience, 50*(3), 442–446.
9. Morgan, R. (1993). *Water and the Land.* Adams Publishing Corp.
10. Rungta. (2004). *Sprinkler Irrigation System.* Electronic Complex, Kushalguda, Hyderabad.
11. Souza, E. A., Coelho, E. F., Santos, M. R., & Melo, D. M. (2021). Agronomic performance of 'BRS Princesa' banana under fertigation and mulching. *Semina: Ciências Agrárias, 42*(3), 979–998.
12. Thomas, J. M., Thomas, T., Stuetz, R. M., & Ashbolt, N. J. (2014). Your garden hose: A potential health risk due to *Legionella* spp. growth facilitated by free-living amoebae. *Environmental Science & Technology, 48*(17), 10456–10464.
13. Tsakiris, G. P. (1985). Infiltration efficiency in sprinkler irrigation. *Advances in Water Resources, 8*(1), 44–47.

Chapter 7

Evaluation and Optimization of Subsurface Irrigation (SDI) System

Contents

INTRODUCTION

Water is applied below the soil's surface using emitters that typically have discharge rates within the same range as drip irrigation, a practice known as subsurface drip irrigation (SDI). Thus, aside from the specific details pointed out, subsurface (SDI) unit is simply all the drip irrigation network buried at a certain depth.

A technique called subsurface textile irrigation (SSTI) was created especially for subsurface conditions, such as heavy clays or desert sands. Utilizing SSTI will drastically cut down on the amount of water, fertilizer, and herbicide used. It will reduce ongoing operating expenses and have a decades-long lifespan with proper maintenance. Plants are healthier and produce much more when water and nutrients are given directly to them. Because water never reaches the surface, it is the only irrigation system that can be used safely or without the need for costly treatment. In contrast to standard, the geotextile moves water along the fabric up to 2 meters from the dripper, so the distance between emitters in the drip pipe is not important.

The depth at which the irrigation laterals are buried varies from 13 to 20 inches, contingent upon the type of soil and crop, climate, and management techniques, among other factors. In addition to preventing runoff, burying drip lines underground reduces the amount of surface soil evaporation. The opportunity to directly supply the crop root zone with the appropriate amount of crop water and nutrients is made possible by SDI. Water and nutrients can be spoon-fed to crops. When handled correctly, the spoon-fed feature of the SDI system has the great potential to reduce or completely eliminate the movement of water and nutrients below the crop rootzone. Before being applied to the field via drip laterals in an SDI system, the irrigation water is filtered at the filter and control station. When used correctly, the SDI has much fewer irrigation application losses than the other irrigation system (wind drift, soil evaporation, deep percolation, and runoff).

Through buried polyethylene, subsurface drip irrigation delivers water and nutrients straight to the crop root zone. To maintain an acceptable level of irrigation uniformity for a range of field lengths, drip lines come in different diameters and thicknesses. When short lateral lengths are needed, drip lines with a smaller diameter are employed. A larger diameter drip line needs to be chosen as lateral length increases in order to keep irrigation uniformity sufficient. The durability of a drip line wall is directly correlated with its thickness. Small-thickness drip lines are typically utilized for transient installations that will be removed quickly, like when high-value crops are being irrigated. When installing driplines permanently, thicker ones are used. Higher operating pressures are also capable of withstanding the thicker drip lines. The diameter and thickness of the drip line are directly correlated with its coast. Along the drip line, tiny holes known as emitters are often spaced every 8 to 24 inches.

Drop by drop, pressure drives water out of the emitters during irrigation. Once in the soil, the water's movement and pattern of wetting will be determined by the soil's physical properties, flow rate, and initial water status. Whereas water tends to move primarily downward in sandy soils, it will tend to move laterally and upward in fine-textured soils. Operating pressure, emitter size, emitter design, drip line diameter and spacing, and emitter spacing all affect how much water can be delivered through a drip system. Today, a variety of drip lines are offered by various manufacturers to meet the unique design specifications for various types of soil, crops, and weather.

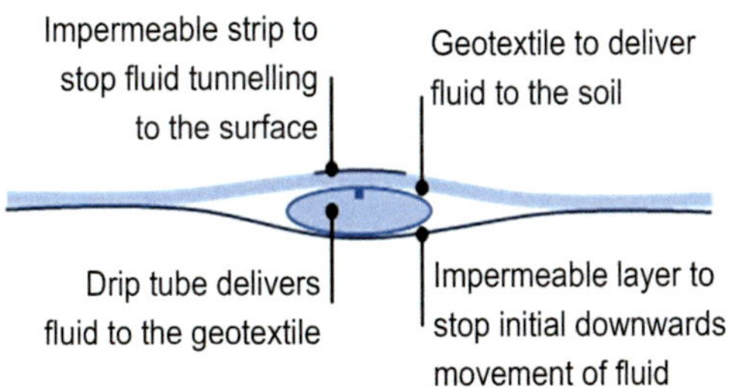

Figure 7.1. *Represents, diagram SSTI is installed 30–50 cm below the surface for agricultural applications.*

Source: Abou Seeda M.A., Hammad S.A., 1Yassen A.A. and Abou El-Nour E.A., Evaluation and Optimization of Subsurface Irrigation (SDI) System: A review, DOI: 10.36632/mejas/2020.10.3.45.

Many different types of plants are irrigated using SDI, including alfalfa, corn, cotton, grass, pepper, broccoli, melon, onion, potato, tomato, and herbaceous crops like lettuce, celery, asparagus, and garlic, as well as woody crops like citrus, apple trees, and olive trees. Although subsurface drip irrigation (SDI) uses less water than drip irrigation, its efficiency may be comparable. When compared to surface irrigation, it may save between 25 and 50% of the water. The features of SDI will be discussed in detail, and some design and management guidelines will be emphasized (Figure 7.1). Currently, rain-fed agriculture accounts for 60% of global food production and occupies 80% of all agricultural lands. However, the IPCC's assessment of climate change poses a serious threat to rain-fed agriculture's sustainability in some areas, endangering not only food security but also the social structure in many nations. Better water availability in turn ensures optimal yields from a given combination of inputs because it directly influences the efficient use of all other inputs.

Therefore, the ideal development of emerging irrigation technologies would be to improve crop water availability in order to make agricultural practices long-term sustainable. Important corrective actions against a changing climate can be taken by switching from rain-fed to irrigated agriculture or from low-efficiency to high-efficiency irrigation techniques. However, farmers' resistance to adopting practices that increase operating costs, like high-efficiency irrigation techniques, may prevent adaptation. Therefore, if new irrigation technologies were created with the least amount of labor, energy, and precious water resources in mind, they would have a higher chance of being adopted. There are several irrigation techniques to choose from, including drip, sprinkler, surface, and subsurface irrigation. According to Ali, the choice of irrigation technique depends on social, economic, and physical factors, which in turn affects the sustainability, economic viability, and efficiency of upland farming systems. Historically, surface irrigation techniques like borders, basins, and furrows have been used to irrigate upland crops in many parts of the world because they are inexpensive and easy to use. Surface irrigation techniques use gravity to move water across the whole field or along furrows. Water seeps into the soil when it flows, giving crops' root zones irrigation water.

Because the level of land leveling determines the uniformity of distribution and application efficiency, high labor costs are associated with land preparation. Subsurface, sprinkler, and drip irrigation techniques have partially supplanted surface irrigation due to rising labor and energy expenses as well as the growing need for dwindling water supplies. Upland crops are still primarily irrigated by surface irrigation, which is still the case globally.

For high-frequency crop irrigation, drip and sprinkler irrigation systems were created using emitting devices and a methodically constructed pipe network. Depending on the kind of irrigation system and water source, water in drip and sprinkler systems is supplied under pressure and frequently passes through a variety of filters. The high efficiency of water application, superior fertilizer application, and increased crop yields associated with sprinkler irrigation surpass those of surface irrigation. Sprinkler irrigation includes solid sets, periodic move or continuous move systems, traveling guns, and boom sprinkler systems. It does, however, have certain disadvantages, including high setup and operating costs because of the high energy and maintenance requirements, as well as a propensity to be negatively impacted by wind conditions. However, drip irrigation overcomes some of these shortcomings because it uses less energy and is not impacted by wind. It supports crop agronomy in addressing the issues associated with irrigated agriculture and boasts unique features that save energy and water. The main problem with drip irrigation systems is that their emitters can become clogged, resulting in poor performance and frequent maintenance needs. Moreover, one reason why farmers are still reluctant to adopt such a promising technology is that it is susceptible to damage from weathering and farm machinery. (Evans et al., 2007)

Subsurface drip irrigation systems have been developed to address the practical issues with drip irrigation that still exist, but their performance has fallen short of expectations because emitter clogging exacerbates the issue of poor water distribution efficiency. In order to improve profitability and sustainability, new irrigation techniques are being developed to irrigate upland crops with the goal of using water more effectively and efficiently while minimizing costs.

7.1. HISTORY OF SUBSURFACE IRRIGATION

More than 2,000 years ago, in China, subsurface irrigation was first mentioned where water was kept in clay containers that were buried in the ground. Water slowly permeated the soil, moistening the roots of the plants. The modern form of SDI emerged in the US around 1959 as a variation on drip irrigation, particularly in California and Hawaii. Polyethylene PE or polyvinyl chloride PVC plastic pipes with punched emitters or holes were used for SDI laterals in the 1960s. The water quality and filtration systems used to determine the low pressure at which these systems operated. By the 1970s, this technique had been applied to crops like avocado, corn, potatoes, pineapple, citrus, cotton, fruit trees, and grass. The primary drawbacks of this system were the intricate upkeep requirements, the uneven distribution of water application, and the clogging of emitters primarily from soil particles and oxide. As an alternative, Israel developed the machinery needed to install SDI in the field and began injecting fertilizers into irrigation systems. Additionally, because there was less clogging, the quality of commercial emitters and laterals increased, and SDI became more respected than other irrigation techniques. Early in the 1980s, a number of factors contributed to the increased interest in SDI, including lower costs for emitters and pipes, better fertilizer application, and field units that were maintained for several years.

7.2. SUBSURFACE DRIP IRRIGATION IN THE WORLD

There is widespread agreement regarding the spread of SDI, but gathering data to support this trend is challenging because most surveys count surface irrigations with SDI as drip irrigations. Only about 27% of the land area in the US is used for drip and subsurface drip irrigation, according to the USDA Farm and Ranch Irrigation Survey. However, as farmers replace their traditional drip irrigation techniques with SDI systems, this percentage is steadily rising over drip irrigation. The potential for SDI to rise would still exist if this framework persisted.

Many references to the so-called low-cost SDI with simple laterals and low emitter pressures are found in developing nations, which emphasizes the SDI potential increase even more. The energy and water savings also improve this.

7.3. MANAGEMENT OF MICRO IRRIGATION IN ARID AND SEMIARID REGIONS

In the meantime, conserving water, enhancing efficiency over the long run, and raising the yield to water consumption ratio rise to the top of the priority list in order to overcome the restricted supply of irrigation water in arid and semiarid areas. When combined with alternative cropping options, micro irrigation technology is thought to be a more water-efficient option that can be used in these areas. The slow application of water on, above, or below the soil surface using bubblers, micro sprinkler systems, subsurface drip irrigation, and surface drip irrigation is known as micro irrigation. Water is applied with emitters or applicators along a water delivery line next to the plant row, or as discrete or continuous drips (also called trickles), tiny streams, or miniature spray troughs. To highlight that only a portion of the soil volume is wetted, micro irrigation is sometimes referred to as localized irrigation in different parts of the world. The following characteristics typically apply to micro irrigation: (i) low rates of application; (ii) long application times; (iii) frequent application intervals; (vi) application of water close to or into the root zone; (v) low-pressure delivery system; and (vi) routine application of water for the purpose of transporting fertilizers and other agricultural chemicals.

Numerous researchers have documented a number of advantages of micro irrigation, including: (a) higher crop yields and quality; (b) reduced non-beneficial use by minimizing surface evaporation and irrigation runoff from hillsides or sloping fields; (c) decreased deep percolation; and (e) enhanced fertilizer and other chemical applications with less leaching and runoff. The economics and mechanics of applying water with micro irrigation systems for specific soils, water qualities, and environmental conditions have shown a number of drawbacks, despite the successes that have been noted and the potential benefits. The drawbacks are as follows: (1) high initial and ongoing expenses; (2) complex and specialized knowledge is needed to set up the irrigation system; (3) possible blockage of emitters as a result of solid and root particle intrusions; and (4) when high salinity water is utilized in arid areas, salt buildup close to the plants. In dry and semi-arid regions, persistent microirrigation use eventually causes changes in the water table and frequently causes issues with salinization and water logging. Groundwater is frequently contaminated by nitrate, which is produced when nitrogen fertilizers are used in agriculture. Nitrate has a negative impact on animals, humans, and the ecosystem. Pesticides, heavy metals, and general plant nutrients are also contaminants linked to irrigation, in addition to nitrates. When wastewater is used without first being treated for irrigation, pharmaceutical compounds become pollutants. (Siemens et al., 2008).

Persistently using saline water for irrigation via surface drip systems could lead to salt buildup near the soil's surface. In traditional surface drip irrigation (DI), salts that accumulate near the surface during watering or rainfall can subsequently move downwards and penetrate the primary root zone.

This procedure may prevent the uptake of water and nutrients, which would impact crop growth and yield. A workable solution was suggested to counteract the osmotic shock that the leached salts imposed on the crop. It was suggested that by continuing to drip irrigate in tandem with the periods of precipitation, more salts would be leached from the main root zone into deeper layers. Conversely, Oron et al. along with Phene et al. have suggested that using a subsurface drip-irrigation (SDI) system to apply salt water can solve this issue. It is expected that under SDI, the risk of damaging the plants' main roots will be reduced because the salt front will be partially driven down into the deeper soil bulk media and to the periphery of the root zone. Furthermore, the inhibiting effects of the salts in the saline water are offset by the improved moisture conditions close to the emitter. Furthermore, Phene et al. stated that because the SDI is buried in the soil, well-managed systems may enhance the benefits of traditional DI systems, particularly in the areas of effective nutrient and water use, salinity control, and deep percolation.

7.3.1. Subsurface Drip Irrigation Method for Improve Crop Quality and Productivity

One type of micro irrigation is subsurface irrigation, which applies water below the soil's surface using buried emitters such as drip tapes, perforated pipes, clay pipes, and tubes. This technique can guarantee that the root zone and crop water requirements receive an adequate amount of water and nutrients on time, potentially increasing crop productivity and quality. Weed growth can also be inhibited by the relatively dry soil on the surface during subsurface application. Increasing in the working life of the system as vandalism, and solar radiation degradation are avoided; ploughing and other cropping practices are facilitated, and the development of fungal disease are diminished. In addition, subsurface irrigation has a special advantage of securing system safety against pilferage and damage by animals and during intercultural operations. Subsurface irrigation gained popularity in the early 1980s, picked up speed in the latter part of the decade, and is still growing today, mostly in areas with limited water supplies, irrigation-related environmental issues, and wastewater irrigation.

Subsurface drip irrigation (SDI) is currently the most widely used type of subsurface irrigation. Camp examined the findings of a few earlier studies that contrasted the crop yield in subsurface irrigation with that of other distinct irrigation techniques. He came to the conclusion that, in every situation—including those involving various crops, soil types, and cultivation conditions—crop yields from subsurface drip systems were on par with or higher than those from the other systems. A few comparative studies revealed that drip irrigation under subsurface cover resulted in slightly higher drip grapevine water uptake and coefficient than surface drip. Miguel and Fransisco discovered that when subsurface drip irrigation was used instead of surface drip irrigation, tomato yields increased by 66.5%. Subsurface drip irrigation (SDI) offers many benefits, but there are also many problems that need to be resolved because the technology and equipment development for it are still in their infancy. Numerous studies have focused on the hydraulic properties of embedded emitters and the variables influencing their flow rate. Consider Xu et al. investigated the impact of operating pressure and emitter type on flow rate through the use of an orthogonal experimental design. Lazarovicth along with others said that soil hydraulic properties could have an impact on SDI dripper discharge rates, unlike surface drip irrigation. In the soil close to a subsurface dripper, positive water pressure backpressure may form during irrigation, adhering to the concepts of point source flow. The amount of pressure that develops is determined by the hydraulic properties of the drip system, the soil texture, the dripper's discharge, and any potential cavity development near the dripper. Reduce the discharge from subsurface drippers by developing positive soil water pressures; this effect is especially noticeable when non-pressure-compensated drippers are operating at low pressures. (Shani et al., 1996)

Furthermore, differences in the hydraulic properties of the soil in a field and in the cavities formed during installation lead to variations in the pressure build-up near each dripper, which in turn results in uneven water application and discharge. As a result, the discharge of drippers should be chosen such that it closely matches the rate at which water is absorbed by the roots. The majority of SDI systems are made to function at pressure heads higher than 7 m. Typically, pumps are required to generate these kinds of pressures. A growing number of people are interested in SDI systems that can operate at much lower pressures, such as 1.5–2.0 m. These systems have the potential to reduce the need for pumps, increase the irrigated area without requiring a larger pump, or decrease the discharge rate without reducing the orifice size. Adoption of low-pressure systems may be aided by the fact that they are typically less expensive than high-pressure systems. These SDI systems could potentially be used in conjunction with low discharge continuous irrigation to maintain a high root zone, soil water content, and hydraulic conductivity—all of which would increase the amount of water available to plants.

Since access to electricity can be a challenge in developing nations, low-pressure SDI systems may be especially helpful there. Nevertheless, soil-induced variations in dripper discharge are especially common in low pressure systems. Subsurface irrigation's ability to provide effective irrigation water use and be properly designed, managed, and operated should all be measured.

Users can quantify and regulate the emitter or dripper discharge, system operating pressure, the impact of backpressure on dripper discharge, the creation of cavities, and the application of irrigation water to meet crop water requirements. The distribution of crop roots and the characteristics of the soil must be taken into account when determining the emitter's proper depth. Understanding crop responses to irrigation and fertigation requires knowledge of root distribution, particularly in light of the small amount of wetted soil volume that forms under subsurface drip. Phene and associates. proved that shallow subsurface drip irrigation applied frequently could increase the root length and rooted soil volume of sweet corn. They showed that regular irrigation kept part of the root zone within the range of ideal matric potential. Whether drippers are positioned on top of or below the crop row, the soil water distribution beneath them largely determines the root length density and water uptake patterns in highly frequented corn. When processing tomatoes are drip-irrigated, the majority of the root system is located in the top 40 cm of the soil profile. In heavy clay soils, the supply of aerated water combined with subsurface drip irrigation can maintain aeration of the root zone and greatly increase vegetable yield, particularly for zucchini and soy beans. The depth of the drip tape in subsurface drip irrigation is determined by various factors such as the crop, soil, climate, and planned cultural practices. Generally, the drip tape depth falls between 0.02 and 7 m. For shallow-rooted horticulture crops, it typically falls between 0.05 and 0.2 m. (Camp, 1998)

The primary goal of reducing soil evaporation and capturing the potential benefit of improving water use efficiency (both in yield and quality) studied on SDI under different system design, installation, application, and operation may require a deeper placement of 0.45 m, even though horticulture is the primary reason for choosing installation depth. For multi-year or long-term use of the SDI system, they recommended installing lateral SDI at a slightly deeper depth for finer-textured soil and at a shallower depth for coarser-textured soil. One of the key elements of any subsurface irrigation system that helps to achieve greater irrigation efficiency is irrigation scheduling. Numerous variables, including soil, crop, climate, water availability, and cultivation techniques, affect when and how much irrigation is needed. As a result, creating an effective irrigation schedule in the context of the current local circumstances is crucial. Numerous techniques are employed for scheduling irrigation in many cultivation crop practices, including the estimated crop evapotranspiration rate, the ratio of irrigation water to cumulative pan evaporation, and soil water depletion. The maximum yield in a subsurface drip system was achieved by applying irrigation at 100% crop evapotranspiration, with the drip tape positioned at a depth of 10 cm.

7.3.2. Irrigation Water Productivity and Water Use Efficiency

Water productivity can be used as an indicator on farms to assess how well an irrigation system is working. The phrase "water productivity" has multiple definitions. It can be understood as the physical ratio of yields to water use (productive or unproductive), or as the relationship between water use and product value.

The ratio of an (agricultural) product (biomass, yield) to the volume of water lost or diverted is known as water productivity, or WP. The population of humans is expected to reach 9 billion by 2050, and 70% of the extra food that will be needed will need to be produced in the next 40 years. In order to overcome this difficulty, modern agriculture places a high value on the "Water Productivity term," which tries to raise yield production per hectare per unit of water used. An increase in water productivity serves as an indicator and a means of coping with the predicted effects of climate change. The most significant development affecting water, according to expert survey respondents in the World Water Scenarios Project of the United Nations World Water Assessment Program (WWAP), is an increase in agricultural water productivity.

In agricultural production systems, "producing more food with the same water resources" or "producing the same amount of food with less water resources" are the main goals of the concept of water productivity. The ratio of the net benefits from crop, forestry, fishery, livestock, and other mixed agricultural systems to the amount of water used in their production process is known as water productivity. It quantifies how well a system uses water to produce goods and services. Water productivity can be simply defined as "crop production" divided by "amount of water used." WP is computed using the following basic equation: WP = Output derived from water use/water input. Depending on the goal of the study, there are several options for how to express the water input.

- Irrigation diversion
- Gross/net inflow
- Evapotranspiration
- Precipitation

The water output or the benefits can also be measured with various terms, for example with:

- Economic value (monetary)
- Physical mass (kilogram)
- Nutritional value (calories)

The ratio of the total dry matter of the harvested (economic) portions of the crop produced to the amount of water consumed is known as water use efficiency, or WUE. Micro irrigation has been shown to significantly increase yield without significantly increasing the amount of water used, improving water use efficiency. Using subsurface micro irrigation for intercropping, or even multiple cropping, of vegetable and melon crops resulted in higher yields, better water use efficiency, and fewer water applications between cropping seasons.

An effective strategy to improve water use efficiency in areas where urbanization and rising water prices have occurred is continuous cropping. The productivity of irrigation water is measured by the yield of plant product per unit of irrigation water used during cultivation, which is known as the irrigation water use efficiency. Drip irrigation water efficiency is largely reliant on evaporation losses from the permanently wet soil below emitters. One strategy to reduce this inefficiency is the introduction of subsurface drip irrigation. Many techniques for reducing evaporation, like mulching, have been used in irrigation.

7.3.3. Subsurface Water Flow

In order to effectively deliver fertilizers, pesticides, and irrigation water to the root zone, subsurface irrigation systems are growing in popularity. Improving irrigation water usage efficiency is a key strategy for dealing with water scarcity in arid and semiarid areas. One subsurface irrigation technique that has been developed and implemented recently in certain arid regions of Indonesia to cultivate both annual and perennial crops is ring-shaped emitter subsurface irrigation. Small-scale farmers in these rural areas can afford to make a ring-shaped emitter out of a regular water hose.

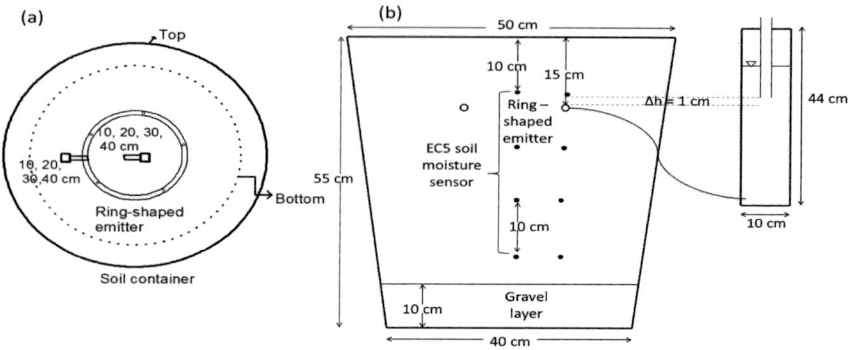

Figure 7.2. *Top view (a) and cross-section; and (b) of the soil container used in the laboratory experiments with a ring-shaped emitter.*

Source: Abou Seeda M.A., Hammad S.A., 1Yassen A.A. and Abou El-Nour E.A., Evaluation and Optimization of Subsurface Irrigation (SDI) System: A review, DOI: 10.36632/mejas/2020.10.3.45.

Small-scale farmers with limited water resources can benefit from a low-cost subsurface irrigation system. In Indonesia, a ring-shaped emitter for subsurface irrigation has been developed and introduced using a regular rubber hose. It is an inexpensive irrigation system built with locally available resources and expertise. A rubber hose with a diameter of roughly 20 cm is bent into the shape of a ring, and five 5-mm holes are drilled into it at even intervals to create the original design's ring-shaped emitter. Subsequently, a permeable fabric is wrapped around the entire ring-shaped hose to allow water to seep into the area surrounding the underground emitter. By modifying the pressure head imposed at the emitter's inlet, water is applied from a water tank that is connected to the emitter. The ring-shaped emitter has been used successfully in practice, but its performance has not been thoroughly assessed.

Furthermore, it could be challenging to find and fix any problems with the ring-shaped hose because it is completely covered in textile. It is crucial to create an emitter that is simple to maintain if you want small-scale farmers in arid or semi-arid areas to adopt ring-shaped emitters for subsurface irrigation. This study suggests changing the covering technique and reducing the number of holes. Numerical simulations can be conducted in addition to experiments because the number of experiments that can be conducted to assess the performance of alternative ring-shaped emitters is typically limited. Thus, the primary goals of this study were to: 1) examine the water movement around a buried emitter in the form of a ring; and 2) assess numerically the impact of altering the emitter's design on the water dynamics of the surrounding soil. HYDRUS, one of the most comprehensive packages for simulating variably saturated water flow in two- or three-dimensional domains, was used to conduct numerical simulations. The soil hydraulic parameters for HYDRUS simulations in a three-dimensional domain were optimized based on experimental data gathered from experiments conducted with the original ring-shaped emitter. The results of the simulation verified that covering the entire emitter is not required and that decreasing the number of holes has no discernible impact on the water availability in silt for model plants like tomato and strawberry. Compared to the original fully covered emitter, partially covering the emitter makes maintenance much easier.

A water hose with a diameter of roughly 20 cm is first bent into the shape of a ring to create the original design ring-shaped emitter. It has 5 mm holes drilled into it spaced evenly apart. Water can then seep through the permeable textile in all directions by covering the entire ring-shaped hose in it. Water is delivered to the emitter's inlet via a tube from a water tank. Although Indonesia has successfully used subsurface irrigation with ring-shaped emitters to cultivate crops, the system's entire design and operation have been entirely empirical. To get the best possible emitter design, it is essential to know the dimensions, spatial extent, and volume of the wetted areas surrounding the emitter. To effectively manage subsurface irrigation with a ring-shaped emitter and wet the plant root zone uniformly, which will increase water use efficiency of water or fertilizer use, and to prevent water losses due to either evaporation in soil surface or deep percolation to groundwater, one must know the precise distribution of soil water content around the emitter. Furthermore, determining the infiltration phase depends critically on the hydraulic properties of the soil. As a result, using a ring-shaped emitter for subsurface irrigation with a high-pressure head of water requires understanding how to control irrigation discharge rates, which affect how much water enters the root zone depending on the hydraulic properties of the soil.

Three theoretical modeling approaches (numerical, analytical, and empirical) were compared by Kandelous and Šimůnek in order to determine the wetting patterns under surface and subsurface irrigation drip irrigation that are crucial for the best possible design. In comparison to other models, they found that the numerical model utilizing HYDRUS-2D produced accurate predictions of the distribution of soil water content. In recent times, the HYDRUS numerical model has been extensively employed to model water flow in various irrigation schemes and management situations. As an instance, Fares et al. investigated HYDRUS-2D to assess drip irrigation effectiveness in three distinct soil types. The findings showed that while the horizontal distance wetted volume was higher in clay soil, the water moved more vertically in sandy soils. Sandier soils are less suitable for drip system performance than clay soils.

Furthermore, Cote et al. examined the patterns of wetting under subsurface drip irrigation (SDI) without taking root-system water uptake into account. They employed four distinct irrigation fluxes and three distinct soil types, including sand, silt, and silt clay loam. They discovered that, if design and management were modified to take into consideration the soil hydraulic properties of medium and low permeability fine-textured soil, subsurface drip irrigation could enhance plant water availability. In addition, water and nutrients are rapidly transferred downward from the emitter in coarse-textured, highly permeable soils. Gärdenäs along with others investigated the impact of root water uptake on the hydraulic properties of the soil during subsurface irrigation. They came to the conclusion that percolation was lowest in fine-textured soils (silt clay soil) and highest in coarse-textured soils (sandy loam soil).

By comparing simulation results with experimental observations of the matric potential for a subsurface drip irrigation system in a sandy loam soil with a 10-cm installation depth, Provenzano evaluated the accuracy of HYDRUS-2D and found satisfactory agreement. Siyal in conjunction with others employed HYDRUS2D to study the effectiveness of the pitcher irrigation system and discovered that, when subsurface water was applied through the pitcher, the simulated and observed soil water contents agreed closely.

As Nagazawa et al. investigated HYDRUS-1D to find out how well the buried porous bottle performed under subsurface irrigation in quartz sand soil using various water pressure head applications. They discovered that, for the most part, the predicted soil water contents agreed well with the observed data. Applying a pressure head lower than the soil's air entry value can also slow down the rate of deep percolation and evaporation from the soil's surface.

In general, they came to the conclusion that HYDRUS 1D, 2D, and 3D is a useful tool for researching and creating irrigation management strategies. The first consideration that needs to be made in order to provide design and management for subsurface irrigation with ring-shaped emitters is to look into the wetting patterns and soil water content distributions surrounding the buried emitter while taking the soil hydraulic properties and irrigation water applications into account. As a result, the objectives are to: (i) investigate how soil hydraulic properties affect the movement of water in the soil surrounding a buried ring-shaped emitter; and (ii) calibrate and validate the HYDRUS 2D/3D numerical model's ability to evaluate the water flow in soil under subsurface irrigation using a ring-shaped emitter designed originally. In order to achieve these goals, the initial laboratory tests were conducted to track variations in the soil moisture content at specific locations when water was applied using the original ring-shaped emitter that was buried. Although the emitter was designed to facilitate plant cultivation, no plants were used in the laboratory tests to minimize potential issues with plant root water uptake. For sand and silt soil types, two distinct water pressure head applications—1 and 5 cm H_2O—are made. As a result, four simulations were taken into consideration to represent the infiltration processes. The experimental soil water content data gathered from the subsurface water application trials using the original ring-shaped emitter was then used to calibrate HYDRUS. Next, additional experimental data were predicted using the calibrated model to ensure its validity.

7.3.4. Yield and Water Productivity of Bell Pepper (Capsicum annum L.)

In arid and semi-arid areas with limited water supplies to support agricultural output, efficient irrigation water use is crucial. Understanding the intricate dynamics of soil water in the root zone under different crop and environmental situations is essential to optimizing irrigation management. In order to improve encouraging ring-shaped emitters for subsurface irrigation and achieve higher water use efficiency, it is therefore necessary to investigate soil water dynamics around buried ring-shaped emitters under such conditions by taking into account several factors, including crop management system as well as root distribution and crop water requirement in agro-climate conditions. Developing water-saving agriculture requires careful emitter design and efficient water management. In order to increase water use efficiency, the number of holes in modified ring-shaped emitter designs was decreased from five to two. The holes were only partially covered to facilitate maintenance and repair. Using the HYDRUS 2D/3D software package, the performance of the original and alternative ring-shaped emitter designs was quantitatively assessed under various soil types. Using soil water stress to the plant roots, the numerical analysis evaluated the impact of various emitter configurations on the assessment of wet volumes surrounding the emitter. As a result, root water uptake was calculated as a proxy for how well the ring-shaped emitter dispersed water to the root zone in a uniform manner.

As a result, the original five holes can be lowered to three or two holes, depending on the type of soil. More water was used efficiently by the emitter with the new hole configuration, which also created the ideal environment for plant roots to absorb water in the root zone. To make repairs and maintenance easier, a permeable textile was used to partially cover the emitter. The quantity of water used for irrigation and the frequency of irrigation determine yield production. In horticulture, bell peppers (*Capsicum annum* L.) are among the crops most vulnerable to water stress. Costa and Gianquinto found that constant water stress greatly decreased the fruit's overall fresh weight, while Antony and Singandhupe came to the conclusion that the overall yield of peppers was low at lower irrigation levels.

Water stress could be caused by an improper irrigation schedule and management for the plant. To meet the crop's water requirements, precision irrigation is therefore essential. Moreover, one of the key elements in maximizing irrigation is the distribution and storage of soil water in the root zone. A more accurate simulation of soil water distribution for plants will be possible with a deeper comprehension of root density, distribution pattern, and root water uptake under subsurface irrigation. Accurate numerical modeling of plant root water uptake and water flow can aid in comprehending the dynamic soil water process during irrigation. HYDRUS 2D/3D has been extensively and successfully used to investigate and evaluate the designing of subsurface irrigation management practices involving a wide range of crops because of its dependability, flexibility, and ability to accommodate various boundary conditions as well as root water and nutrient uptakes. Its user-friendly graphical interface further adds to the model's ease of use.

In order to determine whether subsurface irrigation with a ring-shaped emitter is appropriate for the actual field conditions, an experimental validation of the HYDRUS model that takes crop cultivation into account is required. Taking these backgrounds into account, extensive simulation and experimental studies were conducted while growing bell peppers in glass houses. This chapter's primary goals were to: (1) examine the water dynamics around the buried ring-shaped emitter through experiments and numerical modeling; and (2) assess the emitter's effectiveness in terms of irrigation water productivity and water productivity. In order to verify how well the HYDRUS three-dimensional model replicated soil water movement around a buried ring-shaped emitter, the measured and simulated soil water contents from the cultivation experiments were first compared. The effectiveness of the ring-shaped emitter in supplying the right amount of soil water to the root zone was then assessed by computing the parameters of the root water uptake model. The observed plant growth data can be modeled for more thorough analysis because a new version of the HYDRUS software package was recently released. This version enables simulating plant root growth to characterize the actual of root water uptake. In order to compare the simulation results in terms of water use efficiency, the ratio of actual transpiration to the total amount of irrigation water used during the growing season was finally calculated.

7.3.5. Crop Water Consumption

Water is necessary for crop transpiration. To survive and develop, plant roots either take up or remove water from the soil. The majority of this water evaporates through the leaves and stems and enters the atmosphere instead of staying on the plant. We refer to this process as transpiration. Water from an open water surface evaporates as vapor into the atmosphere during the day, a process known as transpiration. Water on the soil's surface and water on a plant's leaves and stem have the same effects. We refer to this process as "evaporation." A crop's total water requirement therefore includes both transpiration and evaporation. As a result, evapotranspiration refers to the crop's water requirement. The value of evapotranspiration when soil is sufficiently wet is known as the potential evapotranspiration, and it is influenced by air temperature, relative humidity, wind speed, net radiation, and air temperature. In temperate monsoon regions, summer crop evapotranspiration typically falls between 3 and 6 mm per day.

Crop management, environmental factors, and crop growth stages all affect crop evapotranspiration (ETc). The total ETc for the duration of the crop growth period is the crop water requirement, or CWR. The rate of evapotranspiration, also known as adjusted or actual crop evapotranspiration (ETc), must be adjusted to the current conditions when management practices or environmental factors depart from the ideal state. Rainfed or irrigated crops can use the CWR and ETc concepts. The idea of an irrigation water requirement (IWR), which is the net depth of water [mm] needed by a crop to completely meet its specific crop water requirement, must be used in conjunction with the concept of CWR for irrigated crops. The portion of the CWR that is not met by groundwater, rainfall, or soil water storage is known as the IWR. This depth of water is also included in IWR when adding a leaching fraction is required to ensure proper leaching of salts in the soil profile. To account for the effectiveness of the irrigation systems used in practice, IWR must be translated into gross irrigation requirements.

Crop water requirement can be used as a guide in crop management practices to find the balance between the amount of water required at a given growth stage and the amount of extractable soil water available for the crop. Water can move around relatively freely in a wet soil because of its high potential energy, making it easy for plant roots to absorb. However, crop plants find it more difficult to extract water from a dry soil because water is bound by capillary and absorptive forces to the soil matrix. The crop experiences water stress when its potential energy in the soil drops below a certain point, which is typically the lowest amount of soil water that can be extracted by plants. Doorenbos and associates. The depth of water required to meet a crop free of disease's evapotranspiration (ETc) loss of water is known as the defined crop water requirements. Doorenboos et al. suggest a three-step process for calculating ETcrop.

1. The reference crop's evapotranspiration (ETo) indicates how the climate affects crop water requirements. ETo, which is measured in millimeters per day, is the average value over a given time period based on the kind of available meteorological data. The Penman-Monteith equation serves as the foundation for the most widely used method of calculating reference evapotranspiration.
2. The effect of the crop characteristics on crop water requirements is given by the crop coefficient (Kc) which presents the relationship between reference (ETo) and crop evapotranspiration (ET) or ETcrop = Kc. ETo. Values of Kc vary with the crop, its stage of growth, growing season and the prevailing weather conditions. The unit of ETc is in mm day^{-1}.
3. The local impact of climate variations over time, altitude, field size, advection, soil water availability, salinity, irrigation technique, and cultivation methods and practices are some of the factors that affect crop water requirements.

7.3.6. Their Effects on Root Distribution

Different irrigation techniques affect how much water is available at different soil profile depths, which in turn affects how roots develop. All crops need a healthy root system because it makes it easier for the plants to absorb nutrients and water. Determining the amount of irrigation depth and frequency requires taking into account the root zone at different stages of crop growth.

Shallow-rooted plants may need short, frequent irrigations because they can only draw water from a restricted depth. Conversely, deep-rooted plants can draw water from very deep down, but if the subsoil gets too wet, they may become vulnerable to root disease. The variation in root distribution with depth could be one cause of the observed differences in responses of field-grown and pot-grown plants to the soil-water regime. While root density in the field typically varies with depth, it can be uniform in a pot. Furthermore, the characteristics of water uptake and transmission may vary amongst roots located in different layers. For example, deeper rooted plants may be able to withstand more water movement within the plant than higher rooted plants. Digging out the entire plant, shaking off the dirt, or excavating a soil pit and measuring the root system's depth are some methods for determining the planting depth. The two most important factors in figuring out how much water to apply are the rooting depth and the depletion fraction (p), which is the fraction of total available water that can be removed from the root zone before moisture stress occurs. The spatial-temporal root distribution has a significant impact on the distribution of RWU in the soil profile.

More precisely, it is typically believed that RWU is a direct function of RLDs, or root length densities. The dynamics of root distributions must be appropriately taken into consideration in order to depict the interactions between root development and soil water status. Important extra data that can be used to constrain other parameters related to stress and water uptake compensation functions in RWU models is direct measurements of RLD distributions and their temporal dynamics. Root length density is defined as either the length per unit surface area of soil [cm cm^{-2}] or the length per unit volume of soil [cm cm^{-3}]. While the latter is used to calculate the number of roots per unit ground surface or crack surface, the former is used to calculate the root length distribution with depth. Water availability affects how roots grow. Plants enhance their capacity to increase water uptake under unfavorable soil water conditions by promoting extensive root growth and proliferation in the soil matrix. Soil water uptake under varying soil moisture regimes is a function of root growth. Thus, the growth and distribution of roots in the soil matrix are significantly influenced by the spatial and temporal variations of soil water content, soil texture, and soil structure. An essential phase of plant growth is root growth. Numerous aspects of plant roots, including their growth and distribution, water absorption, and adaptive self-regulation, have been the subject of numerous studies.

Water can be drawn out of the soil, irrigation systems, groundwater, and rainfall by plant roots. Water-saving irrigation techniques affect the root system's distribution, maximum root depth, and root absorption of water in orchards. Furthermore, the range of distribution of roots varies according to the water content of the soil.

Groundwater, soil water at varying depths, and irrigation water all have substantially different stable oxygen isotope compositions due to slow chemical reactions, evaporation, and penetration. In particular, it is thought that water absorbed by plant roots is a combination of water from various sources. Thus, by comparing the relationship between the water content of xylem and the isotopic properties of other sources of water, the relative contribution rate of each source of water to plant growth can be quantitatively ascertained. This approach provides the theoretical framework for using the stable oxygen isotope method to evaluate plant water use. For more accurate assessment of water uptake patterns during crop growth, the stable isotope technique is thought to be a more valuable tool than traditional methods like root excavation and soil moisture measurement. This approach harms the root environment very little, making it appropriate for long-term studies. Yano along with others claimed that the impact of precipitation causes variations in the stable hydrogen isotope composition of dry-farmed rice fields in the lowlands. Throughout the growing season and in different soil moisture conditions, trees adjust how much water they absorb. Cao and associates discovered that, depending on the degree of water deficit during different growth stages, cherry trees could absorb soil water and irrigation water at different depths. These findings show that, in water-limiting environments, the ability to adjust and vary the water extraction strategy can enhance root growth and improve a plant's water-use efficiency, which is crucial for irrigation water productivity and final yields. The spatial distribution of roots in soil can be quantitatively evaluated using a variety of techniques, such as excavation of intact root structures, clod method, soil drilling, and profiling. Even though stable isotope methods have many benefits, understanding root growth and distribution in detail is thought to be the most direct—and difficult—way to understand crop growth, particularly for large perennials like fruit trees.

In orchard research, the profiling method is by far the most often utilized approach. Crop root distribution is influenced by numerous variables, such as soil types, attributes, rootstock kinds, and orchard management Atkinson and Wilson, Atkinson). Numerous research teams have documented a significant relationship between the properties of soil moisture and the root distribution of fruit trees. Drawing from the technique of root excavation, Williamson et al. and Parker and Meyer assessed the peach trees' root distribution. Yao and companions investigated how apple tree roots grew, varied, and dispersed under various orchard ground management rules. Liang along with others examined the features of the intercropped apricot root system's spatial distribution. In China's suburbs, the cherry tree (Prunus avium) is a significant species of fruit tree. Therefore, it is imperative to investigate how irrigation affects cherry root distribution, root water absorption characteristics, and water-use efficiency in agricultural water-saving practices. To shed light on how different water sources affect root distribution and water absorption, the majority of previous studies concentrated on field crops, forests, and grasslands.

Few research, notably those examining fruit trees, have examined the effects of varying water sources—particularly irrigation techniques—on root water uptake patterns and subsequent root redistribution. An in-depth analysis of root water uptake and root system distribution in cherry orchards is necessary for the selection of a water source and the estimation of each water source's potential contribution, via irrigation water, to the overall water absorption under various irrigation techniques.

Meng along with others claimed that microbially mediated rhizosphere nutrient transformations and root exploitation of soil heterogeneity are essential to plant resource uptake. Uncertainty exists regarding how these processes alter in response to water-saving irrigation techniques, particularly in organic systems where crops depend on soil ecological processes for plant nutrition and productivity. In a field experiment, the impact of water-saving (SDI) and organic fertilizer application on root exploitation was examined in relation to traditional furrow irrigation (FI) in an organic tomato system. The study focused on how these interventions changed root traits and rhizosphere processes. Distribution and morphology of the roots, rhizosphere-resident C, N, and P-cycling enzyme activity, microbial N-cycling gene abundance in the rhizosphere, and rate of root mycorrhizal colonization under two irrigation schemes were all measured. In comparison to furrow irrigation (FI), tomato plants developed shorter and finer root systems, with higher root densities surrounding the drip line, decreased soil C-degrading enzyme activity, and changes in the abundance of microbial N-cycling genes and mycorrhizal colonization rates in the rhizosphere of (SDI) plants.

Meng along with others found that although subsurface drip irrigation (SDI) produced 66.4% more irrigation water productive than forced irrigation (FI), it also resulted in excessive vegetative growth and a 28.3% decrease in tomato yield. Additionally, they suggest that in order to promote resource uptake under SDI, roots and root-microbe interactions have a high potential for coordinated adaptation to water and nutrient spatial patterns. Nonetheless, discrepancies still exist between plant requirements and available resources, emphasizing the significance of evaluating the temporal dynamics of root–soil–microbe interactions to optimize their capacity to mine resources for creative irrigation systems.

In order to optimize the uptake of soil moisture and nutrient resources—the distribution of which is progressively influenced by irrigation systems—plants can control the development of their root systems. Several tactics, such as spatially targeted root proliferation to mine nutrient- and moisture-rich patches while avoiding resource-poor areas, can be used to achieve this plasticity. Additionally, plasticity develops over time. After soil rewetting brought on by rainfall or irrigation events, plants may experience rapid root regrowth in addition to increased root physiological activity, such as N uptake and hydraulic conductivity. The design of new resource-saving irrigation technologies and the development of efficient root systems for irrigated landscapes will benefit from a greater understanding of the responses of roots to wetting patterns brought about by various irrigation strategies. It is also essential to integrate root developmental patterns with rhizosphere processes mediated by microbes associated with the roots to improve our mechanistic understanding of how plants adapt to resource availability under various irrigation schemes. Microorganisms in the soil are involved in important biogeochemical processes that affect the uptake and retention of nutrients by plants. For example, bacteria and fungi with low nutrient availability release extracellular enzymes that control the depolymerization of soil organic matter (SOM), thereby mediating the soil's overall rate of cycling carbon (C) and nitrogen (N).

Furthermore, quantifying microbial genes involved in N cycling pathways in the rhizosphere, such as nitrification (amoA) and denitrification pathways (nirS, nirK, nosZ), can reveal N transformation rates and consequently N availability to plants. Soil microorganisms compete with roots for mineralized soil nutrients, such as inorganic N. Through extended mycorrhizal hyphae networks in soils, plants also profit from direct associations with arbuscular mycorrhizal (AM) fungi for the uptake of water and low mobile nutrients, particularly phosphorus (P). Different soil moisture patterns and irrigation-induced wet-dry cycles can either accelerate or inhibit these microbial processes, which in turn controls plant nutrient uptake and soil nutrient cycling. However, rhizosphere microbial processes' responses to irrigation have received little attention in earlier research and are frequently taken into account independently of root developmental characteristics. This restricts our ability to understand the mechanisms by which microbes, soil, and roots interact and adjust to the availability of resources under various irrigation techniques.

In a two-year certified organic tomato (*Solanum lycopersicum* L) and maize (*Zea mays* L) rotation system, the comparison of two-irrigation strategies, such as SDI and FI, was started in 2015.

The experimental plot was divided into two halves, with FI on one side and SDI treatment on the other, in order to establish the comparison. In the 2017 organic tomato-maize rotation, the experiment was conducted on three replicated plots during the processing tomato phase (i.e., two years following the start of the irrigation comparison). A single row of processing tomatoes was transplanted at a planting density of 21,000 plants per hectare on a raised bed that was 150 cm wide. One drip line, about 25–30 cm below the soil's surface, was buried in the middle of the bed to implement SDI. During the growing season, FI was applied by surface flood irrigation, with alternating irrigation of the two furrows between each bed. During the 2017 growing season, FI received 11,222 m^{-3} ha^{-1} and SDI received 4,839 m^{-3} ha^{-1}. The irrigation recommendation of Tule, which uses in situ sensors to measure a field's actual evapotranspiration (ET) using the Surface Renewal method, is where the weekly irrigation schedule originated. Tule offers weekly irrigation recommendations based on actual ET from the previous week, plant responses, atmospheric demand forecasts, and an estimate of water demand based on production targets.

Using flow meters, the frequency and volume of each irrigation event for FI and SDI was noted throughout the growing season. Composted poultry manure (9 Mg ha^{-1}) contained approximately 20.0% C, 2.0% N, and 1.4% P was applied prior to tomato planting. The same amount of composted manure was applied to both irrigation treatments, but in different ways: the FI system mixed the compost with the top 20 cm of soil, while the SDI system banded the compost on top of the drip line.

Different application methods of composted manure are part of the long-term design of this experiment and implemented to maximize the overlap between fertilizer distribution and the specific moisture pattern of the two irrigation strategies. Winter cover crop was seeded with a mix of legumes and grass, composing of 90.0 kg ha^{-1} bell bean (Vicia faba L.), 22.5 kg ha^{-1} Lana vetch (Vicia villosa Roth), and 28.0 kg ha^{-1} oats (Avena sativa L.). Winter cover crop was terminated and incorporated into the soil before compost application and planting. Root Distribution and Analysis Digital analysis obtained in both treatments 105 days after transplanting (i.e., 2 weeks before tomato harvest and 3 days after irrigation stopped) was used to characterize the root distribution of tomatoes on fully developed root systems. Using shovels and mattocks, we manually excavated the soil trench on a randomly chosen crop row in each treatment to observe the distribution of roots. We used a sharpshooter spade to smooth the surface of the soil profile. The dimensions of the observation trench were 90 cm in length, 80 cm in width, and 60 cm in depth.

On the soil profile surface, root distribution was noted tangential to the plant stem and perpendicular to the plant row. To extract data on morphological traits and root distribution, a digital processing technique was applied to roots that were taken 90 cm below the surface of the soil profile. To improve contrast and fix distortion, the smartphone app CamScanner was utilized. Prior testing of this technique demonstrated that it produced high-quality images appropriate for the analysis of root traits. Processed images were converted to 8- bit gray images using the Fiji platform Wu et al. and imported into WinRHIZO software (Regent Instruments, Canada) for root analysis. Total root length and root diameters were obtained from each quadrat. Root distribution was calculated as the percentage of root length contained in each quadrat relative to the total root length.

Meng et al. found that compared to furrow irrigation using compost mixed into topsoil, root distribution subsurface drip irrigation using banded compost produced a more concentrated root distribution pattern. Compared to FI, SDI represented a larger percentage of root length at a depth of 20–30 cm and a distance of 10–20 cm from the drip line.

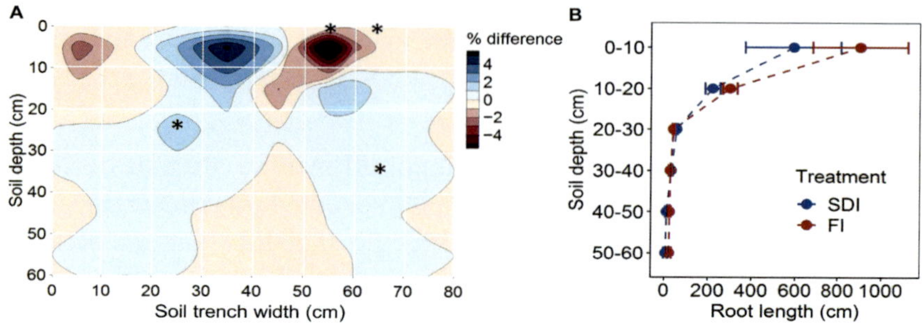

Figure 7.3. *Represents a diagram of root distribution under different irrigation systems subsurface drip irrigation (SDI) and furrow irrigation (FI). (A) root distribution by position; and (B) root length at different soil depths of tomato plants.*

Source: Abou Seeda M.A., Hammad S.A., 1Yassen A.A. and Abou El-Nour E.A., Evaluation and Optimization of Subsurface Irrigation (SDI) System: A review, DOI: 10.36632/mejas/2020.10.3.45.

Similarly, in the SDI system as opposed to the FI system, a larger percentage of root length was discovered at a depth of 30 to 40 cm, 20 to 30 cm away from the drip line. In contrast to SDI, FI caused a larger root distribution at deeper depths and nearer the bed's edge. Compared to the SDI system, a higher percentage of roots were discovered in the top 10 cm of soil, as well as at 10–20 and 20–30 cm from the bed center in the FI. Future research in organic systems is also required to improve the physiological and morphological plasticity of roots, utilize beneficial microbes for better nutrient cycling, and develop irrigation plans that optimize the resource-mining capabilities of roots and rhizosphere processes. With the help of these innovations, agro ecosystems will be better able to make use of biological processes as opposed to expensive irrigation and inorganic fertilizer inputs. They will also become more resilient to future resource shortages and climate change.

7.4. ADVANTAGES AND DISADVANTAGES OF SUBSURFACE DRIP IRRIGATION

A conceptual discussion is held regarding the benefits and drawbacks of subsurface drip irrigation (SDI) in comparison to other irrigation systems. Three groups comprise each category (advantages and disadvantages): i) issues related to water and soil; ii) issues related to cropping and cultural practices; and iii) issues related to system infrastructure. A significant portion of the adaptation and adoption of SDI systems into various cropping systems, geographic locations, soil types, and climates hinges on how potential benefits are weighed against potential drawbacks. Sometimes the benefits of a particular cropping system are limited to a small number, but they are powerful enough to offset any potential drawbacks. Future studies and advancements will most likely address and lessen the drawbacks while expanding the list of possible benefits. Nonetheless, additional applications for SDI can be developed and modified using the current listing. In order to balance benefits and drawbacks, specific instances of SDI application in wastewater, cantaloupe (*C. melo*), almonds (*Prunis dulcis*), alfalfa (*Medicago sativa*), and maize (*Zea mays* L.) are covered.

7.4.1. Advantages of SDI

When utilized appropriately and/or when cropping systems and site conditions permit, the following list of benefits of subsurface drip irrigation (SDI) should be taken into consideration. Furthermore, a feature may be advantageous to some growers while being a disadvantage to others. For instance, SDI offers potential for enhanced cultural practices, but it may also result in fewer options for tillage. These benefits can be further broken down into categories such as problems with soil and water, cropping and cultural practices, and system infrastructure.

7.4.1.1. Advantages Related to Water and Soil Issues

- **More Efficient Water Use:** Deep percolation, surface runoff, and soil evaporation are all significantly decreased or eliminated. Drier soils that have less soil crusting can improve seasonal precipitation infiltration and storage. Depending on the drip line depth, flow rate, and soil limitations, the system may be able to be used in certain situations for a brief irrigation event for germination. Near the end of the cropping season, the innate capacity to apply small irrigation amounts can enable more water-efficient decisions to be made regarding irrigation events. A smaller portion of the soil volume can be wetted in crops that are widely spaced, further minimizing needless irrigation water losses.
- **Less Water Quality Hazards:** Because of deep percolation, there is less nutrient and chemical leaching and less runoff into streams.
- **Improved Opportunities for Use of Degraded Waters:** A more constant and lower soil matric potential can be maintained by smaller, more frequent irrigation applications, which may lessen the risk of salinity. Applying subsurface wastewater can lessen animal and human contact with these waters as well as pathogen drift.

- **Greater Water Application Uniformity:** Increased control over water, nutrients, and salts can be attained through improved in-field uniformities. If enough soil is sufficiently wet, watering widely spaced crops closer to the crop may be beneficial.

7.4.1.2. Advantages Related to Cropping and Cultural Practices

- **Enhanced Plant Growth, Crop Yield and Quality**: A number of crops respond positively.
- **Improved Plant Health**: Less disease and fungal pressure occurs due to drier and less-humid crop canopies. The system can also be used for some types of soil fumigation.
- **Improved Fertilizer and Pesticide Management**: Precise and more timely application of fertilizer and pesticides through the system can result in greater efficacy and, in some cases, reduction in their use.
- **Better Weed Control**: Reductions in weed germination and weed growth often occur in drier regions
- **Improved Double Cropping Opportunities**: Since the system doesn't have to be taken out during harvest and put back in before planting the next crop, crop timing may be improved.
- **Improved Farming Operations and Management:** Events involving irrigation can involve a variety of field operations. Field operations reduce soil compaction and significantly lessen irrigation-induced soil crusting. Compared to surface drip irrigation (DI), SDI frequently reduces variability in soil water regimes and redistribution. Furthermore, weather-related application restrictions like strong winds, low temperatures, and wet soil surfaces don't matter as much. Even with minimal irrigation requirements, necessary fertilization can be administered in a brief irrigation event. The ability to irrigate in freezing weather can be especially helpful in situations where preseason irrigation is applied to maximize the capacity of seasonal irrigation. Additionally, there is less irrigation equipment vulnerable to car damage. Dryer soils are advantageous for hand laborers because they minimize manual strain and injury.

7.4.1.3. Advantages Related to System Infrastructure

- **Automation**: The system is well-suited for automation and sophisticated irrigation control technologies due to its closed-loop pressurized feature, which can minimize variability in application and soil water and nutrient redistribution.

- **Decreased Energy Costs**: Compared to some types of sprinkler irrigation, operating pressures are frequently lower. Energy expenses will be lowered along with any water savings resulting from SDI.
- **System Integrity Issues**: Compared to mechanical-move sprinkler irrigation systems, a sprinkler distribution system (SDI) has fewer mechanized components. Since plastic makes up the majority of the parts, irrigation system corrosion affects them less. Since SDI systems don't have to be taken down and put back up in between crops, they may sustain less damage. Moreover, the possibility of vandalism is diminished.
- **Design Flexibility**: When it comes to matching field size and shape, SDI offers more flexibility than center pivot sprinkler irrigation systems. It is simple and affordable to size the SDI system to match the available water supply. Driplines can be positioned in widely spaced crops to maximize nutrient and water uptake. Compared to surface gravity irrigation, pressure compensating SDI systems have fewer slope restrictions.
- **System Longevity**: Proper design and management can result in a long economic life for SDI installations. Long system life enables investment costs to be spread out over a number of years, making SDI an economically viable option for lower-value commodity crops.
- **Less Pest Damage:** Insects and wildlife may occasionally cause less pest damage to SDI systems than to DI systems. This must be balanced, though, with the possibility that detecting and fixing pest damage to SDI systems may require more work.

7.4.2. Disadvantages of SDI

Similarly, there are situations and circumstances where choosing an SDI system is disadvantageous. These drawbacks can also be broken down into categories such as problems with soil and water, cropping and cultural practices, and system infrastructure.

7.4.2.1. Disadvantages Related to Water and Soil Issues

- **Smaller Wetting Pattern:** On coarse-textured soils, the wetting pattern may be too small, resulting in an inadequate crop root zone. Insufficient irrigation capacity or system failure can be more likely in this scenario, making system reliability and capacity extremely important concerns.

- **Monitoring and Evaluating Irrigation Events**: It is more challenging to assess system performance and application uniformity when it comes to water applications, which may be mainly invisible. Mismanagement of the system can result in either over-irrigation, which causes poor soil aeration and issues with deep percolation, or under-irrigation, which reduces crop yield and quality.
- **Soil/Application Rate Interactions**: Certain soils may not be able to redistribute water at emitter discharge rates that are greater than what is necessary for normal redistribution processes. Under such circumstances, the water pressure in the area immediately surrounding the emitter may be higher than the atmospheric pressure, changing the emitter flows. Unwanted damp patches in the field could result from water unintentionally "surfacing," or from the emitter flow tunneling to the soil's surface. Small soil particles may be carried by the water in "surfacing" issues, creating a "chimney effect" that creates a preferred flow path. Since some of the "chimney" remains above the dripline even after tillage, it might be challenging to remove it completely.
- **Reduced Upward Water Movement**: Depending on installation depth and soil properties, there may be restrictions on using the SDI system for germination. This could be especially problematic for soils that exhibit vertical cracking. Above the dripline, salinity levels may rise, posing a greater risk to young transplants or emerging seedlings.

7.4.2.2. Disadvantages Related to Cropping and Cultural Practices

- **Less Tillage Options:** Primary and secondary tillage operations may be limited by dripline placement.
- **Restricted Plant Root Development:** Smaller crop root zones may increase the importance of fertilization and irrigation in terms of timing and quantity. Even with good irrigation, smaller crop root zones might not be enough to prevent diurnal crop water stresses. For the best yields, nutrient application through the SDI system might be necessary. Applying micronutrients could also become more crucial as the smaller soil volume loses these nutrients more quickly.
- **Row Spacing and Crop Rotation Issues:** Crops with different row spacing may be more challenging to accommodate in SDI systems because they are spatially fixed. Certain crops may demand

extremely close dripline spacing, which may not be feasible from an economic standpoint. When planting annual row crops, extra caution must be used to make sure that crop spacing and orientation are suitable for the location of the dripline.

- **Plant Development Issues:** Under SDI, certain crops might not grow well in particular soils and climates. Peanuts might not firmly root in arid soil. Tree crops could gain from a more extensive pattern of wetting.

7.4.2.3. Disadvantages Related to System Infrastructure

- **Costs:** In comparison to certain other irrigation systems, SDI requires a substantial initial investment. The system frequently has little to no salvage value or no market value. Before approving loans for SDI systems, lenders might need more collateral and a higher percentage of equity. Large-scale investments of this kind might not be justified in places where fuel and water supply is erratic, especially if the outlook for commodity prices is dim. Since alternative irrigation systems often have longer design lives than SDI systems, annualized depreciation costs must rise to account for system replacement.
- **Filtration Issues**: Similar to other micro-irrigation systems, water filtration plays a crucial role in guaranteeing the appropriate functioning and durability of the system. For long-term SDI systems, where a system life of more than 10 years is desired, the problem could, nevertheless, become more urgent. Since there are no opportunities for manual emitter cleaning, SDI may require more sophisticated water quality management than certain surface micro-irrigation systems.
- **Other Maintenance Issues**: Repairs and maintenance must be done on schedule and with consistency. Rodents can make leaks more difficult to find and fix, especially in deeper SDI systems. In addition to using safeguards in system operational and design procedures to limit or prevent further intrusion, the driplines must be watched for root intrusion. Certain perennial crops' roots have the potential to pinch driplines, reducing or eliminating flows. The driplines must be flushed on a regular basis to get rid of any silt and other precipitate buildup that may happen inside of them.
- **Operational Issues:** Compared to some alternative irrigation systems, operation and management call for more constant supervision. Less visual cues about how the system is operating and

how uniformly the applications are used are present. It is necessary to use irrigation scheduling procedures to avoid both over- and under-irrigation. System flowmeters and pressure gauges must be monitored in order to assess whether the system is functioning correctly.

- **Design Issues:** The technology behind SDI is not as advanced as that of some other irrigation alternatives. This is especially true in some areas where growers are less familiar and knowledgeable about these systems. Turnkey systems are getting harder to come by. In certain areas, insufficient contractor capacity may lead to installations being scheduled improperly during wet seasons. Since the majority of the SDI system is underground, fixing design flaws is more challenging. Generally speaking, SDI systems require more components than DI systems. If a vacuum occurs during system shutdown, there is a chance that soil will be ingested, so vacuum breakers and air relief devices need to be present and functioning properly. The system hydraulics will determine the zone size and run duration, just like in any other microirrigation system. In certain soils and at certain depths, soil overburden may compress the dripline, which will have a negative impact on flow. Site Specific Variable Applications are generally not a good fit for SDI systems.

REFERENCES

1. Allison, G. (1982). The relationship between 18O and deuterium in water in sand columns undergoing evaporation. *Journal of Hydrology, 55*, 163–169.
2. Antony, E., & Singandhupe, R. B. (2004). Impact of drip and surface irrigation on growth, yield and WUE of capsicum (*Capsicum annum* L.). *Agricultural Water Management, 65*, 121–132.
3. Atkinson, D. (1983). The growth, activity and distribution of the fruit tree root system. *Plant and Soil, 71*, 23–35.
4. Atkinson, D., & Wilson, S. A. (1980). The growth and distribution of fruit tree roots: Some consequences for nutrient uptake. In *Proceedings of the Acta Horticulturae ISHS: Symposium on Mineral Nutrition and Fruit Quality of Temperate Zone Fruit Trees* (pp. 137–150). Canterbury, UK.
5. Batra, B. R., Inder, M., Arora, S. K., & Mohan, I. (2000). Effect of irrigation and nitrogen levels on dry matter production by okra (*Abelmochus esculentus* L. Moench). *Haryana Journal of Horticultural Sciences, 29*(3, 4), 239–241.
6. Bertrand, G., Masini, J., Goldscheider, N., Meeks, J., Lavastre, V., Celle-Jeanton, H., Gobat, J. M., & Hunkeler, D. (2014). Determination of spatiotemporal variability of tree water uptake using stable isotopes (δ18O, δ2H) in an alluvial system supplied by a high-altitude watershed, Pfyn forest, Switzerland. *Ecohydrology, 7*, 319–333.
7. Bhattari, S. P., Huber, S., & Midmore, D. J. (2004). Aerated subsurface irrigation gives growth and yield benefits to zucchini, vegetable soybean and cotton in heavy clay soils. *Annals of Applied Biology, 144*, 285–298.
8. Black, B. L., Drost, D., Lindstrom, T., Reeve, J., Gunnell, J. D., & Reighard, G. L. (2010). A comparison of root distribution patterns among *Prunus* rootstocks. *Journal of the American Pomological Society, 64*, 52–62.
9. Bryla, D. R., Banuelos, G. S., & Mitchell, J. P. (2003). Water requirements of subsurface drip irrigated faba bean in California. *Irrigation Science, 22*, 31–37.
10. Bufon, V. B., Lascano, R. J., Bednarz, C., Booker, J. D., & Gitz, D. C. (2012). Soil water content on drip-irrigated cotton: Comparison of measured and simulated values obtained with the Hydrus 2-D model. *Irrigation Science, 30*, 259–273.
11. Cai, G., Vanderborght, J., Vereecken, H., Couvreur, V., & Mboh, C. M. (2017). Parameterization of root water uptake models considering dynamic root distributions and water uptake compensation. *Vadose Zone Journal, 17*(1).

12. Camp, C. R. (1998). Subsurface drip irrigation: A review. *Transactions of the ASAE, 41*(5), 1353–1367.
13. Cancela, J. J., Fandino, M., Rey, B., & Martinez, E. M. (2015). Automatic irrigation system based on dual crop coefficient, soil and plant water status for *Vitis vinifera* (cv Godello and cv Mencia). *Agricultural Water Management, 151*, 52–63.
14. Cao, X., Yang, P., Bernard, A. E., & Li, P. (2018). The effects of rainfall and irrigation on cherry root water uptake under drip irrigation. *Agricultural Water Management, 197*, 9–18.
15. Coelho, E. F., & Or, D. (1999). Root distribution and root water uptake patterns of corn under surface and subsurface drip irrigation. *Plant and Soil, 206*, 123–136.
16. Costa, D. L., & Gianquinto, G. (2002). Water stress and water table depth influence yield, water use efficiency and nitrogen recovery in bell pepper, lysimeters studied. *Australian Journal of Agricultural Research, 53*, 201–210.
17. Cote, C. M., Bristow, K. L., Charlesworth, P. B., Cook, F. J., & Thorburn, P. J. (2003). Analysis of soil wetting and solute transport in subsurface trickle irrigation. *Irrigation Science, 22*, 143–156.
18. Dawson, T. E., & Pate, J. S. (1996). Seasonal water uptake and movement in root systems of Australian phreatophytic plants of dimorphic root morphology: A stable isotope investigation. *Ecology, 107*, 13–20.
19. Dexter, A. R. (2004). Soil physical quality: Part I. Theory, effects of soil texture, density, and organic matter, and effects on root growth. *Geoderma, 120*(3, 4), 201–214.
20. Du, T., Kang, S., & Zhang, J. (2011). Water-saving and crop quality improvement of alternate partial root-zone irrigation and application of isotope technology in the research of crop water use. *Chinese Journal of Plant Physiology, 47*, 823–830.
21. Ehlers, W., & Hamblin, A. P. (1991). Root system parameters determining water uptake of field crops. *Irrigation Science, 12*, 115–124.
22. El-Nesr, M. N., Alazba, A. A., & Šimůnek, J. (2014). HYDRUS simulations of the effects of dual-drip subsurface irrigation and a physical barrier on water movement and solute transport in soils. *Irrigation Science, 32*, 111–125.
23. Evans, R. G., Wu, I., & Smajstrala, A. G. (2007). Microirrigation systems. In G. J. Hoffman, R. G. Evans, M. E. Jensen, D. L. Martin, & R. L. Elliott (Eds.), *Design and Operation of Farm Irrigation Systems* (pp. 632–679). American Society of Agricultural and Biological Engineers.

24. Fandino, M., Cancela, J. J., Rey, B. J., Martibez, E. M., Rosa, R. G., & Pereire, L. S. (2012). Using the dual Kc approach to model evapotranspiration of Albarino vineyards (*Vitis vinifera* L. cv. Albarino) with consideration of active ground cover. *Agricultural Water Management, 112*, 75–87.
25. Fares, A., Parsons, L. R., Wheaton, T. A., Morgan, K. T., Šimůnek, J., & van Genuchten, M. Th. (2001). Simulated drip irrigation with different types of soils. *Proceedings of the Florida State Horticultural Society, 114*, 22–24.
26. Ferrara, A., Lovelli, S., Di Tommaso, T., & Perniola, M. (2011). Flowering, growth and fruit setting in greenhouse bell pepper under water stress. *Journal of Agricultural Science, 10*, 12–19.
27. Flanagan, L. B., & Ehleringer, J. R. (1991). Stable isotope composition of stem and leaf water: Applications to the study of plant water use. *Functional Ecology, 5*, 270–277.
28. Ghazouani, H., M'hamdi, B. D., Autovino, D., Mguidiche Bel Haj, A. M., Rallo, G., Provenzano, G., & Boujelben, A. (2015). Optimizing subsurface dripline installation depth with Hydrus 2D/3D to improve irrigation water use efficiency in central Tunisia. *International Journal of Metrology and Quality Engineering, 6*, 1–9.
29. Ityel, E., Lazarovitch, N., Silberbush, M., & Ben-Gal, A. (2011). An artificial capillary barrier to improve root zone condition for horticultural crops: Physical effects on water content. *Irrigation Science, 29*, 171–180.
30. Jarvis, N. J. (1989). A simple empirical model of root water uptake. *Hydrological Journal, 107*, 57–72.
31. Ju, S. H., & Kung, K. J. S. (1997). Mass types, element orders and solution schemes for Richards's equation. *Computers and Geosciences, 23*(2), 175–187.
32. Kandelous, M. M., Šimůnek, J., van Genuchten, M. Th., & Malek, K. (2011). Soil water content distributions between two emitters of a subsurface drip irrigation system. *Soil Science Society of America Journal, 75*, 488–497.
33. Kreij, D. C., van der Burg, A. M. M., & Runia, W. T. (2003). Drip irrigation emitter clogging in Dutch greenhouses as affected by methane and organic acids. *Agricultural Water Management, 60*(2), 73–85.
34. Lila, A. T. S., Berndtsson, R., Persson, M., Somaida, M., El-Kiki, M., Hamed, Y., & Mirdan, A. (2013). Numerical evaluation of subsurface trickle irrigation with brackish water. *Irrigation Science, 31*, 1125–1137.
35. Shani, U., Xue, S., Gordin-Katz, R., & Warrick, A. W. (1996). Soil

limiting flow from subsurface emitters. I: Pressure measurements. *Journal of Irrigation and Drainage Engineering, 122*(5), 291–295.

36. Siemens, J., Huscheka, G., Siebe, C., & Kaupenjohanna, M. (2008). Concentrations and mobility of human pharmaceuticals in the world's largest wastewater irrigation system, Mexico City Mezquital Valley. *Water Research, 42*, 2124–2134.

Chapter **8**

Precision Irrigation Technologies

Contents

INTRODUCTION

When there is not enough natural precipitation, crop production can only occur with artificial watering, and precision irrigation is the best option in this situation. Farmers can avoid using excessive amounts of fuel and water in semi-arid regions and ensure healthy plant growth by implementing precision irrigation technology. Appropriate irrigation system installation also guards against yield losses in agriculture brought on by climatic unpredictability, such as unusual droughts. Cutting-edge innovations make farm management and growers' efforts easier.

8.1. UNDERSTANDING CONCEPT OF PRECISION IRRIGATION

The technique of applying water to crops precisely and targeted maximizes crop yield and water efficiency while reducing water waste. This is known as precision irrigation. As it solves the problems caused by conventional irrigation techniques and ushers in a new era of sustainable agriculture, this inventive technique is a ray of hope. Its potential to transform farming methods and lessen the urgent problems of resource inefficiency and water scarcity underpins its significance. Conventional irrigation techniques, which frequently entail flood or furrow irrigation, are infamous for wasting water and distributing it unevenly. These practices result in overuse of water, which damages the environment and reduces crop yield by causing nutrient leaching, waterlogging, and soil erosion.

The Food and Agriculture Organization (FAO) estimates that 70% of the freshwater extracted worldwide is used for agriculture. A large amount of this is lost as a result of ineffective irrigation techniques. Contrarily, precision irrigation directly addresses these problems. Farmers can precisely ascertain the water requirements of their crops by employing cutting-edge technologies like sensors, weather forecasts, and data analytics. Water waste is greatly decreased by this real-time data-driven approach, which guarantees that water is delivered exactly where and when it is needed. As a result, agricultural productivity is maximized and water resources are preserved, supporting sustainable farming methods.

8.1.1. Types of Precision Irrigation Methods

It includes a range of techniques and tools, each designed to handle particular facets of crop requirements and water distribution. These are a few of the most common varieties of precision irrigation:

8.1.1.1. Surface Drip Irrigation

The precise water delivery system known as drip irrigation is a significant improvement over conventional flood or furrow technique. The fundamental idea is to use a system of tubes, pipes, and emitters to slowly and consistently deliver water to plants' root zones. This targeted application ensures that plants receive the ideal amount of moisture precisely where it's needed by reducing water waste, evaporation, and runoff. Water delivery from drip irrigation systems can be customized to meet the unique needs of individual plants by being automated and adjusted in response to various factors such as crop type, soil composition, and meteorological conditions. Israel is among the most notable examples, as the country has used drip irrigation to transform barren areas into fruitful agricultural centers. Israel's proficiency in drip irrigation, coupled with its dedication to research and innovation, has resulted in significant water savings and elevated crop yields.

Beyond Israel, nations all over the world—from the US to India—have adopted this strategy to solve water scarcity issues and boost agricultural productivity. The Food and Agriculture Organization (FAO) estimates that drip irrigation and other micro-irrigation systems are 30–70% more water-efficient than conventional techniques. Furthermore, studies show that drip irrigation can save water in areas like the Central Valley of California and the vineyards of Australia by up to an astounding 50%. Drip irrigation has been essential in China for increasing crop yields in areas with limited water resources.

8.1.1.2. Micro-Sprinkler Irrigation

Low-volume sprinkler irrigation, or micro-sprinkler irrigation, is a localized water delivery technique that lies in between drip and conventional sprinkler systems. The system uses a network of high-pressure, low-flow emitters to simulate a light rain by dispersing water in the form of tiny droplets. Water is applied carefully to reduce evaporation and drift while giving crops complete coverage. Micro-sprinkler irrigation, which meets the specific water requirements of plants, is more focused on efficiency and precision than traditional sprinklers, which could waste water due to overspray. Worldwide interest in micro-sprinkler irrigation has grown as a potential solution for sustainable agriculture, especially in areas where water is scarce. This technique has been adopted by nations like Spain and India to maximize water resources and raise agricultural output. Farmers can now direct water to the root zone, promoting healthier plants and minimizing water waste, thanks to micro-sprinkler systems.

8.1.1.3. Subsurface Drip Irrigation

The innovative technique known as subsurface drip irrigation, or SDI, provides water directly to the plant root zone beneath the soil's surface. The system is made up of underground pipes that have emitters placed thoughtfully along their lengths. The precise amount of water released by these emitters guarantees a steady and focused supply to the root zone of the plant. By minimizing weed growth, preventing water contact with foliage, and reducing water evaporation, this innovative technique lowers the risk of disease. In arid regions where conserving water is crucial, subsurface drip irrigation is especially beneficial. To increase agricultural yield and reduce water waste, nations all over the world—from Australia to the United States—have implemented SDI systems. In order to deal with scarce water supplies, Israel, a country renowned for developing cutting-edge irrigation methods, has also adopted subsurface drip irrigation.

8.1.1.4. Precision Mobile Irrigation

Automated or robotic irrigation, commonly referred to as precision mobile irrigation, is a development over conventional stationary irrigation systems. With this innovative method, water is precisely delivered to crops through the use of mobile platforms with sensors, GPS, and data analytics. These platforms could be autonomous tractors, drones, or even self-propelled cars. Water can be applied effectively and on schedule thanks to the technology's real-time monitoring of plant health, weather, and soil moisture levels. When it comes to maximizing water distribution and responding to shifting field conditions, precision mobile irrigation offers unmatched flexibility. Precision mobile irrigation has been adopted by nations such as the Netherlands, which are renowned for their inventive farming methods, in order to increase crop yields while reducing water waste.

8.1.1.5. Variable Rate Irrigation (VRI)

The practice of variable rate irrigation, or VRI, differs significantly from conventional irrigation methods. In contrast to uniform irrigation, variable-rate irrigation (VRI) applies water to different parts of a field on a targeted basis according to crop requirements, soil properties, and other factors. Farmers are able to modify the distribution of water in real-time thanks to the incorporation of cutting-edge technologies like sensors and GPS. VRI has been adopted by nations all over the world, from Australia to the United States, as a means of increasing crop productivity while reducing the negative environmental effects of excessive water use.

8.1.1.6. Fertigation

Fertigation is a novel method of crop nutrition that combines "irrigation" and "fertilization" Fertilizers are injected straight into the irrigation system during fertigation, as opposed to conventional methods that apply fertilizers in addition to water. This makes it possible to supply nutrients straight to the root zone of plants, maximizing their absorption and application. Specialized tools, like hydroponic systems, sprinkler systems, and drip systems, make fertilization possible. Plant growth, yield, and general health are all enhanced by this integrated approach, which guarantees that nutrients are given to plants precisely when they are needed. Fertigation has the potential to significantly boost crop yields; these can often range from 20–100%, depending on the crop and the surrounding circumstances. According to UN estimates, runoff and volatilization are two examples of inefficiencies that cause about 50% of fertilizers to be lost after application. Fertigation minimizes losses and maximizes plant nutrient uptake by supplying nutrients straight to the root zone.

8.1.1.7. Remote Sensing and Automated Control

Crop health, weather, and soil moisture are all tracked by technologies such as sensors, weather stations, and satellite imagery. Water delivery is then optimized by automated control systems that modify irrigation based on real-time data. These controllers modify irrigation schedules based on current weather conditions, taking precipitation and evapotranspiration rates into account. Several other pertinent methods that are employed are as follows:

- **Localized Overhead Irrigation:** This method uses overhead sprinklers with low pressure to irrigate particular sections of a field. It works well for crops with different water needs or with asymmetrical field designs.
- **Canopy Sensors:** These sensors gauge crop water stress by measuring light reflection from the crop canopy. The timing and volume of irrigation are modified using this information.
- Soil tension is measured by soil tension meters, which show how hard roots must work to draw water out of the soil. This information can be used by farmers to schedule irrigation.

8.1.2. Benefits of Precision Irrigation

There are several advantages to precision irrigation, which help make agricultural practices more effective and sustainable. A few of the main benefits are as follows:

- **Water Efficiency and Conservation:** By providing water directly to plant roots, it minimizes runoff and evaporation. By ensuring that water is used effectively, this focused strategy protects this valuable resource and advances sustainable water management.
- **Enhanced Crop Yield and Quality:** It promotes healthy plant growth by preserving ideal soil moisture levels and avoiding water stress. Increased crop yield, quality, and general agricultural productivity result from this.
- **Eco-Friendly Practices:** By reducing water runoff and the leaching of nutrients and chemicals into ecosystems, it lessens the impact of farming on the environment. This environmentally friendly strategy improves the quality of the soil, water, and surrounding areas.
- **Energy Savings:** When compared to traditional methods, these techniques use less energy to distribute water. This is consistent with energy-efficient agricultural practices since it results in lower energy consumption and operational costs for farmers.
- **Tailored Nutrient Management:** Farmers can supply nutrients straight to plant roots by combining fertilization techniques with precision irrigation. This focused strategy lowers the chance of nutrient pollution, maximizes nutrient uptake, and minimizes nutrient waste.
- **Effective Weed and Pest Control:** It inhibits weed growth and competition with crops by limiting the amount of water available to weeds. Furthermore, it prevents the growth of pests and diseases by maintaining the foliage of plants drier, which results in more efficient pest management.
- **Adaptability to Variability in the Field:** Variations in crop health, topography, and soil type are common in fields. It adjusts to these variations, making sure that the proper amount of water is applied to every part of the field, thereby encouraging consistent crop growth.
- **Data-Driven Decision Making:** Precision irrigation gives farmers access to real-time insights by utilizing cutting-edge technologies like sensors and meteorological data. Making educated decisions regarding crop management, resource allocation, and irrigation scheduling is made possible by this data-driven approach.
- **Climate Resilience:** Its weather forecasting capabilities allow it to adapt irrigation practices to changing weather patterns brought on by climate change. This helps to guarantee steady crop production and reduce risks.

- **Long-Term Sustainability:** It conforms to sustainable agriculture's tenets. It encourages prudent resource management, maintains the health of the soil, lessens ecological effects, and enhances the general wellbeing of farming ecosystems.

Using these techniques helps farmers by increasing productivity and efficiency, but it also protects the environment and ensures agriculture will continue to exist in the long run in a world that is changing quickly.

8.1.3. Components of Precision Irrigation Systems

Through precision irrigation systems, these elements collaborate to maximize water use, improve crop growth, and support sustainable agriculture. A few of the fundamental elements are:

1. Sensors and Monitoring Devices:
   - Sensors measure soil moisture, temperature, and humidity.
   - Monitoring devices provide real-time data for informed irrigation decisions.
2. Control Systems:
   - Control units process sensor data and automate irrigation schedules.
   - They ensure precise water delivery based on field conditions.
3. Water Delivery Mechanisms:
   - Drip lines, micro-sprinklers, or localized overhead sprinklers deliver water directly to plant roots.
   - Minimizes water wastage and optimizes plant hydration.
4. Weather Stations and Forecasting:
   - Weather stations collect local weather data.
   - Forecasting informs irrigation timing based on climate conditions.
5. Geographic Information Systems (GIS):
   - GIS technology creates field maps with topography and soil type data.
   - Helps tailor irrigation plans to field variability.
6. Automated Valves and Pumps:
   - Valves and pumps regulate water flow within the system.
   - Integrates with control systems for precise water distribution.
7. Data Analytics and Software:

- Software processes sensor and weather data.
- Generates insights for optimal irrigation strategies.

8. Variable Rate Irrigation (VRI) Technology:
   - VRI adjusts water application based on field variability.
   - Customizes irrigation for different areas of the field.
9. Remote Sensing and Imaging Tools:
   - Satellite imagery and drones monitor crop health and water distribution.
   - Identifies areas needing adjustments.
10. Mobile Applications and Remote Access:
    - Mobile apps offer real-time control and monitoring.
    - Enables remote management for timely adjustments.
11. Fertigation Systems:
    - Integrates irrigation with nutrient application.
    - Directs nutrients to plant roots for efficient uptake.
12. Alarms and Alerts:
    - System alerts notify of anomalies or failures.
    - Ensures prompt responses to prevent crop damage.

8.1.4. Challenges and Limitations in Precision Irrigation

With increased crop yield, resource efficiency, and sustainable practices promised, the world is drawn to the era of precision agriculture. But there are obstacles in the way of putting precision irrigation technologies into practice.

8.1.4.1. Initial Investment: Balancing Cost and Benefit

The significant upfront cost associated with implementing precision irrigation technologies is one of the main obstacles. A sizable financial commitment may be necessary for the installation of sophisticated irrigation systems, sensors, data analytics software, and other supporting infrastructure. Even though these technologies have long-term advantages like reduced water usage and higher crop yields, many farmers may be put off by the initial costs, particularly those who live in areas with limited resources. Governments, agricultural organizations, and businesses in the private sector must work together to address this issue by offering financial incentives, subsidies, or reasonably priced financing options that will enable a wider range of farmers to use the technologies. Broader adoption can also be aided by emphasizing the long-term financial returns and providing success stories of individuals who have surmounted the initial investment hurdle.

8.1.4.2. Technical Expertise: Bridging the Knowledge Gap

Acquiring new skills and knowledge is necessary for farmers to effectively incorporate technology into agriculture. It's possible that many farmers, especially those from earlier generations, lack the technical know-how required to efficiently operate and maintain precision irrigation systems. It can be intimidating to comprehend the nuances of sensors, data analysis, and system calibration. Programs for education and training that are adapted to the unique requirements of farmers are needed to address this challenge. Farmers can be equipped with the skills they need to handle its complexities through workshops, online courses, agricultural extension services, and partnerships with nearby agricultural universities.

8.1.4.3. Data Management: Wise Decision-Making

Large volumes of data, such as crop health indicators, weather patterns, and soil moisture levels, are produced by precision irrigation systems. Effectively managing and analyzing this data can be daunting, particularly for smaller-scale farmers who lack access to advanced data analysis tools. Information overload is a real risk, and its potential advantages may be lost in the absence of appropriate data management techniques. User-friendly platforms and software that make data analysis and interpretation simpler are crucial to meeting this challenge. By incorporating machine learning algorithms, cloud-based solutions, and intuitive dashboards, farmers can make well-informed decisions without being paralyzed by the complexity of the data. Furthermore, encouraging partnerships between farmers and data specialists can result in customized solutions that address particular farm requirements.

With its innovative approaches to drip irrigation, micro-sprinklers, subsurface systems, precision mobile irrigation, and variable rate methods, precision irrigation is a promising solution to contemporary agricultural problems. Global adoption demonstrates their transformative power and is backed by data and science. Cooperation is necessary to overcome obstacles like start-up costs, technical expertise, and data management. It connects agriculture and technology to ensure food security while preserving resources, serving as a bridge to a sustainable future.

8.2. ROLE OF TECHNOLOGY IN WATER-SAVING IRRIGATION

The main cause of the world's water problems is agriculture because of its enormous water consumption and major role in water pollution. It might, nevertheless, also contribute to the solution. Thanks to technological developments, we now have the means to protect water quality, increase efficiency, and promote water conservation in the agriculture industry. A number of technologies reduce the amount of water used in water conservation. By delivering water straight to the plant's root zone, drip irrigation systems minimize water losses due to evaporation and runoff. Similar to traditional sprinkler systems, micro-sprinkler systems use less water and enable precise water delivery.

In order to avoid over-irrigation, soil moisture sensors provide current information about the water requirements of plants. They function by determining the soil's water content, which they can then communicate to irrigation systems or farmers directly to automate water conservation and scheduling of irrigation. Water efficiency can also be improved with a combination of technologies, guaranteeing that every unit of water used for agriculture produces the most value possible. Utilizing tools such as GPS and remote sensing, precision agriculture enables farmers to customize nutrient and water applications to the unique requirements of various sections of their fields, maximizing crop yields and input efficiency.

On a bigger scale, drones and satellite imagery can offer comprehensive data regarding crop health and water requirements, allowing for accurate and timely interventions. This increases crop productivity by managing pests and diseases better, using water and nutrients more efficiently, and increasing crop yield. Enhancing water efficiency can also be greatly aided by machine learning algorithms and artificial intelligence (AI). These algorithms can give farmers advanced decision support tools by integrating a variety of data sets, such as crop characteristics, market prices, and weather and soil conditions. These resources can assist in choosing the most effective irrigation schedules, choosing the ideal crop varieties for the circumstances, or even forecasting market trends for tactical planning. Another essential component of sustainable agriculture is maintaining the quality of the water, and technology can make a big difference in this regard. Biofiltration systems are a straightforward but efficient method of treating agricultural runoff because they use microorganisms found in soil or other media to extract pollutants from water. These systems can drastically lower the amounts of pesticides, fertilizers, and other pollutants that agricultural fields release into water bodies.

Another natural technology for preserving water quality is constructed wetlands. They are designed to treat wastewater from agriculture or other sources by removing pollutants using a mix of physical, chemical, and biological processes. They resemble natural wetlands. To eliminate pollutants from agricultural wastewater, advanced treatment technologies can also be used, such as membrane filtration or sophisticated oxidation procedures. These technologies can be useful for specific applications, like recycling wastewater for irrigation or another reuse, even though they are typically more expensive. Additional tactics to preserve water quality that are made possible by technology include conservation tillage and cover crops. With the help of seed drills and other planting technologies, cover crops can lessen nutrient runoff and soil erosion. Farmers can leave crop residues on the field with conservation tillage equipment, improving soil health and water retention while lowering runoff and erosion.

Research and development, farmer education, community involvement, and supportive policies are frequently necessary for the effective application of these technologies. In order to adopt and advance these technologies, governments, agricultural extension services, research institutes, technology companies, and farmers themselves all have important roles to play. To put it briefly, a wide range of technologies are available to protect water quality, increase productivity, and conserve water in agriculture. With the aid of these technologies, agriculture can be changed from contributing to the water crisis to helping to solve it. We can use technology to ensure the sustainable use of our limited water resources, ensuring food security and environmental health for future generations, provided we continue to innovate and work together.

8.3. REMOTE SENSING AND GIS APPLICATIONS

Precision agriculture is a farming management technique that leverages technology to boost productivity, cut down on waste, and enhance efficiency. It involves gathering and analyzing data about soil, crops, weather, and other variables that affect crop growth and yield using a variety of technologies, including GPS, GIS, remote sensing, and sensor-based systems. Afterwards, with the use of this data, decisions regarding crop management, resource allocation, and input application precision can be made with greater knowledge.

8.3.1. Remote Sensing Techniques for Precision Agriculture

Several platforms and sensors are used in remote sensing for precision agriculture in order to gather important data regarding crop and soil conditions at various scales. Important methods for remote sensing are as follows:

8.3.1.1. Satellite Remote Sensing

Large areas are covered synoptically by satellites at regular intervals. Important details consist of:

- Pixel size represents spatial resolution. crucial for field-scale observation. The resolution of commercial satellites is between 0 and 30 meters.
- Spectral resolution is the measurement of the number and location of spectral bands. Crucial for differentiating between crop stress and health.
- Sensitivity to signal intensities is known as radiometric resolution. Essential for quantifying the conditions.
- Temporal resolution: Revisit time. The majority of satellites take pictures every 1–16 days. Crop dynamics are better captured at higher frequencies.

Landsat, SPOT, Sentinel, PlanetScope, and MODIS are a few of the various satellite sensors used in PA. Each has unique specs that are appropriate for specific monitoring requirements.

8.3.1.2. Aerial Remote Sensing

Compared to satellites, aerial platforms such as manned aircraft and unmanned aerial vehicles (UAVs) offer very high-resolution imagery with more control over timing.

- Manned aerial vehicles are small, multispectral, thermal imagery collection aircraft. Resolutions as low as 10 cm are achievable. high access and operating expenses.
- Unmanned Aerial Vehicles (UAVs): A new technology in PA is UAVs with lightweight sensors. Facilitate quick, affordable, high-resolution (1–100 cm) small-area monitoring.

8.3.1.3. Proximal Remote Sensing

Sensors positioned on the ground that are carried by tractors or portable gadgets while in the field. used to measure crop nitrogen, biomass, and chlorophyll in order to modify fertilizer rates in real time. extremely high measurement density.

8.3.2. Geographic Information Systems for Precision Agriculture

With its ability to capture, manage, analyze, and visualize spatial data, geographic information systems (GIS) play a crucial role in precision agriculture. Important elements are as follows:

8.3.2.1. GIS Concepts and Components

A framework for gathering, storing, modifying, analyzing, and presenting geographical data associated with particular locations is provided by GIS. Essential elements consist of:

- **Hardware**: Computers, data loggers, sensors, GPS devices
- **Software**: Databases, analysis tools Data - Geospatial data like remote sensing images, yield data
- **People**: Expertise in geodata analysis, agriculture
- **Methods**: Techniques for data processing, analysis, modelling

Together these allow both spatial data management and complex spatio-temporal analysis.

8.3.2.2. GIS Data Models

Two key GIS data models used are raster and vector models:

- The raster model divides a surface into cells with values that describe conditions in order to represent geographic reality. Imagery from remote sensing is used.
- Geometric shapes and points with specific locations and attributes are used in vector models to depict reality. Used for sampling sites and farm boundaries.

GIS integrates both models for layered geographical analysis.

8.3.2.3. GIS Analysis Applications in Precision Agriculture

Key applications of GIS in precision agriculture include:

- **Yield Mapping:** Using combine harvesters fitted with GPS and weigh cells, this technique measures the variation in yield within a field. Prescription maps are created in GIS by importing georeferenced yield data. permits input optimization to raise low-yielding zones.
- **Soil Mapping:** Using GIS to interpolate soil sample data across landscapes, fine-scale soil type maps are created. reveals trends

in acidity, cation-exchange capacity, and nutrient levels. permits adapting practices to the variability of the soil.

- **Crop Health Monitoring:** Geographically-based crop growth variability associated with soils, pests, and weather is identified through time-series mapping of vegetation indices from satellite data using GIS analytics. Directs the use of pesticides in specific areas. An essential platform for planning and decision-making for accurate field management in precision agriculture systems is provided by GIS integration of multi-source spatiotemporal data.

8.3.3. Applications of Remote Sensing in Precision Agriculture

Several precision agriculture applications, such as the following, heavily rely on remote sensing imagery and data:

1. **Crop Type Classification:** Classification of imagery into landcover maps that identify major crop types and other land uses is made possible by multispectral satellite data. For major crops, >90% mapping accuracy is attained. directs the creation of management plans tailored to each zone.
2. **Crop Growth Monitoring and Yield Prediction:** Plant vigor and phonology are tracked seasonally by vegetation indices that are derived from satellite timeseries. allows for in-season yield forecasting at regional scales with an error of about 8–15% when combined with weather data in crop growth models.
   - **Soil Mapping:** In order to interpolate intricate digital soil maps, predictive soil mapping combines extensive soil sampling with remote sensing-derived elevation, landform, landcover, and geology layers. reveals patterns within the field.
   - **Water Stress Detection:** Optimized irrigation management is made possible by high resolution thermal infrared data, which measures crop water requirements and signals the beginning of soil moisture deficits. Used platforms are satellites and UAVs.
   - **Disease and Pest Detection:** The detection of diseases and infestations before they cause visible symptoms is made easier by the subtle spectral differences detected by hyperspectral sensors. enables quick, focused intervention to reduce yield loss. For prompt, data-driven decision making in precision agriculture systems, the distinct spectral information obtained

from remote sensing provides priceless insights into crop status and field conditions.

8.3.4. Applications of GIS in Precision Agriculture

Key applications of geographic information systems in precision agriculture include:

1. **Yield Mapping and Analysis:** Yield monitor data can be mapped and interpolated using GIS to identify management zones. When coupled with elevation models and soil maps, statistical analysis reveals the variables influencing yield variability for optimization.
2. **Variable Rate Technology**
   - **Applying Fertilizer:** Fertilizer applicators use GIS prescription maps to determine the appropriate rates for each field based on the nutrient levels of the soil. Refrain from overfertilizing.
   - **Irrigation:** GIS-based variable rate irrigation systems are fed by terrain layers, yield, and soil. adapts the spatial application of water to requirements.
3. **Drainage Mapping and Analysis:** Designing effective drainage infrastructure placement that is terrain-appropriate is made easier by the integration of digital elevation models and pipe flow models in GIS.
4. **Infrastructure and Logistics Planning:** In order to minimize expenses and soil compaction, equipment routing, storage locations, and transportation are optimized using GIS networks and proximity analysis.

8.3.5. Integration of Remote Sensing and GIS for Precision Agriculture

Precision agriculture uses information-based technologies to manage field variability in order to maximize yields and optimize crop production. It basically entails appropriate management at the appropriate time and place. In order to evaluate and adapt to field variability, precision agriculture's site-specific management depends on the integration of geospatial technologies such as geographic information systems (GIS) and remote sensing. Remote sensing offers imagery at various resolutions for crop land monitoring, and geographic information systems (GIS) enable spatial modeling and analysis for decision-making. The integration of remote sensing and GIS technologies, their methods for processing and analysis, and their overall contribution to precision agriculture are covered in this section.

8.3.5.1. Geo-Database Development

Having an integrated geo-database that captures the spatial variability in soil, crop growth, and yields across agricultural fields is essential for putting precision agriculture techniques into practice. Such geospatial databases, which enable effective data storage, updating, manipulation, and analysis, can be developed with the help of GIS. When creating a geo-database for precision agriculture, the following factors are crucial.

- Gathering information on soil properties such as texture, organic matter, pH, fertility, and micronutrients through methodical field sampling.
- Recording yield levels through harvesting equipment fitted with yield monitors and GPS
- Stratification of land using past yield maps into zones with comparable yield-limiting factors.
- Integrating remote sensing imagery to extract vegetation indices indicating crop vigor and development stages
- Correlating vegetation indices with yield data to predict spatial distribution of yields
- Establishing a connection between the spatial database and variable rate applicators, which vary the amount of fertilizer or pesticide applied to a field according to yields and expected response.

8.3.5.2. Image Processing and Classification

To extract meaningful landcover and crop information from the raw remote sensing imagery, processing and classification are necessary. Some useful techniques are as follows:

- Radiometric correction to retrieve true surface reflectance by removing sensor distortions and atmospheric effects
- Geometric correction to remove spatial distortions due to sensor optics, platform instability and terrain
- Image enhancement to improve visual distinction between features using filtering, pan-sharpening etc.
- Vegetation indices like NDVI for indicating crop greenness, leaf area, canopy cover and growth stages

- Supervised and unsupervised classification to generate landcover thematic maps exploring spectral clustering algorithms
- Object based image analysis using texture, context and ancillary data besides spectra for classification
- Change detection for identifying changes in vegetation vigor and land use over time

8.3.5.3. Spatial Analysis Models

Several models, workflows, and spatial analysis techniques can be implemented with GIS to facilitate precision agriculture decision-making. Important techniques consist of:

- In order to forecast field variables such as soil nutrient levels and yield at unsampled locations based on surrounding measured values, interpolation using kriging is used.
- Zonal statistics to provide an overview of yield, vegetation indices, etc., for management zones having comparable conditions for crop growth.
- Buffer analysis to pinpoint nearby areas that require special attention because they could affect fields through runoff or leaching.
- Terrain analysis is used to simulate how aspect, slope, and topography affect crop growth, drainage, and erosion patterns.
- Path distance modeling is used to map least-cost routes for field operation planning and optimized logistics.

8.3.5.4. Decision Support Systems

Farm planning and operations can benefit from the development of customized GIS-based dashboards and interfaces as decision support systems. To enable intelligent precision agriculture, these systems combine spatial analysis with specialized knowledge of crop growth models, best practices, and advisories. Some instances are as follows:

- Utilizing yield maps and soil data, variable rate application tools adjust fertilizer inputs among fields.
- Web-GIS and mobile apps providing specific advisories on irrigation, spraying, harvesting etc., considering location and stage specific requirements.
- Crop simulation models and drone imagery analytics combined for better risk management and early growth anomaly detection.

- AI-powered cognitive analytics to draw conclusions from various data sources and give farmers useful information.

8.3.6. Unmanned Aerial Systems for Precision Agriculture

Drones with sensors installed on them, or unmanned aerial systems (UAS), are becoming increasingly useful instruments in precision agriculture. UAS provide ultra-high resolution imagery that can be used to evaluate field variability and crop health at various stages of growth. These are some of the ways they can help with site-specific crop management.

- UAV Platforms, Sensors and Data Acquisition
  - Lightweight fixed-wing, multi-rotor and hybrid VTOL UAVs equipped with visual, multispectral, hyperspectral, thermal sensors
  - Generation of ortho-mosaics and 3D reconstruction for precise measurement and modelling
  - Capacity to provide rapid, cost-effective and flexible imagery on demand without reliance on satellite data

8.3.6.1. Applications

- Monitoring
  - Mapping the vegetation index to evaluate plant vigor, forecast yield, and identify early signs of stress.
  - For variable rate operations and field prescriptions, high resolution model input is required.
- Pest Management
  - Early identification of incidence, type and spread patterns of weed, insects for targeted control
- Irrigation Monitoring
  - Detection of spatial variability in soil moisture levels based on thermal data
  - Deriving irrigation recommendations based on deficit patterns observed
- Image Processing Workflows and Analysis
  - Orthorectification and mosaicking for precise spatial referencing of images
  - Machine learning techniques like neural networks for automated feature identification from UAS data

In order to collect, manage, analyze, and present spatial information, GIS technology integrates hardware, software, and data. GIS facilitates the management and analysis of spatial data in precision agriculture. To establish customized management of inputs and cultivation practices, farmers use GIS overlays of soil survey maps, yield maps from prior seasons, and remote sensing imagery to define management zones. This site-specific management maximizes resource use while boosting productivity. Precision agriculture now has access to a potent decision support system thanks to the integration of remote sensing data and GIS. The integration and analysis of spatial data from multiple sources, such as sensors, satellite imagery, and drones, is made possible by GIS technology. The results support effective management plans that are adapted to regional requirements.

The use of remote sensing and GIS technologies in precision agriculture is anticipated to increase as they develop further. But there are still issues, such as high prices, complexity, poor technical support, and trouble converting data into management plans. Persistent advancements in self-governing systems, data analysis, and analytical instruments could potentially surmount present adoption obstacles. In general, GIS and remote sensing are essential parts of the technological toolkit for precision agriculture, with room for future advancement and increased farm roles.

8.3.7. Big Data and AI in Precision Agriculture

The field of data-driven agriculture has advanced significantly since the introduction of sensing technologies and computer analytics. Big data and artificial intelligence (AI) are used in precision agriculture to facilitate real-time decision-making and advanced analytics. Large, intricate datasets that can be examined to find patterns, trends, and correlations are referred to as "big data." Data from satellite imagery, weather stations, field sensors, equipment, and yield monitors are all included in precision agriculture. Farmers can extract valuable insights from this deluge of data with the aid of advanced analytics. AI, on the other hand, generally refers to computer systems that simulate human intelligence. An AIrf subset called machine learning enables systems to learn from data patterns without the need for explicit programming. AI is used in agriculture to power agricultural robotics, computer vision for drones, automated control systems, and predictive modeling.

8.3.7.1. Applications

- Precision agriculture analytics through predictive analytics leverage AI and big data to uncover insights not visible through the analysis of discrete data sources. Crop models with predictive capabilities aid in projecting future crop growth and yields. Prescriptive models direct the best practices in management.

For example, AI algorithms can forecast expected yields by combining historical yield data, weather forecasts, and satellite images. Early and strategic marketing decisions are made easier as a result. In a similar vein, dynamic irrigation scheduling that saves water and energy is made possible by absorbing soil moisture data from sensors and weather forecasts. Analytics platforms that integrate diverse datasets will gain value as predictive capabilities advance. However, input data quality has a major impact on model accuracy. The majority of small farms lack comprehensive field trial data. Platforms for collaborative data may be able to get past this obstacle.

- Making decisions in real time: Large amounts of crop and machine data can be transmitted in real time via cloud computing services by modern equipment that is equipped with sensors. Farmers can see mapped metrics on fields and monitor operations with dashboard interfaces. Greater speed, accuracy, and confidence in data-driven management are made possible by this visibility.

For example, during harvest, combine harvesters are now sending yield data to the cloud continuously. Operators can monitor this remotely and adjust machine settings in response to variations. Similar to this, maps of crop stress produced by drones and satellites may encourage prompt inspection or intervention in impacted areas. However, because farmers find it difficult to turn raw data into action, real-time analysis applications are still behind. Real-time data utilization will be accelerated by decision support platforms that offer monitoring features in addition to contextual recommendations.

- Automation AI drives automation in a variety of agricultural tasks, including cargo hauling, irrigation, crop scouting, harvesting, and weed control. The resulting accuracy lowers waste, boosts productivity, and lessens the need for labor.

By using computer vision camera systems, smart sprayers are able to identify weeds and only spray vegetation with their targeted nozzles. Sensor-equipped experimental robotic fleets rove fields on their own, collecting data to guide decisions about harvesting, fertilization, and irrigation. AI training data keeps advancing computer vision and autonomous machinery. But concerns about data security and transparency raise red flags for AI in agriculture. Farm policy and incentives that promote quick automation while maintaining shared prosperity are still unresolved issues that require focus.

8.4. AUTOMATION AND SMART IRRIGATION SYSTEMS

A critical issue in contemporary agriculture is the need for effective water management. Innovative and sustainable irrigation techniques are required to replace traditional methods due to climate change and the rapid depletion of water resources. Irrigation systems that are automated and intelligent are two noteworthy strategies that have gained prominence. Both represent the idea of precision irrigation, a sophisticated method of watering designed to reduce water consumption and increase output.

One of the most sophisticated types of precision irrigation is automated irrigation. By ensuring optimal water usage and relieving farmers of the time-consuming manual labor of field watering, these systems allow farmers to concentrate on other crucial duties. An automated irrigation system makes use of a network of sensors and controllers that are set up to track crop water requirements, weather, and soil moisture levels. The irrigation system starts on its own when the sensors identify that the crops need water. By avoiding overwatering, this not only removes the uncertainty surrounding crop watering schedules, but it also encourages water conservation. Irrigation systems that are automated have many advantages. They can increase crop yields, save a significant amount of energy and water, and reduce the labor-intensive nature of irrigation. Additionally, by lowering soil erosion, water runoff, and fertilizer and pesticide leaching into groundwater, they support environmental sustainability.

Smart irrigation systems use cutting-edge technologies like artificial intelligence (AI), machine learning (ML), and the internet of things (IoT) to manage irrigation tasks more effectively than automated systems. An automated irrigation system with extra layers of intelligence and interaction is called a smart irrigation system. Weather forecasts, crop evapotranspiration rates, soil moisture sensors, and other sources of data are just a few of the sources of information that smart irrigation systems gather and process. Advanced algorithms are then used to process this data and make real-time irrigation decisions that maximize crop productivity and conserve water. The watering schedule and volume can be adjusted by the systems in response to changing conditions. With the help of computer or smartphone interfaces, farmers can monitor and manage their irrigation from any location with smart irrigation systems. This feature increases farmers' flexibility and control over their irrigation techniques, which boosts output and efficiency even more.

One cannot emphasize how crucial effective irrigation techniques are in light of the effects of climate change and the depletion of water supplies facing our planet. Irrigation systems that are intelligent and automated represent the potential of technology to bring about a more sustainable future for farming. By making an investment in these precision irrigation methods, we can increase agricultural productivity, protect our limited water supplies, and promote a more sustainable environment. To build a more intelligent and sustainable future for all, the agriculture sector needs to keep implementing and improving these cutting-edge technologies. By using automated and intelligent systems, precision irrigation provides a way to achieve this goal and is revolutionizing the farming sector one drop at a time.

8.4.1. Automatic Irrigation

Using a device to control irrigation structures allows water to flow from bays differently even when the irrigator is not present. This is known as automatic irrigation.

Automation can be used in a number of ways:

- to start and stop irrigation through supply channel outlets
- to start and stop pumps
- to cut off the flow of water from one irrigation area — either a bay or a section of channel, and direct the water to another area.

Although there is no direct human labor involved in these changes, you might need to invest some time in starting the irrigation system and keeping the parts in good working order.

8.4.1.1. Types of Automatic Irrigation Systems

Over time, automatic irrigation systems have developed and are now offered in a variety of configurations. Let's examine the three categories of drip irrigation systems that are automatic.

- **Time Based System:** Depending on the model and controller manufacturer, the farmer can adjust the start time, irrigation quantity (in minutes or seconds), number of cycles, maximum duration, and number of cycles with this system. It is simple to use and program. Cons include, however, the lack of system feedback and disruptions that result in cycles skipping entirely or partially because of power outages.
- **Volume Based System:** The start time, irrigation volume (measured in liters or M3), and number of cycles per day can all be adjusted by farmers. The system provides flow deviation (no flow, low flow, and high flow), precise quantity and flow calculation, and operational flexibility (time or volume based). The system is more expensive

than the time-based system, though, and it needs extra parts like flow meters and water meters with sensors.

- **Sensor-Based System:** This irrigation system uses sensors such as soil moisture sensor, tensiometer, solar radiation, vapor pressure deficit, and evapotranspiration (ET) to determine when to water. It facilitates the use of irrigation when needed and allows for significant resource savings. It is the most expensive automated irrigation system available. The quantity of sensors needed depends on the size of the land. Sensor maintenance is also required. A sensor malfunction may also have an impact on the irrigation system.

8.4.1.2. How Do Automatic Irrigation Systems Work?

Many automated components of automatic sprinkler and drip irrigation systems differ from those of regular systems. These components include an electronic controller that can start the water pump automatically at a pre-set time, high-tech parts on the head unit that allow the system to flush the sand media filter, hydro cyclone, screen, or disc filters automatically without interruption, an automatic fertigation injector that delivers a controlled amounts of liquid fertilizers and chemicals, and valves that open and close automatically to irrigate a large area in segments. Because of the customization options available with these automated irrigation systems, even manually operated drip, surface, and sprinkler irrigation systems can be improved to get beyond current constraints.

A fully automatic irrigation system costs more because of the innovative technology involved in its various devices. However, farmers can begin by gradually upgrading the manually operated irrigation system's current devices to semi-automatic and eventually fully automatic irrigation devices, depending on their financial situation.

- Because manual hydro cyclone, sand media, screen, or disc filters require frequent cleaning, most farmers avoid using them because of the laborious task of regularly opening and cleaning the filter cartridge; instead, they typically just throw away the inside cartridge and turn on the system. Initially, manual filters can be replaced with semi-automatic filters.
- Once they are comfortable with the semi-automatic product's operation, they can swap out the manual valve for a solenoid-operated hydraulic valve and gradually add more automatic products to their irrigation system until the manual system is completely converted to a fully automatic one.

8.4.1.3. Remotely Monitored Fully Automatic Irrigation System

A farmer can also use the system remotely to monitor every aspect of it, including automatically measuring the amount of water needed and the crop's content, checking for blockages in the water's path, and turning off the system through the device application if a problem is detected. Numerous components, including sensors for soil moisture, precipitation, light, temperature, water level, and flow, are integrated into a smart irrigation system. These components regulate the water pump, solenoid valve, filtration and fertigation units, and various irrigation field segments sequentially.

8.4.1.4. Automating an Irrigation Layout

An irrigation layout can be automated at one of two places — in sections of channel or at individual bay outlets.

- **Automation of Channel Sections:** Changes to the channel level are possible in this system because the channel structures are automated. There are no opening or closing mechanisms on the bay outlets. Additionally, every outlet set—like a set of sills—is calibrated to a particular level. For this automation method to work, there must be more fall in the channel system to accommodate changes in the water level between various locations. When another section needs to be irrigated, this change in water level is necessary to stop water from spilling onto bays that have already been watered. This fall is often unavailable on farms, so this automation technique is frequently inappropriate.
- **Automation of Individual Bay Outlets:** Using control over the bay outlets, this automation technique modifies the water flow onto the irrigated areas. The most common application of this automation system is in regions where there is not enough fall to allow for the automation of channel sections. Automation of channel sections and automation of bay outlets can both be configured with the same kind of automatic devices.

Benefits of Automatic Irrigation

The benefits of automatic irrigation are:

- reduced labor
- timely irrigation — plants being watered when needed
- management of higher flow rates
- accurate cut-off of water compared to manual checking
- reduced runoff of water and nutrients

- reduced costs for vehicles used to check irrigation.

Disadvantages of Automatic Irrigation

The disadvantages of automatic irrigation are:

- costs for purchasing, installing and maintaining the equipment
- reliability of irrigation system (due to human error when setting up)
- increased maintenance of channels and equipment to ensure it is working properly.

8.4.2. Smart Irrigation Systems

Water scarcity is an issue that many nations around the world are dealing with. We value both drinking water and irrigation water greatly. Water is needed in large quantities for both drip irrigation in gardens and farms. Water waste is an inevitable result of this process. As such, the importance of smart irrigation systems is growing. It has sophisticated sensors and controls to reduce water waste. Empirical evidence suggests that smart irrigation technology can effectively curtail water waste by 20–40% when compared to traditional irrigation technology. Based on current regional weather data and soil moisture conditions, it can make a scientific determination about whether and how much watering is necessary for plants. The location and timing of irrigation can be changed flexibly. Minimize water waste and boost irrigation effectiveness while preserving the health and quality of the plants.

The practice of conserving water during irrigation through science and technology is known as smart irrigation. It has several controllers, soil sensors, and weather sensors. The controller decides when to open or close the water valve based on data from the sensor, which also tracks the actual ground humidity and the current weather. Acquire the ability to automate irrigation. Scientific assessment of the necessity of water—its quantity, timing, and quality. It works well for managing water usage in landscapes, farms, and lawns, among other places. Technologies like the Internet of Things, mobile Internet, LORA, and remote sensing monitoring are all integrated into the smart irrigation system. Wirelessly operate valve switches from a distance. To conserve irrigation water, give plants the appropriate amount of water based on the type of soil and the local climate. The primary technologies used in the smart irrigation system are computer, sensor, and automated control technologies. in addition to garden irrigation techniques like drip and sprinkler irrigation.

8.4.2.1. Essential Benefits of Smart Irrigation Systems

Here are the 10 essential benefits of the smart irrigation system:

- **Water Conservation:** The capacity of smart irrigation systems to retain water is one of their most incredible features. Conventional irrigation techniques, such as hand watering or filling systems with timers, frequently result in overwatering or underwatering of the sprinkler system. Conversely, smart irrigation systems optimize watering schedules through the use of sensors and meteorological data. By adjusting watering schedules and amounts primarily based on current data, they can make sure that plants receive the proper quantity of water at the appropriate time. This accuracy results in significant financial and water savings, which is critical in areas where water is scarce.
- **Cost Savings:** By conserving water, smart irrigation systems lower water bills. Water costs can be a worthwhile investment for both owners and farmers. Water waste is reduced by smart irrigation systems, which distribute water most efficiently when it's needed. Furthermore, the systems facilitate timely preservation and preventive measures against water loss in the event of leaks or malfunctions within the irrigation network. The initial cost of a smart irrigation technique can be paid off over time by the money saved on water bills, making it an affordable solution.
- **Enhanced Plant Health:** Maintaining plant health requires proper irrigation. Rarities and root rot can result from overwatering. Underwatering can also result in pressure and prevent plants from growing. Intelligent irrigation systems monitor soil moisture levels and modify irrigation schedules to provide plants with the most consistent and maximum amount of water. This accuracy enables the maintenance of robust root systems and encourages vigorous plant growth. Farmers benefit from healthy flowers because they produce more and yield more. It approaches lush, colorful lawns and gardens for owners.
- **Environmental Sustainability:** There are various ways that intelligent irrigation systems support environmental sustainability. They lessen the strain on neighboring water supplies by storing water, which is particularly crucial in areas that are vulnerable to drought. Furthermore, the systems aid in preventing runoff and erosion that could introduce fertilizers and pesticides into waterways, resulting in pollution, by optimizing the use of water resources. Lower fuel and greenhouse gas emissions result from reduced water consumption, which can require up to tons less power to pump and handle water. In general, the promotion of sustainable water irrigation systems is greatly aided by smart irrigation systems.

- **Convenience and Automation:** The level of automation and convenience provided by contemporary smart irrigation systems is excessive. Through internet interfaces or smartphone applications, they can be remotely controlled, enabling users to change their irrigation schedules and view screens from any location. Busy farmers or business owners who might not have the time to manually operate their irrigation systems will benefit from this. Indicators and notifications, such as those indicating low soil moisture levels or system problems, can also be delivered by automated systems, enabling proactive control.
- **Data-Driven Decision Making:** Intelligent irrigation systems gather and evaluate meteorological, soil moisture, and plant requirements data. Users can make well-informed decisions about their irrigation practices thanks to this data-driven approach. Farmers can use the information, for instance, to schedule irrigations to maximize crop yields while consuming the least amount of water. Watering schedules can be tailored by homeowners to the unique requirements of their gardens and lawns. Having access to comprehensive data enables users to assess the effects of their irrigation techniques and make necessary adjustments.
- **Adaptability to Changing Conditions:** Conventional irrigation techniques are challenged by climate change and erratic weather patterns. Smart irrigation systems use real-time data to adjust to changing conditions. They can modify watering schedules to prevent overwatering in response to unforeseen rain or other sudden changes in the weather. Because of their adaptability, plants are guaranteed to receive consistent care regardless of the weather outside. This has the potential to make the difference between a bumper crop and crop failure in agriculture.
- **Scalability and Customization:** Because of their great scalability and customization capabilities, smart irrigation systems can be used in a variety of contexts. Smart irrigation solutions are available to suit the needs of any size agricultural operation or small residential garden. These systems are adaptable to particular plant species, soil types, and landscapes. Drip irrigation systems, for instance, can be used to precisely water individual plants. Sprinkler systems, on the other hand, have a wider coverage area. Customizable irrigation methods and parts guarantee the best possible water use and plant health.
- **Integration with Other Smart Technologies:** A comprehensive and

effective ecosystem can be created by integrating smart irrigation systems with other smart farm or home technologies. They can be linked, for instance, to smart soil sensors that track soil conditions or smart weather stations that deliver real-time weather data. Users can control their irrigation in addition to other home automation features like lighting and security by integrating their irrigation system with these systems. Smart irrigation systems function more effectively and efficiently overall thanks to this integrated approach.

- **Long-Term Resilience:** The long-term resilience of residential and agricultural landscapes can be improved by making an investment in smart irrigation systems. These systems aid in preserving plant health and encouraging sustainable water use, which helps landscapes resist the effects of water scarcity and climate change. For farmers, this translates to decreased susceptibility to harsh weather events and more consistent crop yield. For homeowners, it means having gorgeous, hardy lawns and gardens that can flourish in spite of difficult circumstances.

Many benefits of smart irrigation systems address some of the most urgent and unsettling issues with water management. The systems signify a significant advancement in irrigation technology, supporting environmental sustainability, improving plant health, and conserving water while also lowering costs. They can deliver accurate, data-driven irrigation, which makes them a useful tool for landscape managers, homeowners, and farmers. As water resources become more limited and the preference for untrained irrigation techniques increases, intelligent irrigation systems are essential to guaranteeing the water's sustainable use for future generations.

8.4.2.2. How do Smart Irrigation Systems Work?

In order to automatically modify watering schedules, smart irrigation systems keep an eye on the weather, soil moisture, evaporation, and plant species. Traditional preset irrigation wastes water, but irrigation work can be scheduled by area and time. Managers can remotely control the smart irrigation system via a computer or mobile device. It transmits signals via lora or home Wi-Fi. The water valve does not need to be manually opened or closed by the irrigator.

8.4.2.3. Smart Irrigation System Composition

The smart irrigation system's composition is relatively complex due to its powerful functions and ability to achieve automatic water-saving irrigation. typically consists of a terminal platform, transmission portion, control portion, and monitoring portion. A unit is formed by the various accessories working together. The accessories that go with each part are as follows:

- **Monitoring Part:** Smart irrigation is monitored by a variety of sensors. The primary sensors include a data collector, a soil moisture sensor, a wind speed sensor, an evaporation sensor, a leaf wetness sensor, and a rain and snow sensor.
- **Control Part:** The foundation of automatic irrigation is a variety of control valves. Some examples of valves that facilitate remote switching are ball, pulse, and butterfly valves. Select different water pipes and valves for various irrigation zones.
- **Transmission Part:** At the terminal and control terminal, the command signal and the monitored data signal are transmitted. The lora gateway and data concentrator are the transmission components when utilizing lora wireless transmission. The router and the monitoring host are the transmission components if you use Wi-Fi at home.
- **Terminal Platform:** Typically, the terminal platform consists of a software system for organizing and analyzing data. It has the ability to store, analyze, and provide control commands. The platform software facilitates computer and mobile phone operations.

The platform receives wireless data from a variety of sensors about soil moisture and meteorological conditions, which it then processes, examines, and reports. It determines how much water and land need to be irrigated, then wirelessly commands the control valve at each location. The water valve's opening and closing are managed by the controller. Make sure you're properly irrigated. In order to adjust to changes in rainfall or outdoor temperatures, smart irrigation systems consider site-specific factors like soil composition and sprinkler rates, among others. Adjust the watering time on a regular basis.

8.4.2.4. Smart Irrigation System Composition

- **Soil Moisture Sensor:** One tool that determines whether the soil is dry directly is the soil moisture sensor. The measurement is more accurate when the probe is typically inserted at the same depth as the plant's root. The closest soil controller uses soil sensors from various locations. Regulate irrigation in various areas. Different regions' soil sensors can have their alarm values set to high and low by the management platform. The controller opens the water valve when the soil moisture level drops below the alarm value and closes the water valve when the soil moisture level rises above it. We can select different soil moisture sensors based on the types of soil. When buying sensors, customers just need to make sure they are compatible with their system.

- **Evaporation Sensor and Leaf Wetness Sensor:** Sensors for evaporation are part of weather monitoring equipment. Its value represents the soil surface evaporation. By gathering this climate-related data, irrigation timing can be modified. Encourage the movement of water in nearby areas. An evaporation sensor can cut irrigation water use considerably. The leaf wetness sensor mimics the properties of a leaf, making it similar to a leaf in that it can precisely reflect changes in leaf wetness. We can determine whether the plants are dry and whether to irrigate the soil by keeping an eye on the moisture content of the leaves. In contrast, the evaporation sensor gauges the amount of water that evaporates from the soil's surface. The evaporation of plant leaves is reflected in the leaf wetness sensor.
- **Rain and Snow Sensor:** If it starts to rain while you are watering your lawn, you make sure to stop right away. Rainfall-induced runoff and water and money waste are the results of watering in the rain. One such use is for smart irrigation. The foundation for implementing this function is the rain and snow sensor. The rain and snow sensor is made of metal wires that, when they come into contact with precipitation or snow, conduct electricity. The controller receives an electrical signal, which the management platform interprets and relays to the controller to promptly close the water valve. Rain and snow sensors are incredibly inexpensive—about $25. Additionally, it features an automatic heating feature that ensures normal operation even during the harshest winter months. Annual variations in precipitation and snowfall mean that irrigators can anticipate significant water and financial savings.
- **Wind Speed Sensor:** Watering a lawn in a windy day can lessen the amount of water that seeps into the soil and the uniformity of the irrigation system. Installing wind speed sensors is necessary so that, should the wind speed surpass a predetermined threshold, the irrigation cycle will be stopped. We strategically plan the irrigation cycle and duration based on the local weather conditions, which can save a significant amount of water resources over time.

8.4.2.5. Using Smart Water Technology to Resist Water Shortages

While we strive to lessen our carbon footprint by utilizing renewable energy sources and efficient transportation, growers can also actively lessen their environmental impact. Water can be saved in the following ways with the use of smart water technology:

- **Leak Detection:** Leaks in irrigation systems can cost you as much as 6,300 gallons of water a month. During severe dry spells, these leaks could harm your water supply and cost you money. The simplest method to keep an eye on and cut down on your usage is to make sure that there are no leaks wasting water. You can get instant access to information about your most valuable resource with leak detection technology. This can cut a farm's water use dramatically without requiring high upfront costs or even stopping operations for maintenance.
- **Measurement:** Good data is essential for making well-informed irrigation decisions. Make sure your irrigation systems are equipped with intelligent sensors and controllers to become more intelligent with water management. You can easily ascertain where and how much water is being used thanks to real-time smart sensors. Smart irrigation technology gives growers complete visibility into where their water is being used, allowing them to proactively address any unplanned deviations from their irrigation schedules. It does this by informing you of the amount of water your crops have received.
- **Automation:** Although the idea of irrigation controllers and timers is not new, their lack of widespread adoption can be attributed to their high cost, complexity, and unreliability. Automation is, nevertheless, far more affordable and user-friendly with contemporary systems. You won't ever overwater your crops again if you automate your irrigation because valves will cut off when the exact volumes are reached. Since you won't need to hire someone to operate and supervise the operation of each valve, you'll also save on labor costs. In cases where conserving water is a significant concern, growers can save up to 20% of their total water usage with enhanced irrigation management alone. These crops include wine grapes, strawberries, and almonds.
- **Harnessing Real-Time Data:** For growers, real-time data analytics are more important than ever. Several data solutions, including the following ones, are available on the market to assist in figuring out the best watering schedules for crops:

- Soil moisture sensors
- Plant sensors and pressure chambers
- Weather stations
- Satellite imagery

In order to allow growers to manage and create their irrigation schedules within a single dashboard and easily make any adjustments, without having to jump between apps, smart irrigation systems frequently integrate with these external data sources.

- **Preventive Maintenance:** Smart irrigation technology uses automatic pressure and water flow monitoring to identify issues before they arise. This implies that you'll be aware of possible problems before they turn expensive or even dangerous. In addition, you can take proactive measures to prevent leaks and clogs, which are major contributors to water waste in many farming operations.
- **Optimization:** Growers can now achieve even greater irrigation optimization thanks to data, automation, and sensor technologies. When all these elements work together in the best possible way, you can avoid wasting water by distributing it precisely where it needs to go. In addition to saving farmers time and money, smart irrigation systems can drastically lower the amount of water required for crops while still producing results that are on par with or better. Making every effort to reduce the effects of climate change and conserve water is crucial. Using water more wisely is one way to achieve this. Growers can use less irrigation water without endangering their crops and means of subsistence by implementing smart irrigation systems.

8.4.2.6. Advantages of Smart Irrigation System Using IoT

Using automation, sensors, and the internet, a smart irrigation system with IoT hydrates plants. Modern irrigation systems have many advantages, including:

- It can save water and money by only watering the plants when they need it, based on the soil moisture meter and weather conditions.
- Improve crop quality and yield by providing the plants with the optimal water and nutrients.
- However, reduce labor and maintenance costs by allowing remote monitoring and control of the irrigation system.
- Help protect the environment by preventing water pollution and soil erosion caused by over-watering.

- Moreover, It Makes farming more interesting and fun by using smart devices and data analysis.

Using IoT-powered smart irrigation techniques is a great way to leverage technology to improve the efficiency, sustainability, and fun factor of agriculture.

8.4.2.7. Examples of Smart Irrigation System

Utilizing technology to save water and simplify farming, smart irrigation systems are methods of watering plants. These are a few instances of intelligent irrigation systems.

- Drip irrigation: This system uses tubes and pipes to deliver water straight to the plant roots. We can save water by doing it this way.
- Soil moisture sensors are instruments used to gauge the soil's moisture content. They are able to detect when the plants require watering and automatically activate the irrigation system.
- Weather stations: These are gadgets that gather meteorological data, including humidity, temperature, and precipitation. They can, however, assist in forecasting the amount of water that the plants will require and modifying the irrigation system as necessary.
- Farmers can monitor and manage their irrigation system remotely using this feature on their computers or smartphones. If there is an issue, they can adjust the settings, check the water levels, or shut off the system.
- Data analytics is the process of analyzing sensor and weather station data with computers. It can assist farmers in better understanding the growth of their crops, the amount of water they use, and how to enhance their irrigation techniques.

The ability of these traditional irrigation systems to help farmers save water, improve crop quality, lower costs, and lessen their environmental impact makes them highly valuable for smart agriculture. Put simply, intelligent irrigation functions akin to a clever farm assistant. It saves water and money by using smart devices to determine the ideal watering schedule for crops. This advanced technology improves crop health, protects the environment, and brings a little bit of fun to farming with computers and smartphones. Nonetheless, these clever methods are crucial for farmers in a world where water scarcity and climate change concerns are prevalent. the surroundings and ensuring that there is an adequate supply of food for everyone.

REFERENCES

1. Acharya, S. M., Pawar, S. S., & Wable, N. B. (2018). Application of remote sensing & GIS in agriculture. *International Journal of Current Microbiology and Applied Sciences, 6495*(4), 63–65.
2. Al-Ghobari, H. M., Mohammad, F. S., El Marazky, M. S., & Dewidar, A. Z. (2017). Automated irrigation systems for wheat and tomato crops in arid regions. *Water SA, 43*(2), 354–364.
3. Anand, B. A., Venkata Reddy, H. K., & Sunil Raj, B. A. (2018). Automation irrigation system for floriculture, vegetable, and cereal crops. *International Journal of Current Microbiology and Applied Sciences, 7*(9), 859–866.
4. Anjali, C. (2016). *Development and Evaluation of a Solar Powered Automated Fertigation System* (Ph.D. thesis). Department of Land and Water Resources Conservation Engineering, Kelappaji College of Agricultural Engineering and Technology.
5. Atzberger, C. (2013). Advances in remote sensing of agriculture: Context description, existing operational monitoring systems, and major information needs. *Remote Sensing, 5*(2), 949–981.
6. Baez-Gonzalez, A. D., Chen, P. Y., Tiscareno-Lopez, M., & Srinivasan, R. (2002). Using satellite and field data with crop growth modeling to monitor and estimate corn yield in Mexico. *Crop Science, 42*(6), 1943–1949.
7. Boutraa, T., Akhkha, A., Alshuaibi, A., & Atta, R. (2011). Evaluation of the effectiveness of an automated irrigation system using wheat crops. *Agriculture and Biology Journal of North America, 2*(1), 80–88.
8. Castle, M. H., Luben, D. J. L., & Luck, J. D. (2019). Factors influencing adoption of precision agriculture technologies by Nebraska producers. *National Science Education, 8*(2).
9. Chang, A., Bullock, D., Purvis, A., Solie, J., & McGuire, M. (2015). Introduction to AgBIZ logic GIS data model: An object-oriented data model for representing farm business information. *Journal of the National Association of County Agricultural Agents, 8*(2).
10. Dadhwal, V. K. (2016). Applications of satellite remote sensing and GIS for crop growth and productivity monitoring and simulation. In *Satellite Remote Sensing and GIS Applications in Agricultural Meteorology* (pp. 263–289).
11. Diallo, Y., Hu, G., & Wen, X. (2009). Applications of remote sensing in land use/land cover change detection in Puer and Simao Counties, Yunnan Province. *American Journal of Environmental Sciences, 5*(4), 157–166.

12. Durga, C. V., Ramulu, M., Umadevi, & Suresh, K. (2018). Evaluation of soil moisture sensors and irrigation scheduling in Rabi maize. *International Journal of Chemical Studies, 6*(5), 1789–1792.

13. Franzen, D. W., & Mulla, D. J. (2016). Mapping soil variability: Making a case for directed soil sampling. In *Precision Agriculture Basics* (pp. 61–73). American Society of Agronomy.

14. Franzen, D. W., Hopkins, D. H., Sweeney, M. D., Ulmer, M. K., & Halvorson, A. D. (2002). Evaluation of soil survey scale for zone development of site-specific nitrogen management. *Agronomy Journal, 94*(2), 381–389.

15. Gebbers, R., & Adamchuk, V. I. (2010). Precision agriculture and food security. *Science, 327*(5967), 828–831.

16. Gonzalez-Dugo, M. P., Zarco-Tejada, P., Nicolás, E., Nortes, P. A., Alarcón, J. J., Intrigliolo, D. S., & Fereres, E. (2013). Using high-resolution UAV thermal imagery to assess the variability in the water status of five fruit tree species within a commercial orchard. *Precision Agriculture, 14*(6), 660–678.

17. Guerbaoui, M., El Afou, Y., Ed-Dahhak, A., Lachhab, A., & Bouchikhi, B. (2013). PC-based automated drip irrigation system. *International Journal of Engineering Science and Technology, 5*(1), 221–225.

18. Guery, S., Lea-Cox, J. D., Martinez Bastida, M. A., Belayneh, B. E., & Ferrer-Alegre, F. (2016). Using sensor-based control to optimize soil moisture availability and minimize leaching in commercial strawberry production in Spain. In *International Symposium on Sensing Plant Water Status-Methods and Applications in Horticultural Science* (pp. 171–178).

19. Hedley, C. B., & Yule, I. J. (2009). A method for spatial prediction of daily soil water status for precise irrigation scheduling. *Agricultural Water Management, 96*(12), 1737–1745.

20. Henderson, S. (2017). *Improving Greenhouse Irrigation Through A Sensor-Based Automated Decision Support System* (Ph.D. thesis). University of Guelph.

21. Hong, G. Z., & Hsieh, C. L. (2016). Application of integrated control strategy and Bluetooth for irrigating romaine lettuce in a greenhouse. *IFAC-Papers OnLine, 49*(16), 381–386.

22. Huang, Y., Lan, Y., & Hoffmann, W. C. (2008). Use of airborne multi-spectral imagery in pest management systems. *Agricultural Engineering International: The CIGR Ejournal, X*, 1–14.

23. Kingra, P. K., Majumder, D., Chandra, B., Viswavidyalaya, K., & Singh,

S. P. (2016). Application of remote sensing and GIS in agriculture and natural resource management under changing climatic conditions. *Agricultural Research Journal, 53*(3), 295–302.

24. Kotamäki, N., Thessler, S., Koskiaho, J., Hannukkal, A. O., Huitu, H., Huttula, T., & Jarvenpaa, M. (2009). Wireless *in-situ* sensor network for agriculture and water monitoring on a river basin scale in Southern Finland: Evaluation from a data user's perspective. *Sensors, 9*(4), 2862–2883.
25. Kumar, J. (2017). *Development of Indigenous Sensor Network-Based Irrigation System For Improving Agricultural Water Productivity* (Ph.D. thesis). Indian Agricultural Research Institute, New Delhi.
26. Kumar, N., Yamaç, S. S., & Velmurugan, A. (2015). Applications of remote sensing and GIS in natural resource management. *International Journal of Current Microbiology and Applied Sciences, 20*(1), 1–6.
27. Lark, R. M., & Stafford, J. V. (1997). Classification as a first step in the interpretation of temporal and spatial variation of crop yield. *Annals of Applied Biology, 130*(1), 111–121.
28. Liu, X., Liu, Q., Xie, X., Wan, L., Liu, S., & Wang, H. (2018). Identification of optimal spectral indices for monitoring winter wheat growth in the North China Plain based on radiometric factors. *Remote Sensing, 10*(3), 443.
29. Long, D. S., Engel, R. E., & Schepers, J. S. (2020). Geospatial data collection and management. In *Precision Agriculture Basics* (pp. 59–92). ASA, CSSA, and SSSA.
30. Long, D. S., Engel, R. E., & Schepers, J. S. (2020). Multispectral data acquisition. In *Precision Agriculture Basics* (pp. 93–120). ASA, CSSA, and SSSA.
31. Lyle, G., Lewis, M., Ostendorf, B., & Chisholm, L. (2014). Remote sensing and image classification of crops versus weeds in cotton and wheat. *Precision Agriculture, 15*(4), 394–416.
32. Mee, C. Y., Balasundram, S. K., & Hanif, A. H. M. (2017). Detecting and monitoring plant nutrient stress using remote sensing approaches: A review. *Asian Journal of Plant Sciences, 16*(1), 1–8.
33. Mueller, T. G., Cetin, H., Flemming, J. S., Li, H., Sudduth, K. A., & Wallander, S. (2004). Soil type mapping using electromagnetic induction and inverted Landsat thematic mapper data. *Transactions of the ASAE, 47*(3), 711–717.
34. Mulder, V. L., de Bruin, S., Schaepman, M. E., & Mayr, T. R. (2011). The

use of remote sensing in soil and terrain mapping: A review. *Geoderma, 162*(1, 2), 1–19.

35. Mulla, D. J. (2013). Twenty-five years of remote sensing in precision agriculture: Key advances and remaining knowledge gaps. *Biosystems Engineering, 114*(4), 358–371.
36. Pinter, P. J. Jr, Hatfield, J. L., Schepers, J. S., Barnes, E. M., Moran, M. S., Daughtry, C. S., & Upchurch, D. R. (2003). Remote sensing for crop management. *Photogrammetric Engineering & Remote Sensing, 69*(6), 647–664.
37. Rani, D. S., Venkatesh, M. N., Naga, C., Sri, S., & Kumar, K. A. (2018). Remote sensing as a pest forecasting model in agriculture. *International Journal of Current Microbiology and Applied Sciences, 7*(3), 2680–2689.
38. Rathika, S., Ramesh, T., Priyadharshini, V., Suriya Prasanth, A., & Shanmugapriya, P. (2020). Automation in irrigation: A review. *International Journal of Current Microbiology and Applied Sciences, 9*(12), 2974–2990.
39. Reis, S. (2008). Analyzing land use/land cover changes using remote sensing and GIS in Rize, North-East Turkey. *Sensors, 8*(10), 6188–6202.
40. Ren, J., Chen, Z., Zhou, Q., & Tang, H. (2008). Regional yield estimation for winter wheat with MODIS-NDVI data in Shandong, China. *International Journal of Applied Earth Observation and Geoinformation, 10*(4), 403–413.
41. Rhoades, J. D., Corwin, D. L., & Lesch, S. M. (1999). Geospatial measurements of soil electrical conductivity to assess soil salinity and diffuse salt loading from irrigation. In D. L. Corwin (Ed.), *Assessment of Non-Point Source Pollution in the Vadose Zone* (Geophysical Monograph No. 108, pp. 197–215). American Geophysical Union.
42. Robertson, M. J., Llewellyn, R. S., Mandel, R., Lawes, R., Bramley, R. G. V., Swift, L., & Metz, N. (2012). Adoption of variable rate fertiliser application in the Australian grains industry: Status, issues and prospects. *Precision Agriculture, 13*(2), 181–199.
43. Sandra, M., Jaume, C., Carlos, C., María, J. M., & Maria, H. P. (2019). Using soil moisture sensors for automated irrigation scheduling in a plum crop. *Water, 11*, 2061.
44. Sangeetha, C., Vishnu Moond, R. G. M., Jamu Singh Damor, S. K. P., Pradeep Kumar, & Barinderjit Singh. (2024). Remote sensing and geographic information systems for precision agriculture: A review. *International Journal of Environment and Climate Change, 14*(2), 287–309.

45. Sarma, P. K. (2009). Crop description and spatial information related issues in precision farming. *American Journal of Agriculture and Biological Sciences, 4*(1), 87–91.

46. Sathya, A., Arthi, B., Giridharan, S., Karvendan, M., & Kishore, J. (2016). Automatic control of irrigation system in paddy using WSN. In *IEEE Technological Innovations in ICT for Agriculture and Rural Development (TIAR)* (pp. 115–118).

47. Schumann, G. J., Brakenridge, G. R., Kettner, A. J., Kashif, R., & Niebuhr, E. (2018). Assisting flood disaster response with Earth observation data and products: A critical assessment. *Remote Sensing, 10*(8), 1–19.

48. Shanmugapriya, P., Rathika, S., Ramesh, T., & Janaki, P. (2019). Applications of remote sensing in agriculture: A review. *International Journal of Current Microbiology and Applied Sciences, 8*(1), 2270–2283.

49. Sibley, A. M., Grassini, P., Thomas, N. E., Cassma, K. G., & Lobell, D. B. (2014). Testing remote sensing approaches for assessing yield variability among maize fields. *Agronomy Journal, 106*(1), 24.

50. Sun, W. C., Ishidaira, H., & Bastola, S. (2011). Towards improving river discharge estimation in ungauged basins: Calibration of rainfall-runoff models based on satellite observations of river flow width at basin outlet. *Hydrology and Earth System Sciences, 14*, 2011–2022.

51. Tey, Y. S., & Brindal, M. (2012). Factors influencing the adoption of precision agricultural technologies: A review for policy implications. *Precision Agriculture, 13*(6), 713–730.

52. Thomas, A. (2000). Spatial and temporal characteristics of potential evapotranspiration trends over China. *International Journal of Climatology: A Journal of the Royal Meteorological Society, 20*(4), 381–396.

53. Underhill, S. J., & Raper, K. C. (2016). Integrating health monitoring technology with farm management information systems. *Journal of Agricultural & Food Information, 17*(1), 32–48.

54. Welch, R., & Remillard, M. (1991). Remote sensing/GIS for water resource management applications in the southeast (pp. 282–284).

55. Wójtowicz, M., Wójtowicz, A., & Piekarczyk, J. (2016). Application of remote sensing methods in agriculture. *Communications in Biometry and Crop Science, 11*, 31–50.

56. Wolfert, S., Ge, L., Verdouw, C., & Bogaardt, M. J. (2017). Big data in smart farming: A review. *Agricultural Systems, 153*, 69–80.

57. Zhang, J., Pu, R., Yuan, L., Wang, J., Huang, W., & Yang, G. (2019). Using unmanned aerial vehicle multispectral imagery to detect and map black poplar late blight. *Remote Sensing, 11*(13), 1543.
58. Zhang, N., Wang, M., & Wang, N. (2002). Precision agriculture: A worldwide overview. *Computers and Electronics in Agriculture, 36*(2, 3), 113–132.

Chapter **9**

Water Harvesting Techniques

Contents

INTRODUCTION

Rainwater can be collected and stored using a technique called water harvesting, which can then be used for agricultural purposes. It is stored in a deep well pit or is collected from a surface that resembles a roof. When there is a drought or little rainfall during the monsoon season, farmers use the water that is collected from rivers or streams. In an effort to increase food security and agriculture's sustainable productivity, new methods of collecting water are being developed. Gathering rainwater is one of the oldest methods still in use for home and residential purposes. It helps relieve the pressure on wells and offers portable water. Watering gardens and feeding cattle are two more domestic uses. It also applies to schools and hospitals that receive funding from the public or private sector. It saves money because it doesn't employ an expensive methodology. The act of gathering and preserving water from rainfall or other sources for use by crops, animals, or people is known as water harvesting. Water can be stored for later use in subsurface reservoirs or aboveground ponds, or it can be used immediately, for example, for irrigation.

Precipitation is captured in one place and moved to another as the fundamental idea behind water harvesting. Rooftops, compounds, rock surfaces, hill slopes, and restored impermeable or semi-permeable land surfaces are examples of artificially constructed catchment areas that can be used for this purpose. After gathering, the water can be poured into a percolating tank, cistern, deep pit, aquifer, or reservoir to allow it to seep down and replenish the groundwater. Through rainwater collection and stormwater flow reduction, water harvesting can help lower the risk of floods and soil erosion. When there are few or no other freshwater sources available, it can also aid societies in surviving in arid and semi-arid regions.

9.1. RAINWATER HARVESTING METHODS

Gathering and storing rainwater instead of letting it run off is known as rainwater harvesting (RWH). Rainfall falls on a surface that resembles a roof and is directed to replenish the ground water table by seeping down into a tank, cistern, deep pit (well, shaft, or borehole), aquifer, or reservoir with percolation. As runoff is usually collected from roofs and other area surfaces for storage and later reuse, rainwater harvesting is different from stormwater harvesting. Gardening, cattle, irrigation, home use after appropriate treatment, and home heating are just a few of its uses. Additionally, longer-term storage or groundwater recharge may be the intended uses for the harvested water. Having been utilized for many thousands of years in South Asia and other nations, rainwater harvesting is one of the earliest and most basic ways for households to self-supply water.

Installations can be created at various scales, such as homes, neighborhoods, and communities. They can also be created to serve establishments like public spaces like hospitals, schools, and other settings.

The technology known as a rainwater harvesting system gathers and holds rainwater for use by humans. Simple rain barrels to more complex structures with pumps, tanks, and purification systems are examples of rainwater harvesting systems. Not only can the nonpotable water be purified for human consumption, but it can also be used for car washes, laundry, toilet flushing, and landscaping irrigation. Rainwater harvesting systems can provide homes and businesses with water for use during dry seasons, reducing the demand on municipal systems. Water scarcity is a pressing issue for many densely populated regions. Rainwater harvesting can be a useful method of obtaining this valuable resource, as rainfall is erratic and only a small amount of precipitation worldwide is readily usable by humans. The majority of rainwater that falls on roads, buildings, roofs, and other hardscaping in cities is diverted into storm sewers rather than seeping into the ground. In many places, impermeable surfaces result in urban flooding and produce contaminated, unusable water that is diverted away from sources of potable water. Local groundwater reserves may be exhausted during the dry months, and many places find it difficult to regularly supply enough drinkable water to meet demand. Rainwater harvesting minimizes the amount of fresh water required overall and the burden on stormwater infrastructure when used for nonpotable purposes like gardening and laundry. In big cities, this difference in the supply and demand of drinkable fresh water is substantial. Rain barrels and other rainwater harvesting systems are encouraged and even subsidized by many localities, but some places—especially in the southwestern United States—consider rainwater harvesting to be a matter of water rights and have laws restricting its use.

9.1.1. Uses

The idea of collecting water is not new, even though the term "rainwater harvesting" is being used a lot these days in India. Techniques for harvesting water had developed over centuries before. The process of percolation replenishes groundwater reserves naturally. However, because of haphazard development and the swift urbanization of the area, the amount of exposed soil surface has significantly decreased, which has decreased rainwater percolation and depleted the ground water supply. The technique known as "rainwater harvesting" involves using artificial techniques to increase the natural infiltration of rainwater into foundations. Rainwater harvesting is the deliberate collection and storage of rainfall to meet water demands for drinking, household use, and irrigation.

9.1.1.1. Domestic Use

Rainwater harvesting on rooftops is a useful technique for supplying drinking water, household water, livestock water, small-scale irrigation, and groundwater replenishment. Rainwater harvesting for irrigation, laundry, and toilets has already proven successful in Kenya. Kenya has made the regulation of its agriculture sector a top priority ever since the country's 2016 Water Act was established. A portion of Australia also uses collected rainfall for drinking and cooking.

9.1.1.2. Agriculture

When it comes to urban agriculture, rainwater collection in cities lessens the effects of flooding and runoff. Building temperatures can be lowered by more than 1.3°C when urban "green" rooftops and rainwater catchments are combined. The United Nations Sustainable Development Goals for healthier and more sustainable cities, as well as for the security of food and water, could be effectively met by combining rainwater harvesting with urban agriculture. In order to use water more efficiently, particularly in an urban setting, the available technology needs to be remodeled. The risk of losing part or all of the year's harvest due to soil or water scarcity can be considerably decreased by collecting and storing rainwater runoff for later use, as demonstrated by missions to five Caribbean countries. Furthermore, there would be a reduction in the hazards linked to soil erosion and flooding during periods of heavy rainfall. Because they can capture runoff and lessen the effects of soil erosion, small farmers, particularly those who farm on hillsides, stand to gain the most from rainwater harvesting. Rainwater harvesting is a common and affordable method of obtaining clean water, particularly in arid countries. Soil ridges that catch and stop rainwater from flowing down hills and slopes are built to improve irrigation in arid areas. Enough water is gathered to support crop growth even during dry spells. Rooftops can be used to collect water, and large-scale rainwater storage structures like dams and ponds can be built to store rainwater so that crops can be adequately irrigated even on days when there is little to no rainfall.

9.1.1.3. Industry

Germany's largest rainwater harvesting system is found at Frankfurt Airport. Approximately one million cubic meters of water are saved annually by the system. In 1993, the system cost one and a half million dinars, or $63,000 USD. This system uses the 26,800 square meter roof top area of the new terminal to collect water. Six tanks totaling 100 cubic meters are used to store the water in the airport's basement. The primary uses of the water are air conditioning system cleaning, plant watering, and toilet flushing. At The Velodrome in the London Olympic Park, rainwater harvesting was implemented to improve the building's sustainability. The park's estimated need for potable water dropped by 73%. Despite this, it was decided that the park's blackwater recycling program was a more effective use of financial resources to increase sustainability than rainwater harvesting.

9.1.2. Technologies

Historically, there was only one goal for stormwater management with retention basins. Still, the current detention capacity remains intact thanks to optimized real-time control that enables this infrastructure to serve as a rainwater harvesting source. This reduces the flow of wet weather while guaranteeing that water is available for later use. It is used in the EPA headquarters to remove stored water before storm events. As a result, during combined sewer overflow incidents, less water is released in total and more water of higher quality is released. Generally, the purpose of building check dams across streams is to improve surface water percolation into subsurface strata. By dislodging the subsurface layers and using ANFOR explosives, which are used in open-pit mining, it is possible to artificially increase the rate of water percolation in the check dam's water-impounded area exponentially. In order to use the available surface water for use during the dry season, local aquifers can be swiftly replenished.

9.1.2.1. System Setup

Systems for collecting rainwater can be as simple as installing them yourself, or they can be as complex as automated rainwater harvesting systems requiring sophisticated setup and installation. Since every building's terrace outlet is connected by pipe to an underground water storage tank, the fundamental rainwater harvesting system is more of a plumbing project than a technical one.

More specifically, the area where rainfall collects, like a building roof, needs to be big enough to keep the water flowing in a sufficient amount. The size of the water storage tank ought to be sufficient to hold the collected water. Rooftop systems, surface water collection, pumping rainwater that has already seeped into the ground or collected in reservoirs and stored in tanks (cisterns) are just a few of the low-tech techniques used to collect rainwater for low-tech systems.

9.1.2.2. Rainwater Harvesting by Freshwater-Flooded Forests

Growing freshwater-flooded forests allows for the collection of rainwater without sacrificing the revenue generated by the utilized, submerged land.

Using locally available rainwater to meet annual water needs without requiring significant capital expenditure is the primary goal of rainwater harvesting. This would make clean water easier to find for irrigation, industry, and residential use.

9.1.2.3. Rainwater Harvesting by Solar Power Panels

A reliable source of high-quality water that is closer to populated areas is getting harder to find and more expensive for users. Rainwater is a significant renewable resource on any land, in addition to solar and wind energy. Every year, in every part of the world, a vast area is covered with solar PV panels. The majority of the rainwater that falls on solar panels can also be collected, and because rainwater has a very low salinity, it can be produced using straightforward filtration and disinfection techniques to produce drinking water free of bacteria and suspended matter. By using rainwater to create value-added goods like bottled drinking water, solar PV power plants can remain profitable even in regions with heavy rainfall or cloud cover thanks to the increased revenue from the production of value-added drinking water. In India, groundwater levels have been found to rise significantly through the use of cost-effective rainwater collection in wells that have already been dug.

9.1.2.4. Other Innovations

The Groasis Waterboxx is an example of low scale technology, in this case to assist planting of trees in arid area. It harvests rainwater and dew.

9.1.3. Advantages

In developed nations, rainwater harvesting is often utilized to augment the primary water supply, as well as to provide an independent source of water during regional water restrictions. It helps prevent flooding in low-lying areas, supplies water during droughts, and lessens the demand on wells, which may allow groundwater levels to remain stable. By raising the level of dried borewells and wells, rainwater harvesting increases the amount of water available during dry seasons. There is a ready supply of surface water for many uses, which lessens reliance on subsurface water. By reducing the salinity, it enhances the quality of the ground. It is environmentally benign and produces no pollution. It is reasonably priced and efficient. Because rainwater is largely free of salt and other impurities, it also contributes to the availability of drinkable water. By lowering the demand for clean water in water distribution systems, generating less stormwater in sewer systems, and reducing stormwater runoff that contaminates freshwater bodies, the use of rainwater harvesting in urban water systems has a significant positive impact on both the wastewater and water supply subsystems. The development of life cycle assessment and its costing methodologies to evaluate the degree of environmental impacts and financial savings that can be achieved by installing rainwater harvesting systems has been the subject of a substantial body of work.

9.1.3.1. Independent Water Supply

During water restrictions, rainwater harvesting offers a stand-alone water supply. Rainwater harvesting is an essential source of clean water in places where it is expensive or hard to obtain clean water. Though collecting rainwater can also lower a household's water bills or usage levels overall, it is typically collected in developed nations to be used as a supplement rather than as the primary source of water. If consumers take extra precautions before drinking, rainwater is safe to consume. Water that is boiling helps to destroy germs. To prevent water contamination, adding an additional supplement to the system, like a first flush diverter, is also a standard practice.

9.1.3.2. Supplemental in Drought

Rainwater gathered in earlier months can be utilized during dry spells. When rainfall is infrequent but erratic, having a rainwater harvesting system installed can be essential to collecting precipitation when it does fall. Rainwater harvesting is a common and affordable method of obtaining clean water in arid countries. Soil ridges that trap and stop rainwater from flowing downhill are designed to improve irrigation in arid environments. Enough water is gathered to support crop growth even during dry spells. Rainwater can be collected from roofs and large volumes can be stored in tanks built for that purpose. Rainwater harvesting also lessens the need for well water, allowing groundwater levels to be further maintained as opposed to being depleted.

9.1.3.3. Life-Cycle Assessment

Throughout a system's life, its environmental effects are assessed using a process called life-cycle assessment. A new measure known as the demand to supply ratio (D/S) was created to address the functional parameters of rainwater harvesting systems. It determines the optimal building design (supply) and function (demand) in relation to the environmental performance of rainwater harvesting for toilet flushing. Given the theory that rainwater collection reduces stormwater entering the combined sewer network, which otherwise needs to be treated, as well as potable water, it follows that buildings connected to a combined sewer network will have lower environmental emissions than those not.

9.1.3.4. Cost-Effectiveness

Standard RWH systems can give impoverished developing areas access to water, but depending on the technology employed, the average cost of an RWH setup can be high. NGOs and government assistance can help impoverished communities by supplying the tools and training required to create and manage RWH setups. According to some studies, rainwater harvesting's affordability and environmental friendliness make it a widely applicable solution for water scarcity and other multiple uses. Building new, large-scale, centralized water supply systems—like dams—has limited uses, especially in developing nations or underdeveloped areas, and is likely to harm regional ecosystems in addition to posing external social costs. However, numerous studies have shown that installing rainwater harvesting systems not only provides local communities with a sustainable water source but also offers a number of other benefits, such as flood protection and runoff control, even in impoverished areas. Rainwater harvesting systems that do not necessitate significant construction or ongoing upkeep by an outside expert are more environmentally friendly and are likely to provide longer-term benefits to the local population. Therefore, there is a greater likelihood that rainwater harvesting systems that are installed and maintained by locals will be adopted and utilized by a larger number of people. Rainwater harvesting can have lower investment costs when in-situ technologies are used. Since in-situ technologies require less material to build, they may be a viable option for rainwater harvesting in rural areas. They can offer a dependable supply of water that can be used to increase agricultural yields. Water can be collected for home use using above-ground tanks, but those living in poverty may not be able to afford such units.

9.1.4. Limitations

In nations that exhibit drought-like conditions, rainwater harvesting is a popular technique for storing rainfall. Different criteria and techniques have been developed by numerous research projects to choose appropriate locations for rainwater harvesting. A model builder in ArcMap 10.4 was developed, and appropriate locations for possible dam construction were found and chosen through research. 1. To ascertain whether the location is suitable for collecting rainwater, the model took into account a number of factors, including slope, runoff potential, land cover/use, stream order, soil quality, and hydrology. When precipitation is below average in arid urban regions like the Middle East, harvested water from RWH systems may be minimal. Since RWH collects water for domestic use and irrigation, it is beneficial for developing areas. To guarantee safe drinking, the collected water must be sufficiently filtered.

9.1.4.1. Quality of Water

Rainwater may need to be properly analyzed and used in a way that ensures its safety. For instance, harvested rainwater in the Gansu province is disinfected using solar energy by boiling it in parabolic solar cookers before consumption. Low-cost disinfection options for treating rainwater that has been stored for drinking are offered by these so-called appropriate technology methods. Although rainwater can be a clean source of water in and of itself, often superior to groundwater or water from rivers or lakes, the process of collecting and storing it can leave the water contaminated and unfit for human consumption. Rainwater collected off roofs may contain mosses and lichens, wind-blown dust, pesticides, inorganic ions from the sea (Ca, Mg, Na, K, Cl, SO_4), dissolved gases (CO_2, NOx, SOx), and excrement from people, animals, and birds. In Europe, high concentrations of pesticides have been discovered in rainwater, peaking during the first downpour following a dry spell. By redirecting the initial run-off water flow to waste, the concentration of these and other contaminants is greatly decreased. Additional methods to achieve better water quality include withdrawing from the last tank in a series and utilizing a floating draw-off mechanism (instead of one from the tank's base). In the industry, prefiltration is a standard procedure used to maintain the integrity of the system and guarantee that the water entering the tank is clear of sizable sediments.

A water supply system should, in theory, match the end user's water quality. Nonetheless, potable water of superior quality is utilized for all purposes in the majority of the developed world. This method has needless negative effects on the environment and wastes money and energy. A sustainable water management strategy may include supplying rainwater that has undergone initial filtration for non-potable water uses like laundry, irrigation, and toilet flushing. Mosquitoes carrying pathogens may also find a home in rainwater cisterns. Therefore, it's important to make sure that female mosquitoes are unable to enter the cistern in order to lay their eggs. Fish that consume larvae can be chemically treated or added to the cistern.

9.1.5. Design Recommendations

Non-pressurized systems, like rain barrels, in which pipes are run from rain gutters into a tank, are the most basic type of rainwater harvesting system. Referred to as "dry systems," those constructions prevent the accumulation of water in the pipes following a rainstorm, thereby preventing the establishment of mosquito and other insect breeding grounds. When the pipelines are not set up to flow straight into the tanks, "wet systems" are required. Plumbing from the gutter descends underground and rises through a riser into the tank in locations where the tanks are situated somewhat distant from the collection surfaces or where multiple tanks are present to service multiple buildings. To ensure that stagnant water does not remain in the lengthy pipe runs, these systems are frequently pressurized.

Effective rainwater harvesting systems make sure that all of the openings, including the pipes, are free from insects, especially when they are wet systems. All tank inlets should have wire mesh screen covers to further help keep debris out of the tank. Tanks should be constructed of nontoxic and noncorrosive materials, and collection surfaces—mostly roofs—should be made of nontoxic materials. Lead-based paints and membranes should be avoided in particular. A tank's floor should be at least 10 cm (4 inches) above the tank outlet taps or draw-off pipes to prevent drawing out any sludge that may have accumulated in the water supply. All systems should have their internal surfaces cleaned on a regular basis, even though some have a sump pump and washout pipe to remove sludge. Catchments should also be maintained free of lichens, moss, dirt, and other debris buildup. Removing branches from trees that hang over those catchment surfaces is a good idea. For optimal operation, yearly tank inspections and routine cleaning of gutters, tank inlets, and screens are required. It is ideal to test the water on a regular basis to keep an eye on its quality.

9.1.6. Quality

As rainwater descends through the atmosphere, it picks up dust and other pollutants as well as soluble and insoluble materials from the surfaces it falls on. Plants, fungi, and other organic materials can be contaminants, as can inorganic materials like chemicals, metals, dissolved minerals, and water-soluble paints. Rainwater collected from dirty surface runoffs is unfit for drinking or cooking, even though it doesn't need to be extremely pure for use in gardens or agriculture. The quality of the water in the rainwater storage tank can be enhanced by removing the first flush of rainwater from the roof, gutters, and other collection surfaces. Rainwater collection needs to be purified before it can be used for domestic purposes. The removal of bacteria, organic material, and chemicals that form films on surfaces or sink to the bottom of the tanks as sludge can be accomplished through flocculation, settling, and biofilm skimming. Incoming raw water can also be treated with a liquid alum solution to bind fine suspended particles into larger particles that can be filtered and settled to remove. This gets rid of the drinking water's unwanted color, turbidity (cloudiness), and aluminum. If the prefiltered water is meant for drinking purposes, it can then be treated with chlorine or another chemical to eliminate harmful substances, or it can go through solar water disinfection. Fluoride, calcium hydroxide, and potassium permanganate are additional chemicals used in water purification.

9.1.7. Why Is Rainwater Harvesting Important?

Rainwater harvesting is beneficial for a number of reasons, the primary one being that we are losing ground on indoor water conservation and need to start searching outside for new opportunities. The advantages we've seen with our indoor water fixtures as a result of government regulations and fixture manufacturers' inventiveness are displayed in the graph below. As you can see, there isn't much more that we can do to increase the efficiency of our interior fixtures. The law of diminishing returns refers to this phenomenon. Is a 0.2-gallon toilet the next? Probably not. Where, then, will the next revolution in water conservation occur? We think our services will be needed in those areas.

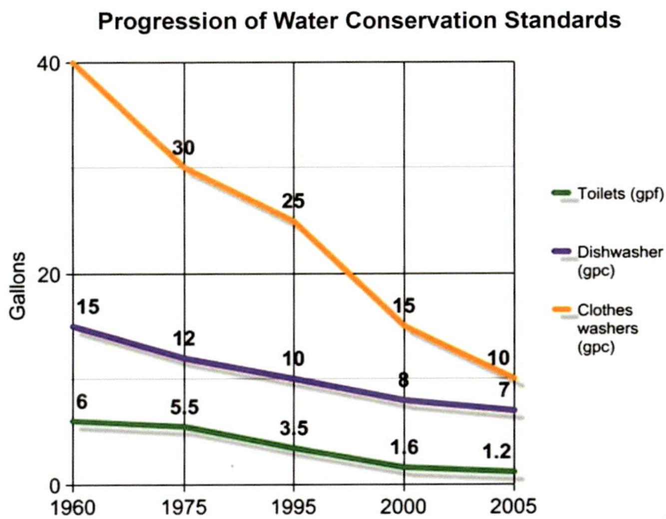

9.1.7.1. What Are the Uses of Collected Rainwater?

Rainwater can basically be used anywhere tap water is used. Considering the nation's growing population and chronic water scarcity, it is foolish and wasteful to use drinking water for lawn irrigation and toilet flushing. One way to green your home and reduce your environmental impact is to collect rainwater.

There are basically three areas where rainwater can be used:

- Irrigation use
- Indoor, non-potable use
- Whole house, potable use

Here are some ideas for specific uses of rainwater:

- Hand water your lawn and garden

- Connect rainwater collection system to irrigation/sprinkler system
- Wash your vehicles
- Wash your pets
- Refill your fountains and fish ponds
- Refill your swimming pool
- Replace the use of tap water with rainwater to wash your driveways and sidewalks (if you don't use a broom)
- Use it for all indoor non-potable fixtures (toilets and clothes washer)
- Use it for all potable needs when properly filtered and disinfected
- Use it for industrial processes instead of municipally treated water

9.2. STORAGE AND DISTRIBUTION SYSTEMS

A distribution system's water storage is a crucial part. The major goal is to supply enough water to balance or average the daily demands on the system. Because storage offers sources of supply both during the day and at night, when utility workers are normally not at work, it increases operating conveniences. It balances out the need for pumping, can save pumping-related power expenses, offers a water supply during blackouts, acts as fire storage, and has additional advantages. This text's chapter will look at water storage systems, different kinds of water storage, how water storage impacts water quality, and a general rundown of how water storage structures are maintained and operated.

9.2.1. Why Water Storage Is Needed

Early in the morning and late at night are when there is the greatest demand for water. Water is needed in the morning when patrons take showers, use the restroom, and cook. Furthermore, before the weather gets too hot, the majority of irrigation systems are used in the early morning. People are usually returning home in the evenings from work and school, which causes an increase in the demand for water. Midday and overnight are the least used times. In the absence of storage, the pumping capacity would be roughly twice that of the typical need. Uniform pumping is made possible by adequate storage. Minimal operator intervention is also possible during non-working hours thanks to storage. In the event of an unforeseen spike in demand, employees would need to be on hand to run the system in the absence of storage. To adapt to variations in demand, a utility would need a variety of pumps in different sizes. Pumps and motors suffer greater wear and higher electricity costs when they are cycled on and off frequently.

Having water on hand to supply and store during emergencies and blackouts is an additional advantage of water storage. Should a water utility rely exclusively on pumps to fulfill demand, water service would be disrupted during any power outage. Service during these periods can be maintained without interruption thanks to water tanks. The water supply is brought into the community being served by long transmission water mains in certain distribution systems. Since the water main would need to be shut off for any repairs to be made to these pipes, the customers' access to water would be disrupted. For community water systems, fighting fires is also essential. The demands placed on water utilities to put out fires must be met in addition to their operational obligations. As much as fifty (50%) of the total storage may go toward firefighting. Moreover, during main line interruptions, power outages, and peak customer demand, firefighting requirements must be satisfied. While there is a fire, proper pressure must also be kept.

9.2.2. Water Storage and Water Quality

Both positively and negatively influencing water quality are possible with water storage. Certain periods of time must pass before the water is used by the first customer, according to a set of rules known as the Surface Water Treatment Rule (SWTR), which is a component of the Safe Drinking Water Act. Tanks of water can be used to hold water for a certain amount of time, which is necessary to let the chlorine react with the water. Combining supplies from different sources can also happen in the water storage sector. In order to provide sufficient blending of the impaired source, unimpaired sources may also be pumped into a storage tank that contains an impaired source of supply. It's also possible for water storage tanks to lose their quality. Though they offer many advantages, water storage tanks can deteriorate the quality of the water they hold. In storage tanks, if the water is not changed, chlorine residuals can decrease and the water can become stagnant. Water quality deterioration and stagnation can be prevented by cycling tanks, or allowing the level to rise and fall within one.

9.2.3. What Type and Size of Storage Is Needed?

There are many different shapes and sizes of water storage tanks. The size, topography, and layout of a distribution system are some of the factors that affect the type and capacity of storage within the system (i.e., is the system concentrated in a small area or is it spread out widely). The requirements for water storage are designed using these and other factors. It's important to comprehend a few terms linked to water storage.

- **Average Day Demand (ADD):** This is the total demand for water during a period of time divided by the number of days in that time period. This is also called the average daily demand.
- **Maximum Day Demand (MDD):** This is the highest total demand over a 24-hour period within a given year
- **Peak Hour Demand (PHD):** This is the maximum demand over a one-hour period within a given year
- **Float on the System:** This is a method of operating a storage facility. Daily flow into the system is approximately equal to the average day demand. When customer demands are low, the storage facility will be filling and when demands are high the storage facility will be emptying.

Locating water storage facilities inside a distribution system has a direct bearing on system hydraulics. When a pumping station and a water storage tank are close to each other, there could be an excessive head loss (pressure) through regular-sized pipes to the furthest part of the distribution system. Such pressure loss can be mitigated by adding more transmission mains. Enough pressure is usually obtained at the far ends and close to the pumping stations if a storage tank is positioned at the furthest end of a service area. Greater main sizes are not required with this kind of setup. But in order to replenish the tank during off-peak hours, there needs to be sufficient capacity at the remote site. Moreover, lower pressures may occur in the middle of the distribution system if there is a large distance between the pumping facility and storage. Ideally, storage structures should be placed close to the area experiencing the least amount of pressure. Littler diameter water mains can be used because the flow from the tank is divided into two directions, and this usually gives enough available pressure to cover the entire service area. Having several smaller tanks rather than one larger tank is more ideal for hydraulic purposes. Pressures in the distribution system can now be more evenly distributed and stabilized as a result. When long-distance pumping is necessary and peak demand is present, head losses also rise. Most of the time, pumping stations can run at or close to average day demand conditions when storage tanks are placed close to the center of a distribution system.

9.2.4. Types of Water Storage Structures

Water storage facilities come in different varieties. They are shaped differently and constructed from a variety of materials to meet a range of purposes. This section only covers storage structures found in distribution systems, even though there are other storage structures for storing raw water and within treatment plants. Some of the more popular water storage facilities found in distribution systems are as follows:

- **Elevated Storage Tanks:** Elevated storage tanks are frequently utilized in areas with comparatively flat topography. These are above-ground tanks held up by a pedestal or tower made of concrete or steel. Designed to float on the system, the majority are composed of steel. They may overflow and not provide enough pressure if they are not built tall enough.
- **Hydroneumatic Storage Tanks**: Small systems that have sufficient pressure are utilized with these tanks. To supply more water than the pump can handle when needed, they are kept partially filled with compressed air. Even in the event of a pump failure, these systems can supply water for a brief period of time.
- **Standpipes**: These tanks are taller than they are in diameter and are built right on the ground. They are frequently employed to balance storage close to a supply source, such as a well field. They can also be applied to offer more protection against fires.
- **Above Ground Storage Tanks:** Along the US west coast, these are the most widely used types of storage tanks. They are situated where the terrain permits them to be built on hillsides and have the capacity to store enormous amounts of water. They need a sizable amount of land, which is their primary drawback.

9.2.5. Components of Elevated and Above Ground Storage Tanks

These tanks have very similar components. This section will review the major components and address any differences between the two styles of storage structures.

9.2.5.1. Inlet and Outlet Pipes

Above-ground storage tanks may have separate or common inlet and outlet pipes, but elevated storage tanks typically have a common inlet and outlet pipe. These pipes' functions are to supply water and permit it to leave the tank. To facilitate water circulation within the tank, distinct inlet and outlet piping configurations are used. An elevated storage tank's common pipe, also known as a riser, normally runs up the center of the support structure that houses the tank. In most cases, the bottom of an above-ground storage tank is where the inlet and outlet piping enters the tank. The inlet and outlet connections are normally located at the opposite ends of the tank when the inlet and outlet are configured separately.

9.2.5.2. Overflow Pipe

Every kind of tank has an overflow pipe installed. In the event that the water-level controls malfunction, it is intended to let water escape the tank and into the atmosphere. They should never be connected directly to a storm drain or sewer because they are typically designed to empty into a catch basin. In order to keep animals out of the pipe, they should be screened or have a weighted flap that keeps them well spaced from the area they are emptying into.

9.2.5.3. Drain Connection

Periodic inspections of all tanks are required. Tanks are frequently drained for inspection, cleaning, and repairs, though some inspections can be done with water in the tank. By stopping pumps from starting up to fill the tank, water levels can be decreased. Only the height of the outlet pipe can be reached by lowering them, and a tank should never be completely empty while it is in use. If the water is lowered to the maximum level while the can is in use, the remaining water can be drained by opening a different drain pipe.

9.2.5.4. Monitoring Devices

Like other water distribution facilities, water storage tanks are frequently outfitted with monitoring equipment. There is another chapter that goes into more detail about these surveillance tools. But there will also be a quick discussion of the here. The water level in storage tanks is one of the most crucial things to keep an eye on. As a result, the majority have either a site gauge that is physically fixed to the outside of the tank or level sensors that allow tank levels to be transmitted to distant locations. These gadgets frequently come with alarms for both high and low water levels.

9.2.5.5. Valves

Along the inlet/outlet piping leading into the tank, a valve must be installed in order to isolate the tank from the distribution system. The tank can then be taken out of service for upkeep and repairs by closing this valve. Occasionally, a valve known as an altitude valve will be installed in a tank. This valve's purpose is to shut off the tank's ability to overflow.

9.2.5.6. Vents

Tanks typically have air ventilation installed at their tops to let air out when the tank fills and air in when the tank's level falls. These air vents need to be properly screened with a minimum mesh size of ¼" and big enough to keep the tank from collapsing.

9.2.5.7. Access Hatches

It is also necessary to allow access inside a tank. On top of storage tanks, there are usually one or more access hatches. They make it possible for workers to access the tank for upkeep and inspection. In order to keep surface water runoff out of the tank, these hatches need to be built correctly with rims beneath the cover. In addition, manways are located at the bottom of the tank to provide access after it has been emptied and taken out of commission for upkeep and inspection.

9.2.5.8. Ladders

Both the interior and exterior of tanks must be accessible. Usually, ladders offer this kind of access. Spiral staircases are sometimes used in place of ladders in above-ground water storage tanks. Typically, three (3) distinct ladders are included with elevated storage tanks. The first one goes from the ground to the balcony surrounding the tank and then up the tower's leg. The second ladder ascends to the top of the tank roof from the balcony. For inside access, the third ladder travels the length of the tank's interior. To prevent unwanted access, outside ladders should be installed between six and eight feet above the ground, or they should have a locked metal shield around the bottom.

9.2.5.9. Cathodic Protection

Internal tank walls are prone to corrosion, particularly in the upper parts that aren't submerged in water all the time. This internal corrosion in coated steel tanks can be lessened by cathodic protection. In order to prevent rust and corrosion, cathodic protection reverses the current flow that tends to dissolve iron from the tank surface. Instead of the tank walls corroding, direct current (DC) electrodes are used. It is possible to suspend the electrodes from the tank roof in warm climates. The electrodes have to be immersed in cold weather. The anodes should be inspected yearly and have a maximum lifespan of ten (10) years.

9.2.5.10. Tank Coatings

Tank interiors and exteriors should be properly coated because water is the "universal solvent" and steel can oxidize and deteriorate. Interior coatings need to be resilient to high humidity, heat, chlorine, fluctuating water temperatures, alternating wet and dry times, ice abrasion, and mineral-rich water. Similar conditions must be met by exterior coatings in order to preserve their good looks. All coatings used inside must have NSF approval.

9.2.5.11. Operation and Maintenance

Every three to five years, all water storage structures should undergo a thorough inspection, according to the American Water Works Association's recommendation. Periodically draining, cleaning, inspecting, repairing, and painting elevated and above-ground storage tanks is recommended. The interior surfaces need to be thoroughly cleaned, either by sweeping and scrubbing or by using a high-pressure water jet. In addition to completely or partially recoating the tank, all dirt and debris should be removed. Before being put into service, every tank must be cleaned and disinfected. This covers both newly built structures and tanks removed from service for maintenance. There are three fundamental ways to clean storage tanks:

- The first approach entails adding water to the entire tank and maintaining a disinfectant residual level of 10 mg/L. The detention period is six (6) hours if the water is disinfected prior to entering the tank, and twenty-four (24) hours if the tank is filled and then disinfected.
- The second method involves spraying the interior walls with a disinfectant solution concentration of 200 mg/L
- The third method requires six (6) percent of the tank to be filled and disinfected to a residual of 50 mg/L. The tank is then completely filled and held for twenty-four (24) hours.

Next, a sample and analysis of the tank's total coliform bacteria are required. Further disinfection is necessary if the results are positive until two (2) consecutive samples are negative. It is also necessary to sample and examine newly coated and new tanks for volatile organic compounds. Water storage tanks should also undergo routine inspections. It is important to regularly check the overflow piping, vents, hatches, ladders, and locks for damage and vandalism. When considered unsafe, ladders should be replaced and kept in good condition. To stop surface water from leaking into the tank, cracks and holes in the roof and access points should also be inspected.

9.2.6. Water Distribution System

A component of the water supply network, a water distribution system transports potable water from wells or a centralized treatment plant to consumers in order to meet their needs for firefighting, commercial, industrial, and residential use. The term "water distribution network" refers to the section of a water distribution system that serves bulk water consumers or demand nodes, which are collections of many consumers. The World Health Organization (WHO) refers to a network of pipes that transport water from water treatment plants to service reservoirs as a "water transmission system" and refers to a network of pipes that typically have a loop structure as a water distribution system. The latter is used to supply water to consumers from the service reservoirs and balancing reservoirs.

9.2.6.1. Components

Pipelines, storage facilities, pumps, and additional accessories make up a water distribution system. Water is moved throughout a distribution system via pipelines referred to as water mains that are installed within public rights of way. To connect water treatment facilities and service areas, large diameter water mains known as primary feeders are utilized. Main feeders and distributors are connected to secondary feeders. In addition to providing water to individual fire hydrants, distributors are water mains that are situated close to water users. A small diameter pipe known as a "service line" is used to connect a water meter at the user's location to the water main via a tiny tap. To cut off the water supply to the user's location, a curb stop, also referred to as a service valve, is situated on the service line next to the curb. When the system needs to store clean drinking water in response to varying demands, storage facilities, also known as distribution reservoirs, provide the necessary space for water treatment. These reservoirs also help to equalize operating pressure and provide storage for clean drinking water (balancing reservoirs). Additionally, in the event of a power outage, they can be momentarily employed to meet firefighting demands. Distribution reservoir types include the following:

- **Underground Storage Reservoir or Covered Finished Water Reservoir**: A sizable, fully covered ground-excavated reservoir or an underground storage facility. To stop ground water intrusion, these reservoirs' bottoms and walls may be lined with impermeable materials.
- **Covered Finished Water Reservoir**: A sizable ground-excavated reservoir without a top cover that is equipped with sufficient lining or measures to stop ground water intrusion and surface water runoff. Because the water in this kind of reservoir won't be further treated before being distributed, it's more vulnerable to pollutants like animal and human activity, bird waste, algal blooms, and airborne deposition.
- **Surface Reservoir (Also Known as Ground Storage Tank and Ground Storage Reservoir)**: A raised water reservoir. Some popular varieties are the composite elevated storage tank, which is a steel tank on top of a large-diameter concrete column; hydropillar elevated storage tanks, which are steel tanks on top of large-diameter

steel columns. Spheroid elevated storage tanks are another type of tank. Storage and multi-story office space are two possible uses for the area inside the large column beneath the water tank. The area's aesthetic appeal is a major consideration when utilizing water towers in the water distribution system.

- **Water Tower (Also Known as Elevated Surface Reservoir)**: A hybrid water tank combining the features of a water tower and a ground storage tank. Useful storage is the upper portion of the storage space, which would be at a height comparable to an elevated water tower, while supporting storage is the lower portion.
- **Standpipe**: The standpipe enables water storage from the bottom of the tank to the top, which sets it apart from an elevated water tower.
- **Sump**: Water is not directly distributed from this facility; rather, it serves as a backup water storage facility. It is usually constructed in a circle, below ground, with a dome over the top. When a service reservoir needs water, the water from a sump will be pumped there.

Usually, storage facilities are found in the middle of the service areas. The water mains to the service locations are shorter because of their central location. As a result, less water is lost through friction during transportation across a water main.

9.2.7. Maintenance

9.2.7.1. Internal Corrosion Control

When metal pipe surfaces and connections corrode in distribution systems, the quality of the water declines. Color, taste, and odor changes in water caused by pipe corrosion can be harmful to human health. Releasing trace metals like lead, copper, or cadmium into the water can have negative health effects. Children's physical and mental development can be delayed by exposure to lead. Liver and kidney damage can result from prolonged copper exposure. Multiple organs may suffer harm from high or prolonged exposure to cadmium. The water becomes rusty or red due to iron pipe corrosion. Metal taste may result from corrosion in iron and zinc pipes. Adjusting the pH level, adjusting the carbonate and calcium to create calcium carbonate as a pipe surface coating, and applying a corrosion inhibitor are some methods for controlling internal corrosion. A type of corrosion inhibitor would be phosphate products, which build up as films on pipe surfaces. The possibility of trace metals leaking into the water from pipe materials is decreased as a result.

9.2.7.2. Hydrant Flushing

The regular release of water from fire hydrants or dedicated flushing hydrants is known as "hydrant flushing," which is done to remove iron and other mineral deposits from water mains. Testing to see if water is supplied to fire hydrants at a pressure suitable for fighting fires is another advantage of using fire hydrants for water main flushing. Customers may notice rust-colored water after flushing their hydrants because the process stirs up iron and mineral deposits.

9.2.7.3. Water Main Renewals

Water mains will deteriorate in structural, hydraulic, and water quality aspects after many years of service. Numerous factors can contribute to structural deterioration. Pipes made of metal corrode both inside and outside, thinning or deteriorating the pipe walls. They might eventually crack or leak. The deterioration of reinforced steel and cement matrix can affect cement-based pipes. Joint failures can occur in any type of pipe. Sedimentation, scaling, and biofilm formation are examples of deteriorating water quality. Hard deposits that form on the inside walls of pipes are known as scaling. This condition, known as tuberculation, may result from the combination of calcium in the water and pipe corrosion. When solids collect in the pipes, usually in the spaces between scaling build-ups, this process is known as sedimentation. Water will become discolored due to the settled solids being stirred up by a sudden change in water flow velocity, such as turning on a fire hydrant. Since the rougher the interior wall is, the more difficult it is for a disinfectant to kill the bacteria on the pipe wall's surface, biofilms can form in highly scaled and therefore rough-surfaced pipes where bacteria are allowed to grow. Deterioration affecting pressures and flows due to hydraulics may result from other types of deterioration that block water flow.

There are a lot of factors to take into account when deciding which water main renewal method to use. One of the methods for pipeline rehabilitation or open-trench replacement can be used for this. Pipe lining, slip lining, and pipe bursting are a few techniques for pipeline rehabilitation. One advantage of using an in-situ rehabilitation method is that it is less expensive because the entire water main pipeline does not need to be excavated. To reach the current water main, only tiny pits are dug. However, since the water main will not be accessible during the rehabilitation, a temporary water bypass system must be constructed to function as the affected area's water main. It is important to carefully design a temporary water bypass system, also called temporary bypass piping, to guarantee that the project area's customers have access to a sufficient supply of water. Water is transferred into a temporary pipe from a feed hydrant. In order to allow cars to cross the temporary pipe, a cover or cold patch should be installed where the pipe crosses a driveway or a road. The temporary pipe can have temporary service connections installed for residences. One popular method for creating a temporary connection is to attach the garden hose to the temporary service connection. For fire safety, the temporary pipe should also include temporary fire hydrants. When a water utility plans a water main renewal project, it is advised that it collaborate with property owners to replace lead service lines as part of the project because water main work can disrupt lead service lines, which can lead to elevated lead levels in drinking water.

9.3. WATER ALLOCATION AND INTEGRATIVE MANAGEMENT OF PRECISION IRRIGATION

In order to minimize the negative effects on the environment while meeting the specific needs of individual plants or management units, precision irrigation is the effective coordination and precise application of water. Precision irrigation, as a comprehensive coordinated performance, addresses the growing issue of water scarcity and the need for more effective irrigation management by facilitating the development of more adaptable and reactive operational systems as well as the optimal management and best operation of agricultural irrigation systems. A wide range of participant component systems are included in precision irrigation, both at the on-farm plot and scheme levels, to correspond to various management and monitoring realities. Precision irrigation has the potential to be used as an efficient water-allocation support tool for irrigation scheduling and cropping patterns in an irrigated area. Following are anticipated direct benefits of precision irrigation's integrative management effectiveness and water allocation mechanism.

- **Irrigation Collaboration:** Sophisticated techniques for designing irrigation systems and tools for optimizing their performance make it possible to develop water allocation for the best irrigation system collaboration possible. These tools can simulate and physically prototype water allocation variations in an agricultural field and can represent changes in water allocation over time and space in a way that is compatible with farming management practices.
- **Irrigation Implementation:** Data about important aspects of the agricultural production system can be instantly gathered when stage irrigation objectives are used as components of water allocation. When planning water allocation schemes, it is possible to modify the irrigation system's corroborative behavior to account for unusual cultivation conditions or changes in the environment. This helps determine the best time to irrigate.

Water resources in the irrigated soil have been appropriately redistributed in the last few years thanks in large part to the water allocation mechanism of precision irrigation. Furthermore, in common application scenarios like moisture infiltration measurement, adaptive irrigation scheduling, and dynamic groundwater distribution, the integrative management effectiveness of irrigation practices becomes a critical benchmark for precise calibration of irrigated agriculture. Even though no prior study has offered a comprehensive and flashy review of the water allocation mechanism and integrative management effectiveness, it is unfortunate that such significant topics have not been focused on and discussed before. This poses a significant barrier to the promotion of irrigation techniques and the improvement of irrigation efficiency during agricultural crop production.

In order to show these important subjects' originality and distinctiveness for precision irrigation, it is imperative that a thorough analysis of them be proposed. This conditional requirement is met in the following ways: first, a systematic overview of the water allocation mechanism is presented; second, the effectiveness of precision irrigation's integrative management is outlined; and third, this paper concludes with a discussion of the prospects for precision irrigation's future development and a set of focused comparisons of the performance properties of water allocation techniques with respect to the factors influencing the irrigation environment.

9.3.1. Water Allocation Mechanism of Precision Irrigation

Precise and effective strategies for distributing water resources to boost crop yield and agricultural productivity in real-world irrigation applications are known as high-efficiency water allocation mechanisms. In addition to preventing uncontrolled water or soil withdrawals, appropriate irrigation water allocation can minimize water loss and precisely measure the amount of water used. It can also help overcome topographic restrictions. Significant benefits, such as increased water use efficiency (WUE), reduced energy requirements, increased field yields, and improved quality of harvested crops, have resulted from research advancements and equipment developments in the water allocation mechanism, as illustrated in Figure 9.1. Environment-oriented water allocation monitoring and information management skills are prerequisites for achieving resource recycling and environmental maintenance. These skills are known as irrigation parametric coefficients or influential process factors. The mechanism of water allocation determines the ideal cropping pattern and irrigation schedule in irrigated agriculture, as well as facilitating the soil/water balance. Three significant issues come up in light of the extensive and general research on this subject.

- Figuring out how to create a general-purpose, efficient water allocation plan that allows field soil to experience a rise in WUE and a quickening of moisture infiltration for their performance benchmarking. In order to standardize and enable their widespread application—which seems to be crucial right now—more research should be done on the internal structure and underlying principles of water allocation mechanisms.

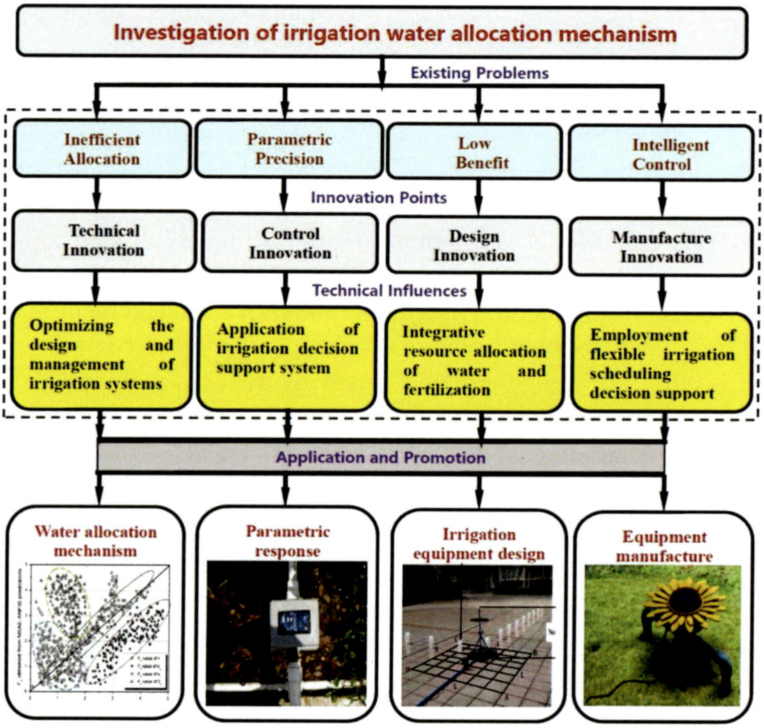

Figure 9.1. *Research progress and equipment development in the water allocation mechanism.*

Source: https://www.mdpi.com/water/water-12-03135/article_deploy/html/images/water-12-03135-g001-550.jpg.

- Figuring out the best way to incorporate the water allocation components that are required for a particular precision irrigation system. The schedule principle and control architecture in water allocation need to be synchronized at the same time because irrigation efficiency needs to be continuously monitored and greatly improved. The instantaneous investigation of irrigation turn duration, based on the adaptive control of moisture infiltration or the coordinated behavior of the water allocation system, makes the task more difficult because it necessitates omniscient water resource allocation for the irrigation scenario management.

- Establishing a comprehensive set of water allocation scenarios that accurately reflect the range of irrigation needs that may arise and the anticipated resource allocation goals. Right now, there isn't nearly enough relevant research on this subject, and what is available does not fully address the water allocation mechanism.

The ability of the water allocation system to cooperate with other facility components, as indicated by the water research literature, is what is meant by the technical adaptivity of precision irrigation. This ability is linked to the physical connections between various irrigation system components. Given its adaptability, a decision support system (DSS) model was created to gather environmental, crop, and water data in real time via a wireless sensor network. This was done in order to preserve the efficiency and consistency of precision irrigation, which can be maintained from the intricate and varied schedule-based systems by taking weather, soil, and water data into account. The need to predict the hourly requirements of soil moisture content (MC) and soil evapotranspiration using the Penman–Monteith method or the Food and Agriculture Organization (FAO) model drove the development of a radial basis function type of neural network. Likewise, a weather-dependent irrigation control mechanism based on fuzzy logic was created and combined with an interoperable decision support system (DSS) to provide notifications for appropriate water allocation through an interface with a Global System for Mobile Communications (GSM) modem. Studies have suggested that a number of factors, including water supply volume, infrastructure for water delivery, groundwater use, water allocation strategy, infrastructure size, and irrigation resource consumption costs, can be used to assess the effectiveness of precision irrigation. Thus, it is important to use energy-intensive irrigation systems sparingly and only when they allow for a decrease in the variability of water supply and risk associated with the distribution of water resources.

In order to assess the beneficial effects of indoor water allocation mechanisms on grain yield and crop productivity, a greenhouse irrigation experiment was carried out. The results showed that thin shallow moist dry irrigation (TSMDI) significantly increased water productivity and allowed for a notable reduction in irrigation water consumption. This suggests that local data on root growth patterns are essential for comprehending cultivar adaptations and helping to ensure accurate deficit irrigation. The IGR ratio, which measures the ratio of irrigation water needed to groundwater recharge, was evaluated using the deterministic distributed water balance model known as mGROWA. This model was created by the Forschungszentrum Jülich in collaboration with the State Agency for Mining, Energy and Geology to identify water balance variables, such as groundwater recharge in sizable regions under current and potential future climate conditions. The results indicated that the areas that were identified as vulnerable coincided with the field regions where high irrigation quantities from groundwater resources had been previously documented.

In terms of how well surface irrigation performs in terms of efficient water allocation, research has shown that soil texture, plot size, and farmer practices substantially impact these factors. Through a review of the variables affecting the water distribution system. As irrigated agriculture necessitates high flows to meet diverse technological demands during seasonal peak times and high-performance reliability from water delivery networks, the application of satellite imagery to assist in accurate water allocation has caught on. Proper allocation strategies utilizing limited freshwater resources depend on the monitoring of crop water consumption, also called actual evapotranspiration (ETa), using satellite imagery and remote-sensing-based monitoring algorithms. Utilizing satellite imagery to estimate crop evapotranspiration (ETc) and derive the normalized difference vegetation index (NDVI) for accurate water allocation was motivated by this scenario.

Based on domestic high-resolution satellite images, an intelligent water allocation method for farmland irrigation and drainage system was developed, taking into account both the spectral features and the geometric features of the system. This was done because the dynamics of maize, alfalfa, fruit trees, and poplars can be analyzed using greenness and wetness extracted from satellite remote sensing data. The results of their experimental analysis showed that the precision of irrigation and drainage systems can be precisely maintained by using the large-scale, fast extraction of water required for large-scale satellite imagery accuracy. To characterize the land surface in terms of spectral indices, such as NDVI and surface temperature (Ts), water allocation data from multiple bands was combined and used to map the seasonal cultivated area based on satellite images. The ability of the Crop Water Stress Index (CWSI) to estimate water stress and schedule irrigation based on satellite thermal infrared data was assessed.

The Penman–Monteith method was used in the previous few years to conduct pertinent water allocation tests for crop cultivation, e.g., By combining the high-resolution NDVI and land surface temperature (LST) datasets with the low-resolution Soil Moisture Ocean Salinity (SMOS) satellite soil moisture data, the disaggregation based on physical and theoretical scale change (DISPATCH) algorithm was proposed as a way to downscale soil moisture satellite data. A comparison of the variations in water allocation could also be made using satellite imagery; Landsat Thematic Mapper (TM) and Spectral Mixture Analysis (SMA) were used in some investigations to evaluate the risk of land degradation and changes in water allocation. It was further verified by others that the maximum likelihood approach and high-resolution remote sensing images work well for allocating water.

9.3.2. Integrative Management Effectiveness of Precision Irrigation

The efficiency of precision irrigation's integrative management can be defined as the operational or maintenance impact that results from utilizing every part of the irrigation system to attain the goal of ideal agricultural irrigation. Coordinated use of management tools, such as remote sensing, on-site monitoring systems, schedule models, and others, is necessary for the successful implementation of integrative irrigation management. Lead to the cooperative development of crop cultivation in the end, depending on current soil/plant water requirements and realistic meteorological conditions. A precision irrigation system needs to take three key actions in order to operate as best it can:

- Definition of a generic model for the efficacy of integrative management. The current irrigation techniques, which are based on particular flow models, are not well matched with general management strategies. Therefore, in order to enable precise verification for system collaboration, it is necessary to define the generic management effectiveness of both the irrigation system's working principle and collaborative behavior;
- Effectiveness of integrative management driven by models. The goal of this task is to create a fully functional irrigation schedule that complies with earlier management models in order to demonstrate the validity of integrative management effectiveness. Expertise in irrigation in the area of resource distribution model transformations should be provided for this task;
- Official confirmation of integrative management's efficacy. In order to demonstrate the special qualities of an irrigation system that are beneficial for controlling agricultural productivity, this task aims to assess the viability of applying formal verification to integrative management effectiveness. Refinement of the proposed control methods to achieve irrigation management objectives may be necessary, given the anticipated management results, based on information feedback from the water/soil resource management.

For conjunctive irrigation management, a novel model of Linked Simulation Optimization (LSO) matched to the Clonal Selection Algorithm (CSA) was developed from the previously mentioned analysis, based on the generic model definition step for integrative management effectiveness. As they have shown, when appropriate irrigation conditions are available, the management effectiveness would aid in enhancing irrigation operations. Based on the model-driven integrative management effectiveness investigation step, genetic algorithms in the Doroodzan network of agricultural irrigation can be used to improve the current on-farm irrigation management methods for maize, water equity, and crop productivity. In a similar vein, some studies revealed that integrative irrigation treatments—that is, full drip (DT), full border (BT), deficit drip (DDT), and deficit border (DBT) irrigation treatments—tended to raise field yield and WUE, confirming their positive effects on crop returns.

Because it significantly affects crop productivity, soil salinization is a major problem in irrigated areas. It is a pervasive issue that poses a serious threat to food security and global food production as well as a major abiotic limitation. Consider the value of using spectra of saline soils and associated vegetation that were obtained in the field to map and characterize the spatial distribution of soil salinization brought on by irrigation. Higher soil salinity concentrations in deeper water tables are linked to lower evaporation, according to macroscopic numerical simulations. Although the primary factor influencing soil salinity distribution is groundwater level, it is still unknown how the spatial distribution of soil salt and groundwater level respond to each other.

Partial root drying irrigation, or PRD irrigation, is a water-saving technique that is used to control groundwater nitrogen contamination. This study examines the effects of applying integrative management of precise root drying irrigation on yield quality, drainage water, nitrogen leaching, nitrogen uptake, and nitrogen efficiency indices while varying the nitrogen application rates of ordinary furrow irrigation (OFI), PRD irrigation as the variable alternate furrow irrigation (VAFI), and fixed alternate furrow irrigation (FAFI). In accordance with the quantity of water available under VAFI, less N fertilizer should be applied. This conclusion is supported by the use of probability distribution functions (PDF) in high-efficiency irrigation equipment management to regulate the applied water depth. An essential reference for effectiveness in allocating water resources during irrigation was furnished by the management analysis.

One essential intelligent feature of integrative irrigation management is decision-making, which can be defined as a ranking and choice made up of recursive irrigation decisions that can be applied in various management architecture layers. These decisions include irrigation conditional preparation, soil parametric monitoring, irrigation logic triggering, irrigation decision design, irrigation effectiveness evaluation, and irrigation decision application. Since agricultural catchment hydrology and water quality are simulated using an eco-hydrological model known as the soil and water assessment tool (SWAT), which is utilized globally. Their integrative managed results confirmed the encouraging and beneficial impact of integrative irrigation management on crop water use by demonstrating that deficit control strategies and wise decision-making could be used to significantly reduce water consumption with only a slight decrease in crop yield.

Alghory and Yazar assessed the crop water stress index (CWSI) in accordance with the conventional irrigation management strategies since integrative irrigation management is typically based on the coordinated measurement and instantaneous control of soil water content or meteorological parameters for modeling evapotranspiration. The relationships that were discovered may be used to forecast how yield will react to water stress. The need to increase crop productivity globally and the growing competition for water resources make efficient use of irrigation water a critical concern. In light of the suggested soil water balance and crop water satisfaction models, an irrigation management decision-making tool was consequently created to optimize water use. On the other hand, arbitrary irrigation management plans frequently lead to significant fertilizer leaching and excessive water use. The relationship between water stress, the current environmental conditions, and species-specific stress-tolerance thresholds can be characterized to enable high-efficiency integrative management effectiveness. Stem psychrometers are used to measure plant water status (PWS) directly.

Precision irrigation aims to boost crop yield while using the least amount of water and causing the least amount of pollution to the environment. In developing nations and on small farms, this is necessary. However, when searching for appropriate irrigation technologies in developing nations like China, Brazil, and India, small-scale farmers, regular agricultural enterprises, and technical engineers face difficulties due to the high-tech nature of precision irrigation developed in developed nations. A few of the difficulties are the degree of technology, the ability to process data in real time, and the expense of setting up and keeping up irrigation applications. It is vital to create straightforward, affordable, and data-driven strategies for water allocation and integrated management in order to support socioeconomic development, since small-scale farms account for around 80% of irrigated agriculture in developing nations and are expected to becomc increasingly significant in the global agricultural production chain in the future. By precisely achieving a soil/moisture equilibrium, precision irrigation reduces water inputs while embracing the idea of sustainable intensification of agricultural production. In order to visualize and compile the data collected by the pH and moisture sensors utilized in a smart agriculture system, an irrigation application mechanism is needed. By reducing direct user interaction with the irrigation control system, this application enhances convenience and allows for easier control of the irrigation process when needed. A low cost method of monitoring, gathering, and processing soil and irrigation data is available to farmers who use this irrigation application mechanism.

Because small-scale farms and developing nations face unique socioeconomic challenges, it is anticipated that balanced precision irrigation technologies will be applied. Here, rather than relying primarily on empirical and visual analyses, precision irrigation relies on fuzzy decision-making to inform the parametric calibration of crop cultivation and soil management. This tendency suggests that developing nations should use construction and maintenance technologies suitable for precision irrigation and drainage projects. To get free information about the water and soil qualities of farms that use irrigation, encourage small-scale farmers and users in developing nations to use the Internet. In order to facilitate agricultural irrigation, geographic information systems (GIS) have been used. Therefore, high-speed data/information connectivity systems (computers/Internet) need to be developed in rural areas. In the interest of farmers and agricultural enterprises, internet control and related intelligent techniques have been incorporated into the irrigation apparatus; thus, the costs of systems managers could be markedly reduced. The best crop for a given area can be determined by this system, which can also be used to automatically identify areas that are suitable for arable land. When small-scale farmers use modern farming concepts and high-tech irrigation equipment, appropriate irrigation solutions allow them to become agricultural experts. Alternatively, service provider entrepreneurs can sell irrigation technologies in rural areas of developing countries. Following the application of precision irrigation, a predetermined number of irrigation tests and data analyses are needed. Proposed strategies for precision irrigation adoption should take into account issues with land fragmentation, a dearth of highly advanced technical centers, software designed specifically for precision irrigation, and the generally low economic standing of developing-nation farmers.

A greater number of advantages than disadvantages are produced for realistic agricultural production conditions when all these recently suggested techniques are integrated into precision irrigation applications. The precision irrigation methods that are now in use present chances for substitute techniques to boost the efficiency of agricultural irrigation in those developing nations while reducing the adverse effects on the workplace. The successful reduction of agricultural production costs and the enhancement of precision irrigation technique to better suit the needs of farmers and users in developing countries will require close collaboration between farmers, agricultural enterprises, farm associations, community groups, machinery manufacturers, research and extension agencies, and other public and private agencies.

To identify distinct irrigation characteristics and confirm the efficiency and dependability of integrated management. We identified the most effective irrigation settings that both characterize the soil moisture distributed along the furrow and result in significant reductions in overall water use. In order to enhance management effectiveness, it is important to take into account both the cut-off length of the furrow and the rate of water discharge. Study results included effects of irrigation management techniques on plant and soil water status, photosynthesis fluxes, and chamber transpiration. Water stress increases lower the physiological parameters of an irrigation system because of stomatal and non-stomatal limiting factors, according to their effectiveness verification results, with the exception of the regulated deficit irrigation (RDI) treatment. A risk management analysis model that measures crop production risk owing to variations in the amount of moisture infiltrated into the cultivation area and the total amount of precipitation, or the uncertainty of agro-climate factors like average temperature, total sunshine, total precipitation, and average solar radiation, which represent the climate change, irrigation water quality, surface water, and groundwater level, respectively. Under different conditions of agricultural production, some other issues that are currently unresolved, like comparisons of management effectiveness and dynamic integrative coordination between irrigation components, might be addressed.

In order to boost irrigation production, they discovered that, in accordance with the goals and the framework for implementing contemporary integrated water resources management (IWRM), it is necessary to enhance public involvement, strengthen administrative skills, and give irrigation management organizations enough funding. Extensive data indicates that irrigation infrastructure is contributing significantly to the advancement of precision irrigation technology. Additionally, irrigation infrastructure enhances the precision of water allocation measurements, firmly preserving the balance of agricultural water supply. The contribution of regional irrigation systems to farmers' enhanced drought response capabilities. They discovered that irrigation infrastructure helps to improve farm production capacity in adapting to drought, concurrently increasing mean yield and lowering risk exposure in crop cultivation, along with its infiltration effectiveness in mitigating the drought risk during agricultural crop production. Therefore, these descriptions make it possible to clearly identify the beneficial effects that irrigation infrastructure, management organizations, and real-time irrigation water resources have on the integrative management of precision irrigation and the mechanism for allocating water.

The majority of agricultural irrigation applications are covered by Figure 9.2, which provides a brief summary of the aspects of precision irrigation's integrative management effectiveness. These aspects include the application's characteristics, importance and necessity, optimization design, allocation mechanism, intelligent controlling, and allocation influence. Each of these facets offers a distinct perspective on the methodical examination of precision irrigation's efficacious integrative management.

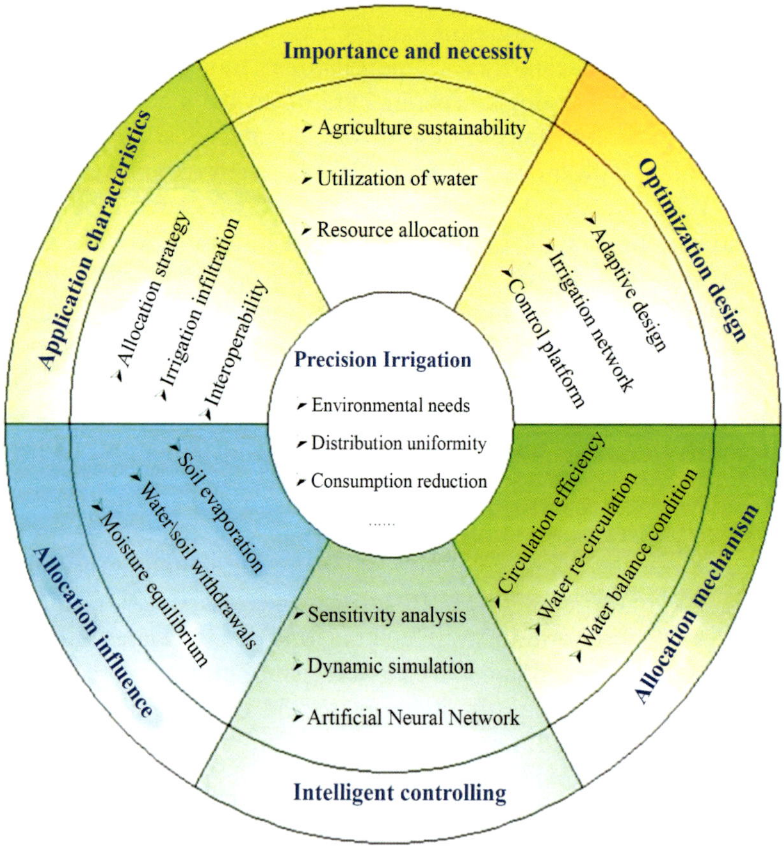

Figure 9.2. *The inclusive subjects discussed in integrative management effectiveness of precision irrigation.*

Source: https://www.mdpi.com/water/water-12-03135/article_deploy/html/images/water-12-03135-g002-550.jpg.

REFERENCES

1. Adeosun, O. O. (2014). Water distribution system challenges and solutions. *Water Online.*
2. Amos, C. C., Rahman, A., Karim, F., & Gathenya, J. M. (2018). A scoping review of roof harvested rainwater usage in urban agriculture: Australia and Kenya in focus. *Journal of Cleaner Production, 202,* 174–190.
3. Aragones, D. G., Calvo, G. F., & Galan, A. (2021). A heuristic algorithm for optimal cost design of gravity-fed water distribution networks: A real case study. *Applied Mathematical Modelling, 95,* 379–395.
4. Chen, X. (2007). Rainwater harvesting benefits farmers in Gansu. *People's Daily Online.*
5. Devkota, J., Schlachter, H., & Apul, D. (2015). Life cycle based evaluation of harvested rainwater use in toilets and for irrigation. *Journal of Cleaner Production, 95,* 311–321.
6. Johari, A. (2019). Mumbai could save water through its mandatory rainwater harvesting rule–But no one seems to care. *Scroll.in.*
7. Kinkade-Levario, H. (2007). *Design for Water: Rainwater Harvesting, Stormwater Catchment, and Alternate Water Reuse* (p. 27). Gabriola Island, B.C.: New Society Publishers. ISBN 978-0-86571-580-6.

Chapter **10**

Use of Treated Wastewater in Irrigation

Contents

INTRODUCTION

An inventive and sustainable way to manage water resources, especially in areas where water is scarce, is to use treated wastewater for irrigation. The availability of freshwater is being affected by climate change and population growth, placing increasing pressure on conventional freshwater sources. Reclaimed or recycled water, commonly referred to as treated wastewater, provides a practical substitute for irrigation, easing the burden on potable water supplies.

After treated wastewater is thoroughly cleaned to get rid of impurities and pathogens, it can be used for a variety of non-potable purposes. Primary, secondary, and tertiary treatments are among the stages that are commonly included in the treatment process. Large solids are removed by primary treatment, bacteria and organic matter are dealt with by secondary treatment, and the water is further polished by tertiary treatment, which also removes any remaining contaminants and pathogens. High levels of purification can be attained by sophisticated treatment technologies, ensuring that the water is suitable for irrigation. The use of treated wastewater for irrigation has a number of advantages for the economy and environment. Because less freshwater is needed, natural water bodies and ecological balance are preserved. This method can increase crop yields in agricultural settings by providing a steady and dependable source of water, particularly in arid and semi-arid areas. Furthermore, nitrogen and phosphorus—two important nutrients found in treated wastewater—can act as natural fertilizers, negating the need for synthetic fertilizers and cutting down on agricultural expenses. Although treated wastewater has many benefits, its use in irrigation needs to be managed and observed carefully to guarantee both efficacy and safety. In order to safeguard the environment and public health, rules and regulations are put in place that address issues with pathogen concentrations, chemical residues, and the possibility of soil salinization. Reclaimed water systems that are properly maintained and tested can reduce hazards by guaranteeing that the water quality stays within safe ranges for irrigation.

Implementing treated wastewater for irrigation successfully depends heavily on public opinion and acceptance. Building trust and addressing concerns about potential health risks require openness and education. It is possible to dispel doubts and encourage broader adoption of reclaimed water by proving its efficacy and safety through pilot projects and open reporting. Utilizing treated wastewater for irrigation can help address the problem of water scarcity while also having positive effects on the environment and the economy. The use of reclaimed water is expected to become more crucial to sustainable water management techniques as treatment technologies improve and public awareness rises.

10.1. POTENTIAL OF TREATED WASTEWATER IN AGRICULTURE

Treated wastewater has a great deal of potential in agriculture and presents a viable way to improve sustainable farming methods and deal with water scarcity. Treating wastewater offers a practical substitute that can boost agricultural productivity while protecting limited freshwater resources, as the population grows, the climate changes, and over-extraction put increasing strain on conventional freshwater sources.

Reclaimed water, also known as treated wastewater, is put through a rigorous purification procedure to get rid of pollutants and pathogens so that it can be used in agriculture. Primary, secondary, and tertiary stages are usually included in the treatment process. Primary treatment gets rid of big particles, secondary treatment deals with organic matter and bacterial contaminants, and tertiary treatment further purifies the water to guarantee high quality. Reclaimed water can be safely and effectively used for irrigation thanks to advanced treatment technologies that can achieve even higher purification levels.

Wastewater that has been treated has various advantages in agricultural settings. It offers a dependable and steady supply of water, which is especially helpful in areas that are vulnerable to drought or water scarcity. Farmers are able to better control irrigation and maintain consistent crop yields thanks to this consistency. Furthermore, the presence of vital nutrients like phosphorus and nitrogen in treated wastewater can improve soil fertility and lessen the need for artificial fertilizers. The provision of both water and nutrients has the potential to reduce expenses for farmers and encourage the adoption of more environmentally friendly farming methods. Environmental sustainability is further supported by the application of treated wastewater in agriculture. Reusing wastewater minimizes the release of untreated or partially treated wastewater into natural water bodies, which can cause pollution and ecological harm, and eases the burden on freshwater resources. Furthermore, well-managed reclaimed water systems can sustain nearby ecosystems and help maintain natural water cycles.

To guarantee efficacy and safety, treated wastewater adoption in agriculture must be carefully managed. Guidelines and regulatory frameworks are necessary to handle possible health risks like the presence of pathogens or chemical residues. To safeguard both crops and consumers, regular monitoring and adherence to safety regulations are essential. Transparent communication and education about the safety and advantages of reclaimed water are essential for overcoming skepticism. Additionally, public perception and acceptance of treated wastewater can be a barrier. With the ability to improve soil fertility, decrease environmental impact, and provide a consistent and sustainable water source, treated wastewater has the potential to revolutionize agricultural practices. Reclaimed water integration into agricultural systems is expected to become increasingly important in addressing global water challenges and promoting sustainable food production as treatment technologies improve and regulatory frameworks solidify.

10.1.1. Treated Wastewater Irrigation

Without water, life cannot exist and industry cannot function, making water a vital natural resource. Water is vital to a nation's development because the establishment of a permanent community requires a consistent supply of fresh water. However, the amount of pure freshwater available worldwide is steadily declining. Many nations have more water demand than supply, and as the world's population grows and water demands rise, freshwater shortages are becoming more prevalent. The primary application of freshwater is thought to be irrigation. 80% of freshwater used globally comes from land irrigation, and by 2030, that percentage will rise to 15%. This will lead to water shortages in areas like the Middle East and North Africa, for example. Furthermore, it was mentioned that almost 11.8 billion people will reside in a region with water shortages by 2025.

As a result, using alternate water sources is crucial. It is projected that more than 40% of the world's population will experience water scarcity or stress in the ensuing decades, which will have a significant effect on water security. As a result, recycling wastewater has many benefits for the farming industry. Reusing wastewater is regarded as one of the most crucial strategies for managing water scarcity. "Water that has received at least secondary treatment and basic disinfection and is reused after flowing out of a domestic wastewater treatment facility" is the definition of treated wastewater (TW). Wastewater's properties differ significantly depending on where it comes from, so research is necessary before treating and recycling it. Across the world, there is concern about the safety of TW that is recycled for crop irrigation. Reusing TW represents another renewable water source, which makes it a viable alternative to increase water resources, particularly in arid regions. In addition, the importance of water reuse was emphasized in the 2017 World Water Development Report. Because there is an abundance of wastewater, the use of TW is therefore encouraged in nations that experience water scarcity. For example, China is thought to be the world's biggest producer of wastewater. Approximately 68.5 billion tons of this water, or the Yellow River's yearly stream volume, were released from industrial and municipal sources in 2012. China generates about 108.16 billion m^3 of wastewater annually, of which 34.33 billion m^3 come from domestic sources and 73.83 billion m^3 come from industrial sources. Approximately 5 billion m^3 of sewage water were collected annually in Egypt. As a result, Egypt's water resources could increase by 5 billion m^3 thanks to the TW. Wastewater is regarded as an additional limitless, dependable, and dependable source of water and nutrients, so treating it and using it for irrigation makes sense. This is especially true in arid areas. Therefore, wastewater reuse is regarded as a long-standing and crucial practice in many water-scarce countries across the world. Applications for wastewater reuse in agriculture and landscaping, groundwater recharge, industry, street cleaning, firefighting, and ecological and recreational purposes are all possible.

Reusing wastewater for irrigation in agriculture is, nevertheless, more acceptable than reusing it in other fields. TW irrigation has grown in significance recently, especially in arid areas. While many nations have extremely stringent regulations, the majority of countries do not have any laws governing the reuse of wastewater. The secondary TW can be used as a fertigation source without facing any major restrictions. Reusing wastewater has reduced the amount of waste released into ecosystems in addition to reducing the use of freshwater. It has also improved the soil by adding nutrients and organic matter (OM). There are both financial and environmental advantages to using TW as an irrigation source because it may lessen or even completely eliminate the need to fertilize the soil with pricey chemical fertilizers. Wastewater has been identified as a valuable resource for boosting agricultural productivity at a low cost because it contains nutrients and organic matter that are beneficial to plants. However, there are risks associated with reusing wastewater in agriculture. For instance, its use increased the salinity of the soil and introduced pollutants and microbial microorganisms. Not only can this water increase the risk of bacterial, viral, and parasitic illnesses in crops that use it for irrigation, but it can also harbor pathogens that pose a threat to human health. Reusing TW will not only solve the issue of water scarcity for agricultural growth, but it will also address the health risks and pollution caused by the careless disposal of untreated sewage water. Wastewater that has not been adequately or completely treated can have negative effects on the environment, public health, and economy. As a result, proper procedures must be followed when treating and using wastewater, especially since improper treatment could put the public's health at risk.

The different degrees of conventional treatment are:

- **Preliminary**: Remove the large solid materials from the crude wastewater that are conveyed by sewers that could hinder the discharge or cause damage to equipment, such as wood, rags, fecal material, and heavier grit particles.

- **Primary**: Remove the suspended solids (SS) and floating substances.
- **Secondary**: By removing or reducing the residual settleable solids and floating materials from primary treatment, the secondary treatment process seeks to reduce the chemical oxygen demand (COD), SS, and a number of other harm parameters. The amount of dissolved oxygen (BOD) required by aerobic biological organisms in water to decompose organic matter present in a sample of water at a particular temperature for a predetermined amount of time is known. The COD is a measure of the oxygen quality needed to chemically stabilize carbonaceous organic matter.
- **Tertiary and/or Advanced**: Removal of nutrients and heavy metals (HM), which are not removed by the previous treatment. Additionally, decreasing the microbiological constituents by using some options such as chlorination, ultraviolet rays, and ozonation in disinfection operation.

All things considered, the detrimental effects of TW can be greatly mitigated by choosing an appropriate irrigation system, a suitable cropping schedule combined with appropriate and efficient irrigation management, regular monitoring of the quality of the water, soil, and plants, and taking cautious and preventative measures against pathogens.

This article aims to shed light on the significance of TW as a substitute freshwater source for irrigation. The effects of TW irrigation on soil characteristics, fertility, plants, and overall health are also covered. A review of the irrigation systems that TW uses is part of this.

10.1.2. Characteristics of Wastewater

Wastewater can be broadly categorized based on its physical, chemical, and biological properties. They added that a complex mixture of minerals and organic matter (OM) in various forms, such as colloidal, floating suspension, and large and small particles, make up the wastewater's liquid portion. Among the toxic substances found in wastewater are arsenic, cadmium, chromium, lead, copper, zinc, mercury, and so forth. Pesticides, carbohydrates, fats, proteins, synthetic detergents, medications, and complex nitrogenous organic matter products are some of the organic materials found in this water. These toxic substances have dangerous consequences for overall health. Direct proof of detrimental effects on human health is still up for debate, though. The wastewater contains microplastics, polymer fibers, polyethylene terephthalate, polyethylene, polypropylene, and polystyrene. Plant growth, both vegetative and reproductive, may be negatively impacted by microplastics. Several studies have reported the presence of pharmaceutically active compounds (PhACs) in TW. These substances could not be effectively removed by activated sludge-based conventional wastewater treatment. Many of these chemicals were therefore later found in soils that had been irrigated with TW. They also stated that there was very little ecotoxicological risk associated with PhACs in soil. But as demonstrated by, this is heavily reliant on the PhACs. If these compounds build up in the soil for a long time, they may have an impact on soil microorganisms. If they reach crops and then the food chain, they may also pose a risk to human health. Table 10.1 lists the physical characteristics, chemical and biological components, and sources of wastewater. Furthermore, Table 10.2 lists the significant contaminants of interest for wastewater treatment.

Table 10.1. *Physical, Chemical, and Biological Characteristics of Wastewater and Their Sources*

Characteristic	Sources
Physical properties	
Color	Domestic and industrial wastes, natural decay of organic materials
Odor	Decomposing wastewater, industrial wastes
Solids	Domestic water supply, domestic and industrial wastes, soil erosion, inflow infiltration
Temperature	Domestic and industrial wastes
Chemical constituents:	
Organic	
Carbohydrates	Domestic, commercial, and industrial wastes
Fats, oils, and grease	Domestic, commercial, and industrial wastes
Pesticides	Agricultural wastes
Phenols	Industrial wastes
Proteins	Domestic, commercial, and industrial wastes
Priority pollutants	Domestic, commercial, and industrial wastes
Surfactants	Domestic, commercial, and industrial wastes
Volatile organic compounds	Domestic, commercial, and industrial wastes

Other	The natural decay of organic materials
Inorganic	
Alkalinity	Domestic wastes, domestic water supply, groundwater infiltration
Chlorides	Domestic wastes, domestic water supply, groundwater infiltration, water softeners
Heavy metals	Industrial wastes
Nitrogen	Domestic and agricultural wastes
Acidity	Domestic, commercial, and industrial wastes
Phosphorus	Domestic, commercial, and industrial wastes natural runoff
Sulfur	Domestic water supply, domestic and industrial wastes
Toxic compounds	Industrial wastes
Gases	
Hydrogen sulfide	Decomposition of domestic wastes
Methane	Decomposition of domestic wastes
Oxygen	Domestic water supply, surface-water infiltration
Biological constituents:	
Animals	Open watercourses and treatment plants
Plants	Open watercourses and treatment plants
Bacteria	Domestic wastes, surface water infiltration, treatment plants
Archae	Domestic wastes, surface-water infiltration, treatment plants
Protista	Domestic wastes, treatment plants
Viruses	Domestic wastes

Source: https://www.researchgate.net/publication/351994621/figure/tbl1/AS:1029536116400131@1622471946914/Physical-chemical-and-biological-characteristics-of-wastewater-and-their-sources-72.png.

Table 10.2. *Important Contaminants of Concern in Wastewater Treatment*

Contaminants	Reason for Importance
Suspended solids	Suspended solids can lead to the development of sludge deposits and anaerobic conditions when untreated wastewater is discharged into the aquatic environment.
Biodegradable organics	Composed principally of proteins, carbohydrates, and fats, biodegradable organics are measured most commonly in terms of BOD (biochemical oxygen demand) and COD (chemical oxygen demand). If discharged untreated to the environment, their biological stabilization can lead to the depletion of natural oxygen resources and the development of septic conditions.
Pathogens	Infectious diseases can be transmitted by the pathogenic organisms in wastewater.
Nutrients	Both nitrogen and phosphorus, along with carbon, are essential nutrients for growth. When discharged to the aquatic environment, these nutrients can lead to the growth of undesirable aquatic life. When discharged in excessive amounts on land, they can also lead to the pollution of groundwater.
Refractory organics	These organics tend to resist conventional methods of wastewater treatment. Typical examples include surfactants, phenols, and agricultural pesticides.
Heavy metals	Heavy metals are usually added to wastewater from commercial and industrial activities and may have to be removed if the wastewater is to be reused.
Dissolved inorganic solids	Inorganic constituents such as calcium, sodium, and sulfate are added to the original domestic water supply as a result of water use and may have to be removed if the wastewater is to be reused.

Source: https://www.mdpi.com/2073-4441/13/11/1527#B72-water-13-01527.

The use of wastewater carries certain chemical risks because it contains HM, OM, salt, nutrients, and toxic compounds. Wastewater contains a greater variety and concentration of chemicals, including oil, coarse particles, chemicals that are chemically contaminated with alkalis, and various acids. High concentrations of calcium, sodium, potassium, phosphate, chlorine, sulfur, bicarbonate, ammonium salts, and HM are examples of inorganic constituents. Many of the micronutrients required by the plant, including copper, iron, manganese, zinc, boron, molybdenum, cobalt, and nickel, are present in TW. Wastewater's chemical makeup has the potential to harm the environment in a variety of ways.

In addition to adding taste and odor to water supplies and lowering oxygen levels in streams, soluble organic materials can also contain harmful substances that could have an impact on the food chain and public health. The physical characteristics of wastewater are those attributes that are detectable by human senses. The solid content of wastewater is one of its most crucial physical properties since it determines the color, clarity, and aesthetics of the water. In a wastewater treatment plant, temperature and odors are examples of other physical parameters that are rarely changed. The primary reason wastewater temperature matters is that it has an impact on the aquatic and biological life in the receiving body of water. Elevated temperatures cause the water's solubility of dissolved oxygen to decrease. The unwanted circumstances of the receiving water's increased turbidity and silt load. As water is treated and used, a sizable amount of dissolved solids may be added. The aquatic life and viruses that can be found in water are referred to as biological characteristics of water. These features have a major impact on the water's quality. For instance, algae produce taste and odor. Certain kinds of algae because sand filters to clog, while others result in unsightly slimy growths on equipment, tanks, and reservoir walls. Additionally, some microalgae release potent toxins that are detrimental to living things. Numerous diseases can be caused by microbiological life found in water, including bacteria, viruses, and protozoa. Wastewater is thought to provide the perfect conditions for the growth of bacteria, viruses, and protozoa. Not all sewage is harmful, but some pathogenic microorganisms are present. The majority of pathogenic bacteria can be eliminated from sewage by biological oxidation systems, which take place during secondary treatment, according to several studies.

By measuring both COD and BOD, characteristics of biological wastewater can be determined. Typically, organic waste needs oxygen to decompose biologically quickly and efficiently. Thus, the oxygen demand increases with the amount of organic pollutants present. Therefore, higher COD and BOD values in wastewater suggest a higher pollutant content. The BOD/COD ratio is frequently used to determine whether wastewater is biodegradable. Following treatment of the wastewater, there is a significant drop in the concentrations of BOD and COD because of a reduction in the amount of biodegradable organic matter in TW. The BOD value is a reflection of the amount of organic pollution in water and a signal that contaminated water or effluent may be consuming oxygen. It is measured in milligrams of oxygen consumed per liter of a sample over the course of five days of incubation at 20°C. Industrial wastes and raw sewage wastewater are the main sources of OM that have an impact on the BOD concentration. Furthermore, they mentioned that BOD values for uncontaminated water are usually 2 mg L^{-1}, while raw sewage has a BOD value of roughly 600 mg L^{-1}. On the other hand, treated sewage effluents have BOD values that vary depending on the level of treatment, from 20 mg L^{-1} to 100 mg L^{-1}. BOD levels in industrial wastes can reach 25,000 mg L^{-1}. The COD test is frequently used to determine how sensitive organic and inorganic compounds in water bodies and the effluent from sewage and industrial plants are to oxidation. It is an indirect indicator of the organic compounds present in the water. It is measured in milligrams (mg L^{-1}) of oxygen per liter of water. For a lot of industrial wastes, COD is a quick and helpful variable to measure. The characteristics of the wastewater will determine the COD to BOD ratio. This ratio is frequently employed as a biodegradation capacity indicator.

10.1.3. Reuse of Treated Wastewater

TW is shorthand for treated municipal wastewater meant for beneficial uses after it has been brought into compliance with certain water quality standards. Approximately 0.4 trillion m^3 of wastewater are released globally annually, contaminating 5.5 trillion m^3 of water. As a result, treating and reusing these massive volumes of wastewater should concern all nations. Wastewater production will increase due to rising domestic water demands brought on by population growth, rising living standards, and the expanding industrial sector. This will encourage the reuse of TW globally. In addition to the risks to public health and the environment, this issue also has technical, institutional, sociocultural, and sustainability issues. Wastewater treatment and its application will therefore be crucial. Accordingly, for instance, China's wastewater treatment rate—that is, the ratio of total wastewater (TW) to all wastewater discharged—rose to 86% in 2014, which is three and a half times higher than it was in 1999. Worldwide, TW irrigation has grown by roughly 41% in Australia and by roughly 10–29% annually in the US, Europe, and China. Table 10.3 lists all of the wastewater that is produced, collected, cleaned, and used as irrigation in various nations. By lowering SS and OM and eliminating wastewater's chemical and biological components that could be hazardous to crops and public health, the primary treatment goal is to provide TW with an acceptable and secure level of risk for the environment and public health.

Table 10.3. *The Wastewater Generation, Collection, Treatment, and Reuse for Irrigation of Crops in Some Countries in Relation to the Total Cultivated Area*

Country	Total Area (1000 ha)	Agricultural Area (1000 ha)	Total Agricultural Area (%)	Generated Municipal Wastewater (10^9 m^3 $year^{-1}$)	Collected Municipal Wastewater (10^9 m^3 $Year^{-1}$)	Treated Municipal Wastewater (10^9 m^3 $year^{-1}$)	Treated Wastewater Used for Irrigation (10^9 m^3 $year^{-1}$)
Australia	774,122	47,307	6.11	–	–	2	0.28
Brazil	851,577	86,589	10.1	–	–	3.1	0.008
China	960,001	122,524	12.7	48.51	31.14	42.37	1.26
Germany	35,738	12,074	33.7	–	5.287	5.213	5.183
India	328,726	169,360	51.5	–	–	4.416	–
Italy	30,134	9121	30.2	3.926	–	3.902	0.087
Jordan	8932	322	3.6	–	0.115	0.113	0.103
Pakistan	79,610	31,252	39.2	3.06	–	-	–
South Africa	121,909	12,913	10.5	3.542	2.769	1.919	–
Turkey	78,535	23,944	30.4	4.297	-	3.483	–
UK	24,361	6279	25.7	4.089	4.048	4.048	–
USA	983,151	157,205	15.9	60.41	47.24	45.35	–
Canada	998,467	50,846	5.09	6.613	5.819	5.632	–
Sweden	44,742	2608	5.82	0.671	–	0.436	–

Source: https://www.mdpi.com/2073-4441/13/11/1527#B72-water-13-01527.

Since agricultural irrigation is currently the world's largest user of total water yield, there are a lot of opportunities for reuse of water in both industrialized and developing nations in the future. About 44 countries use 15 million m^3 of treated wastewater (TW) per day for crop irrigation. That wastewater is used to irrigate 20 million hectares across almost 50 countries. Reusing TW can have positive effects on the environment and human health, but it can also have negative effects because pollutants can persist in the waste after treatment and easily find their way into all living things through the food chain. Increasing health risks and environmental pollution are thought to be the primary drawbacks of reusing wastewater. Nonetheless, TW's detrimental effects on the environment and public health can be mitigated to a manageable degree through efficient irrigation management. Reusing wastewater could be dangerous for your health if it isn't properly pre-treated (i.e., insufficient reduction of pathogens and heavy metal entry into the food chain). Thus, all risks have to be within reasonable bounds. Wastewater can be broadly classified into three categories: SS, colloidal materials, and dissolved materials, in addition to water, which represents the bulk volume. When using TW for irrigation, the primary factors that need to be considered for efficient management are: SS, since filtration may be necessary, particularly when using a micro-irrigation system; nutrients to modify fertilization; salinity to determine the leaching fraction and choose the right cropping pattern; and pathogens to take precautionary measures and choose an appropriate irrigation system.

10.1.4. Treatment Methods and Safety Considerations

Strict safety regulations and efficient treatment techniques are essential for the productive application of treated wastewater in agriculture. Wastewater treatment aims to remove contaminants and pathogens that could endanger human health and the environment, as well as purify water to a level that satisfies safety regulations for its intended use—in this case, irrigation.

10.1.4.1. Treatment Methods

- **Primary Treatment:** The goal of this first step is to filter out large solids and particulate matter from wastewater. It involves physical procedures like sedimentation and screening, in which particles separate from the water and leave behind a clearer liquid. Primary treatment does not significantly address dissolved organic matter or pathogens, even though it lowers the load of suspended solids.
- **Secondary Treatment:** The biological processes that break down organic matter are the focus of this stage. Organic pollutants in wastewater are broken down and consumed by microorganisms, which then produce simpler materials. Aeration tanks, which introduce air to promote microbial growth, or trickling filters, which allow wastewater to pass over microbially covered media, are commonly used in secondary treatment processes. Both suspended solids and biochemical oxygen demand (BOD) are greatly decreased at this stage.
- **Tertiary Treatment:** Tertiary treatment is used to reach higher purification levels. Advanced filtration, chemical treatments,

and disinfection procedures are all part of this step. To get rid of any leftover pollutants, pathogens, and nutrients, methods like membrane filtration, activated carbon adsorption, and ultraviolet (UV) disinfection are employed. Tertiary treatment guarantees that the water is safe for irrigation and satisfies strict quality standards.

- **Advanced Treatment:** Advanced oxidation techniques and reverse osmosis are examples of treatment methods that can be used for applications that demand the highest level of purity. These technologies ensure exceptionally high-quality water by eliminating a wide range of contaminants, including pharmaceuticals and endocrine-disrupting chemicals.

10.1.4.2. Safety Considerations

- **Pathogen Removal:** Effective pathogen removal is critical to prevent health risks associated with treated wastewater. Regular testing for indicators such as *E. coli*, fecal coliforms, and viruses is essential to ensure that the water is free from harmful microorganisms. Advanced treatment technologies, such as UV disinfection and ozonation, enhance pathogen removal and reduce health risks.
- **Chemical Residues:** Treated wastewater may contain residual chemicals from industrial processes or pharmaceuticals. Monitoring for specific contaminants and employing advanced treatment methods to remove these residues is necessary to protect soil health and avoid potential impacts on crops and groundwater.
- **Nutrient Management:** While treated wastewater can provide beneficial nutrients, excessive levels of nitrogen and phosphorus can lead to soil and water pollution. Proper management practices, such as balancing nutrient application and monitoring soil health, are important to prevent nutrient overload and maintain environmental sustainability.
- **Regulatory Compliance:** Adhering to local and international regulations is crucial for ensuring the safety of treated wastewater for agricultural use. Regulations specify permissible levels of contaminants and outline best practices for treatment and monitoring. Compliance helps protect public health and supports the responsible use of reclaimed water.

A multifaceted strategy combining cutting edge treatment techniques and stringent safety precautions is needed for the integration of treated wastewater into agricultural practices. Treated wastewater can be a dependable and sustainable resource for irrigation, supporting water conservation and sustainable agricultural practices, provided that it is effectively purified and safety regulations are followed.

10.2. DEFINITION OF WASTEWATER REUSE IN AGRICULTURE

Water is considered to be reused if it is used twice or more before being reabsorbed into the natural water cycle. Reusing treated wastewater has many advantages; these include recharging or replenishing groundwater basins, industrial processes, agricultural and landscape irrigation, and residential toilet flushing. Reusing water can lessen the amount of water diverted from delicate ecological systems and is defined as relying on the use of surface and groundwater sources. Furthermore, reusing water can lower the amount of nutrients that wastewater flows into streams, which will reduce pollution.

Reusing wastewater in agriculture refers to using treated wastewater again for crop irrigation. Because of the need for a controlled supply to make up for water shortages brought on by seasonality or the erratic availability of other water sources for crop irrigation throughout the hydrological year, this type of reuse is regarded as an effective tool for managing water resources. Even though using wastewater is a long-standing practice, it hasn't always been handled correctly or adhered to quality standards. As a result, knowledge about the use of wastewater has changed throughout human history.

Prehistoric societies used domestic wastewater in agriculture to get rid of waste from urban settlements during the Bronze Age (3200–1100 BC). The most popular method was soil irrigation with wastewater, which has since undergone various stages of development. The utilization of public latrines that flushed wastewater through a sewer system and into a storage chamber by the ancient Greeks is the earliest evidence of wastewater reuse. Furthermore, domestic wastewater was used by the Greek and Roman civilizations around the peripheries of their major cities, Athens and Rome.

Wastewater was hauled out to the fields and applied as crop and orchard fertilizer. Farms in Germany, Scotland, and England began using wastewater directly on their fields between the years of 1550 and 1700. Many rapidly expanding cities in Europe and the US started using wastewater for soil irrigation around 1800. For instance, the practice was accepted as lawful in Boston, London, and Paris and was seen as a practical way to handle and dispose of massive amounts of wastewater. The first major city to use wastewater to irrigate peri-urban fields was Paris. The practice of irrigating four distinct peri-urban areas with wastewater reached its peak in Paris in 1872. Australia also began to implement the disposal of municipal wastewater during this same time frame. Melbourne saw the establishment of the first wastewater-irrigated field in 1897.

The 19th century saw disastrous outbreaks of cholera and typhoid fever due to the transportation and ultimate disposal of raw wastewater onto exposed peri-urban fields. The Public Health Act of Great Britain, which established the "discharge of rainwater in the river and of wastewater on the soil" as the primary principle, was one of the landmarks in sanitation brought about by these epidemics. Furthermore, a number of sanitary conferences on hygiene and demography were organized as a result of the global sanitary movement that was supported by powerful European nations. In addition, the International Office of Public Hygiene was founded in order to carry out border sanitary controls. The emergence of subterranean sewage systems during the mid-1800s is thought to have been a reaction against the unsanitary conditions brought on by the significant industrialization and urbanization that was taking place at that time. Up until the early 20th century, however, large American and European cities continued to widely implement wastewater disposal plans in agricultural fields. Due to the high water demands of this industry, interest in the use of wastewater for agriculture (indirect irrigation with raw wastewater) increased during the 1990s in many parts of the world.

The hazards that wastewater reuse posed to the environment and public health at this time made it a global concern. In order to protect public health and enable the sensible use of wastewater and excreta in agriculture and aquaculture, the World Health Organization (WHO) drafted the document "Reuse of effluents: methods of wastewater treatment and health safeguards" in 1973. This first recommendation was developed using a minimal risk approach and without the benefit of epidemiological studies. A comprehensive review of all available epidemiological studies was carried out in 1986, resulting in a body of evidence that indicated the need to reevaluate the 1973 guidelines. These factors led to an update of the guidelines in 1989, which included new health data (risk assessments, for example) and more details on what constitutes a reasonable risk for society given the state of a given disease in a given nation. The guidelines set limits on wastewater's microbiological quality so that it could be used for irrigation. The WHO guidelines, however, did not include guidelines for surveillance; as a result, the formulation of these guidelines was based on goals and precautionary measures for health. These factors were taken into account when the World Health Organization created its wastewater guidelines in 2006.

The WHO's 2006 recommendations for the safe use of excreta, graywater, and wastewater serve as a tool for the proactive management of wastewater in agriculture and offer decision-makers precise instructions on how to apply wastewater in various local contexts. The main goal of the guidelines is to assist in the development of government regulations and standards pertaining to the use and handling of wastewater while taking into account the unique characteristics of each nation. Included in these guidelines was a crucial microbiological analysis for risk assessment, which entails gathering information on pathogens found in fields, wastewater, and irrigated crops. The guidelines also include health risk management and prevention estimates for wastewater use that are based on person-per-year and disability-adjusted life year (DALY and PPY).

A number of guidelines pertaining to the use of wastewater in agriculture have also been developed by the United Nations Food and Agriculture Organization (FAO). The guidelines for wastewater quality suitable for agricultural use were released in 1987. These recommendations linked salinity, infiltration, and toxicity parameters of particular ions to the extent of water use restriction. The recommended guidelines for the "agricultural reuse of treated waters and treatment requirements" were released by the FAO in 1999. The guidelines categorized the type of agricultural reuse according to the type of crop that receives irrigation (Table 10.4).

Table 10.4. *FAO Guidelines for the Agricultural Reuse of Treated Water*

Type of Agricultural Reuse	Type of Treatment	Quality Criterion
Agricultural reuse in crops that are consumed and not processed commercially.	Secondary Filtration—Disinfection	pH = 6.5–8.4
		BOD < 10 mg/L
		<2 UNT
		<14 NMP *E. coli*/100 mL
		<1 Egg/L
Agricultural reuse in crops that are consumed and not processed commercially.	Secondary—Disinfection	pH = 6.5–8.4
		BOD < 30 mg/L
		SS < 30 mg/L
		<200 NMP *E. coli*/100 mL
Agricultural reuse in crops that are not consumed.	Secondary—Disinfection	pH = 6.5–8.4
		BOD < 30 mg/L
		SS < 30 mg/L
		<200 NMP *E. coli*/100 mL

The Environmental Protection Agency (EPA) established in 1992 that crops exposed to specific trace elements found in irrigation wastewater have harmful effects. The EPA broadened its focus on indirect potable reuse and industrial reuse issues in 2004. This included the inclusion of numerous updated case studies, fresh data on treatment and disinfection technologies, emerging chemicals and pathogens of concern, economic analysis, user rates and funding options, public acceptance and involvement, research endeavors, and information sources.

The Guidelines for Wastewater Reuse were revised in 2012 by the US Agency for International Development (USAID) and the US EPA. Facilitating the development of wastewater reuse through a compilation of global experiences was the primary goal of the update. The updated analysis of regional variations in water reuse, developments in wastewater treatment technologies, community involvement best practices, global water reuse practices, and factors supporting the growth of safe and sustainable water reuse were all included in the 2012 guidelines. In order to provide technical updates of the guidelines, technical revisions, information on standards, and case studies, over 300 experts in the field of wastewater reuse worked together. The EPA and USAID suggested a maximum safe concentration level of trace elements present in irrigation water based on quality considerations. Regulations in many nations around the world have been formulated using the guidelines established by the EPA, FAO, and WHO.

Within the framework of the international AQUAREC project, over 3300 wastewater facilities were registered globally between 2000 and 2006. Agriculture was the main wastewater user, and the various wastewater facilities were distinguished by varying water treatment quality levels and use types. Japan and the United States had the most reuse facilities (1800 and 800, respectively), followed by the European Union and Australia (450 and 230, respectively). It was estimated that there were about 100 wastewater treatment facilities in the Mediterranean and Middle East, compared to 50 in Latin America and 20 in Sub-Saharan Africa.

According to the FAO, 20 million hectares across 50 countries, or about 10% of the world's irrigated land area, receive untreated or partially treated wastewater. Jiménez and Asano did note, however, that the estimated area irrigated by wastewater varies by nation as well as between treated and untreated conditions (Figure 10.1). According to Bixio and Wintgens, 963 mm^3/year of untreated wastewater are reused in agriculture across the European continent. About 400 m^3/s of raw wastewater are released into the environment in Latin America and are then utilized to irrigate various crops.

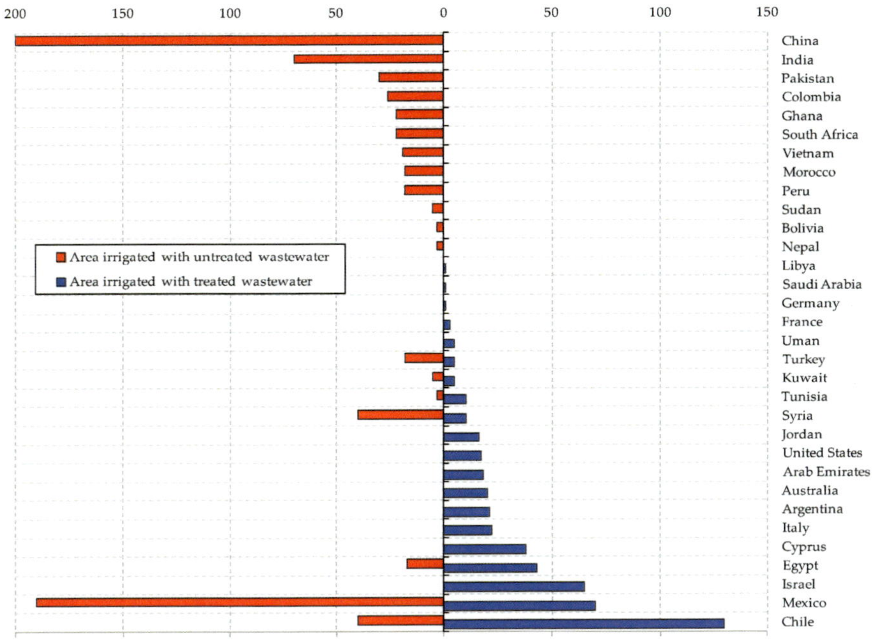

Figure 10.1. *Reuse area in agriculture by country (thousand ha).*

Source: https://www.researchgate.net/profile/Nicoleta-Ungureanu/publication/328772419/figure/fig1/AS:690052078243842@1541532644644/Areas-of-wastewater-reuse-in-agriculture-by-country-data-refers-to-thousand-hectares.ppm.

10.2.1. Reuse of Wastewater in Agriculture

Water has a fixed amount and changes form in response to changes in the atmosphere. Water is never destroyed; its distribution is the only thing that changes. Its circulation is a constant process. It is important to maintain both its quantity and quality. It needs to be handled carefully because it will become the scarcest commodity in the future. 30% goes to industry, 10% to homes, and 70% is used for irrigation—including all water pumped from underground and diverted from rivers. India used 84% of its water for agriculture, 4% for drinking, and 12% for other uses like industry.

Thus, effective management is crucial in the agricultural sector. Allocating household and industrial effluents for removal or disposal without posing an unacceptable risk to human health or the environment is typically the main goal of wastewater treatment. As in slow-rate land treatment, using wastewater for irrigation is actually an efficient way to dispose of wastewater. It can be fully untreated wastewater from industry or municipalities, wastewater that has been mechanically cleaned, or wastewater that has been partially or fully treated by biological means. The efficacy and functionality of wastewater-soil-plant or aquacultural systems are significantly influenced by the quantity of treated effluent used in agriculture and horticulture. When using irrigation, the quantity of effluent needed will vary depending on the crops being grown, the soil's characteristics, and the effluent distribution system that is chosen. Reduced crop constraints and irrigation techniques that prioritize human health can reduce the amount of pre-application wastewater treatment.

Approximately 16% of the world's population lives in India. It makes up only 2.5% of the world's land area and 4% of its water resources. It was discovered that the nation's total usable water resources, or 1123 BCM (690 BCM from surface and 433 BCM from ground), accounted for 28% of precipitation. Currently, 688 BCM, or about 85%, of the water used for irrigation is diverted; by 2050, that amount is expected to rise to 1072 BCM. In India, groundwater is a major source of water for irrigation. About 213 BCM of India's yearly groundwater recharge is used for irrigation, and another 18 BCM is used for residential and commercial uses. The amount of water required for residential and commercial uses will rise to nearly 29 BCM by 2025. Thus, it is predicted that the amount of water available for irrigation will decrease by up to 162 BCM.

By the end of 2050, India's population is predicted to exceed 1.5 billion people based on the current annual population growth rate of 1.9%. The average annual freshwater availability per capita in India has decreased as a result of population growth, general development, and changes in the country's standard of living. According to reports from 1951, 2001, and 2010, India's annual freshwater availability per capita was 5177 m^3, 1869 m^3, and 1588 m^3, respectively. In order to meet the nation's water needs, an effective and efficient water resource management system that incorporates a wastewater treatment process and recycling system is desperately needed. Figure 10.2 displays the water needed/required by the different sectors according to the project. The state of wastewater and effluent disposal is concerning due to treatment plant capacity shortages and rising rates of sewage production.

In order to minimize environmental pollution, wastewater treatment plants are typically designed with the reduction of organic and suspended solid loads as a primary consideration. Although pathogen removal has rarely been regarded as an objective, it is now necessary to give priority to this issue when it comes to the reuse of effluents in agriculture, and processes should be chosen and planned accordingly. Although technically possible, treating wastewater to remove components that could be toxic or detrimental to fish, crops, or aquatic plants (macrophytes) is typically not economically viable.

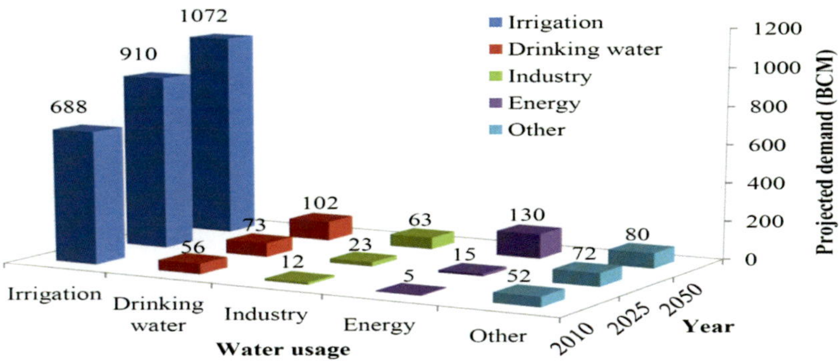

Figure 10.2. *Projected water demand by different sectors.*

Source: https://www.researchgate.net/profile/Bl-Meena/publication/335566315/figure/fig2/AS:828697791721472@1574588358247/Projected-water-demand-by-different-sectors-CWC-2010.png.

There is a daily pattern to the short-term variations in wastewater disposals observed at public and municipal wastewater treatment facilities. Early in the morning, when water consumption is at its lowest, as well as when there is little to no sanitary wastewater infiltration, is when disposal is typically lowest. When wastewater from the morning's peak water consumption reaches the treatment plant, there is typically a first peak of flow in the late morning, followed by a second high flow in the evening. The size of the community, the length of the sewers, and the relative magnitude of the peaks all vary from one country to the next. The ratio of peak flow to average flow is significantly higher in small communities with small sewer systems than it is in large communities. The daily variations in flow from a public or municipal treatment unit make it impractical, in most cases, to irrigate directly from the treatment unit, even though the extent of peaks is satisfied as wastewater passes through a treatment plant. It has long been understood that reclaimed water is valuable for crop irrigation, especially in areas with limited freshwater resources. Although storage has additional advantages, some kind of flow equalization or short-term storage of treated wastewater is necessary to provide a reasonably continuous/uniform supply of treated wastewater for effective irrigation.

Wastewater is composed of 99.9% water and 0.1% of other materials like suspended, colloidal and dissolved solids. Due to the scarcity of water, demands for water from agriculture and domestic and industrial urban areas frequently clash severely. Typically, the only way to resolve this conflict is through the agricultural use of wastewater: after proper treatment, urban wastewater must be used for crop irrigation, with fresh water usage coming first in cities. If this order of use for water resources is not followed, there may be significant restrictions on the growth of both agriculture and urban areas, which would have a negative impact on the growth of the national economy. Furthermore, it's essential to understand the effluent's fertilizer value. The effluent would supply all of the nitrogen and a large portion of the phosphorus and potassium normally needed for agricultural crop production. Moreover, the organic matter and other important micronutrients in the effluent will offer further advantages.

10.2.2. Environmental Toxicity of Untreated Wastewater

Larger applications for treated wastewater include irrigation, firefighting, toilet flushing, and groundwater recharge. Wastewater treatment plants (WWTPs) serve as a primary hub for the accumulation of various hazardous substances, microbes, and heavy metals. It gathers wastewater from various sources, such as residential, commercial, industrial, or hospital wastewater, as well as urban runoff. It has been observed that, in contrast to groundwater (GW) irrigation, grass and crops irrigated with sewage and treated wastewater are higher in heavy metals. Nevertheless, heavy metals—which are categorized as toxic elements—are iron, copper, lead, mercury, and cadmium. Exposure to excessive levels of these heavy metals may pose health and environmental risks. Drinking water is one of these heavy metals' main sources. This could be the result of sewage spilling into a river or groundwater contamination from contaminated soil. Table 10.5 lists the maximum levels of heavy metals that are allowed in drinking water, their effects on human health when they are exceeded, and the pathways through which heavy metals are exposed to the human body.

Table 10.5. *Total Permissible Limits of Heavy Metals in Drinking Water and Diseases Associated with the Surplus Amount*

Heavy Metals Polluting the Water Quality	Permissible Limits in Drinking Water According to WHO (mg/L)	Permissible Limits in Effluent Water According to WHO (mg/L)	Diseases Associated with the Excess Amount	Exposure Routes

Arsenic	0.01	5.0	Skin, lung, bladder, kidney cancer, skin manifestations, gastrointestinal disorders, neurological effects, hormone disruption and infertility, psoriasis	Inhalation and ingestion
Cadmium	0.005	0.003	Psychological disorders, gastrointestinal disorders, central nervous system complications, immune system deficiencies, DNA impairment, cancer, Itai-Itai disease, osteoporosis, respiratory disease	Ingestion of contaminated food and water and, to a significant extent, through inhalation and cigarette smoking
Chromium	0.1	0.05	Gastrointestinal ulceration, nausea and vomiting, fever, diarrhea, toxic nephritis, liver damage, gingivitis, bronchitis, pneumonia, lung cancer	Inhalation and ingestion
Iron	1.0	2.0	Genetic disorder, hemorrhagic necrosis	Ingestion

Lead	0.01	0.05	Hypertension, miscarriages, premature and low births, renal impairment, brain injury, abdominal pain	Inhalation through the nose and ingestion through drinking water and soil
Mercury	0.006	0.001	Down's syndrome, affects the reproductive system, speech defects, memory loss, tremors and muscle incoordination, deafness, vision complication	Inhalation, ingestion and dermal contact
Copper	2.0	0.25	Insomnia, anxiety, agitation, restlessness, fatigue, jaundice, dizziness	Ingestion
Nickel	0.07	0.02	Lung embolisms, asthma, respiratory failure, heart disorders, dizziness, increased possibilities of cancer	Inhalation and ingestion

Reusing wastewater for irrigation or releasing it directly into rivers can have short-term effects such as microbiological and heavy metal contamination, as well as pathogenic interactions with crops and soil. Long-term effects include soil salinity, which increases with repeated use of untreated wastewater. When wastewater is used improperly for irrigation, it poses a risk to public health and the environment. Irrigation using a variety of wastewater kinds, i.e., heavy metals are transferred to the soil by sewage liquid sludge, municipal and agricultural wastewaters, industrial effluents, and sewage, which causes an accumulation of heavy metals in crops as a result of incorrect practices. This has been recognized as a major entry point for heavy metals into aquatic resources. It was found that soil irrigated with treated wastewater from large-scale sewage treatment plants had higher concentrations of heavy metals than normal ground water, with the exception of Cd.

Put another way, using wastewater for irrigation raises the risks to human health while degrading crop quality. Children who consume too much copper experience nausea, headaches, vomiting, liver and kidney damage, and anemia. Increased levels of arsenic can cause osteopenia or osteoporosis as well as kidney and bone cancer. Musculoskeletal disorders are caused by cadmium, while mercury directly damages the nervous system.

10.2.3. Benefits of Agricultural Reuse

The economy, the environment, and human health all benefit from the use of treated wastewater in agriculture. In light of the depletion of surface and groundwater resources brought on by climate variability (CV) and climate change (CC), this use reflects an alternate approach that is being adopted in various regions that are struggling with water shortages and expanding urban populations with rising water demands. Because wastewater-sourced water isn't always treated before it reaches surface channels, it can also affect the availability of water resources, as can associated aquifer pollution.

The associated reduction in strain on freshwater resources is one of the most well-known advantages of using wastewater in agriculture. Wastewater thus provides an alternative source of irrigation, particularly for agriculture, which uses 70% of the water available worldwide. Reusing wastewater also boosts agricultural output in areas with limited water supplies, which improves food safety. Approximately one-ninth of the world's population, or 805 million people, are hungry. Nonetheless, the likelihood of halving the number of undernourished individuals is supported by a declining trend in hunger, according to the FAO's most recent estimates. To be successful, though, a comprehensive strategy that combines public and private investment in order to boost agricultural productivity, as well as improving and expanding the availability of water resources and safeguarding marginalized communities, must first be adopted.

Reusing agricultural wastewater may also save money by avoiding the need to extract groundwater, depending on the local conditions. It is important to keep in mind that the energy needed to pump groundwater can account for as much as 65% of the expenses associated with irrigation. Furthermore, because wastewater naturally contains nutrients, fertilizer costs can be reduced.

This creates a closed, environmentally beneficial nutrient cycle that prevents the inadvertent return of macro- (particularly nitrogen and phosphorous) and microelements to water bodies. Wastewater can potentially contain macro- (N, P, and K) and micronutrients (Ca, Mg, B, Mg, Fe, Mn, or Zn) depending on the nutrients present. Reusing wastewater has in fact been shown to increase crop yield and decrease the need for fertilizers in farming. As a result, both the conditions of eutrophication in water bodies and farmer costs associated with agrochemical use would decrease.

Reusing wastewater in agriculture would also have the benefit of preventing water pollution. Reductions in wastewater discharge contribute to better receiving water bodies' source quality. Reusing agricultural wastewater also preserves groundwater reservoirs by replenishing them with better-quality water. Furthermore, using wastewater more frequently could help install and improve treatment facilities to generate effluent of the right quality for irrigation, which would benefit sanitation projects financially. Low-cost wastewater treatment systems may also be a practical option in places where the climate and geography permit, provided that certain technological solutions meet the goal of agricultural reuse.

Utilizing wastewater in agriculture facilitates the payment of economic instruments by international actors, which helps release capital resources. Given that treating wastewater before it is released for human use is the highest priority, treating wastewater for reuse in agriculture also has an implicit economic benefit. Reusing wastewater lowers the cost to the municipality of looking for water sources by employing more costly methods in certain nations. Reusing agricultural wastewater can support appropriate investment policies and financing mechanisms for pollution prevention and control based on regulatory considerations.

10.2.4. Toxic Contaminations in Wastewater Impacting Human Health

There could be major health risks if untreated wastewater is released into a river. As was previously mentioned, sewage from homes and businesses is largely composed of organic materials and pathogenic microorganisms. These microbes can spread a number of diseases, including typhoid, dysentery, diarrhea, vomiting, and malabsorption. Furthermore, the pharmaceutical industry is crucial to the control and release of biologically hazardous substances. There is another class of pollutants present in the untreated wastewater that are harmful to people.

These hazardous contaminations are divided into two main categories:

- microbial contamination and
- Chemical contamination.

10.2.4.1. Chemical Contamination

Many diseases are primarily caused by different kinds of chemical compounds that are released from factories, tanneries, workshops, irrigated lands, and home wastewaters. Organic materials, hydrocarbons, volatile compounds, pesticides, and heavy metals are a few examples of these contaminants. Lung infections, eye irritation, chronic dermatoses, and skin cancer are among infectious diseases that can result from exposure to these contaminants. The majority of them are unmanageable and non-biodegradable. As a result, they can live in water bodies for a very long time and can easily build up in our food chain. There are numerous pharmaceutical personal care products (PPCPs) and surfactants on the market that might include harmful substances like ethinylestradiol, nonylphenol, estrone, and estradiol. These substances are known to disrupt hormones, and even minute concentrations of these substances within the human body can pose serious risks. Also, there have been numerous reports from all over the world about the presence of perfluorinated compounds (PFCs), which are naturally toxic, in wastewater. PFCs also pose a serious risk to human health because they can lead to immunotoxicity, neurotoxicity, carcinogenesis, pre-eclampsia, birth abnormalities, and decreased human fertility.

10.2.4.2. Microbial Contamination

Researchers have found that the microbial pollutants in untreated wastewater pose a significant risk to human health. Serious health consequences such as campylobacteriosis, diarrhea, encephalitis, typhoid, giardiasis, hepatitis A, poliomyelitis, salmonellosis, and gastroenteritis are caused by this diverse group of microorganisms. A few types of bacteria similar to *P. aeruginosa*, Vibrio cholerae, *Salmonella typhimurium*, and *G. intestinales*, species of Legionella., *E. Shigella sonnei* and *E. coli* have been linked to acute human illness and the spread of waterborne diseases. These previously mentioned microbes can escape into the environment through public water supply systems, animal husbandry, or municipal sewage water networks, and then find their way into the food chain.

10.3. WASTEWATER IMPACT ON AGRICULTURE

It is commonly known that the agricultural sector uses the most water—nearly 70% of all water used worldwide. The fact that wastewater is used to irrigate an estimated 20 million hectares of land worldwide points to a significant irrigation source. On the other hand, the majority of wastewater used for irrigation is raw. Wastewater, whether partially treated or untreated, is mostly used for irrigation in developing nations. Untreated wastewater frequently contains a wide range of chemical contaminants, including heavy metals, fertilizers, textile, leather, paper, sewage waste, chemical wastes from industrial discharges, pesticides, and textile and leather waste. The World Health Organization (WHO) has issued a warning about the serious health risks associated with directly using wastewater for irrigation. Living close to wastewater sources and areas irrigated with untreated wastewater puts communities of farmers, agricultural workers, their families, and consumers of crops irrigated by wastewater at risk for health problems. A large range of organic compounds are also present in wastewater. Certain substances are hazardous to embryos and can cause cancer. Figure 10.3. illustrates the untreated wastewater irrigation pathway and related health effects.

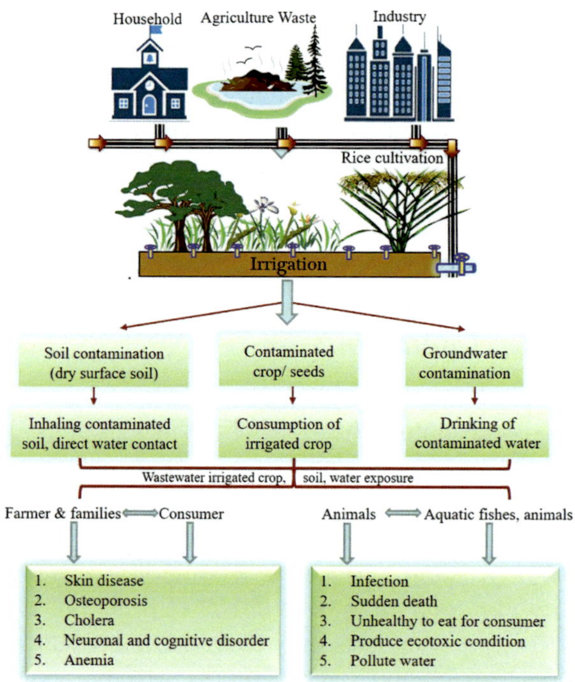

Figure 10.3. *Exposure pathway representing serious health concerns from wastewater-irrigated crops.*

Source: https://www.researchgate.net/publication/351455628/figure/fig2/AS:1022148428722201@1620710584412/Exposure-pathway-representing-serious-health-concerns-from-wastewater-irrigated-crops.png.

Conversely, untreated wastewater is dumped into the existing waterbodies in developing nations because there aren't enough treatment facilities in those areas. Biodiversity is lost when wastewater is directly used for irrigation or agriculture, as this stunts the growth of native grasses and plants. The first indications that diseases are linked to the reuse of agricultural wastewater. When raw wastewater is applied to crops for irrigation, the soil becomes more salinized, which causes land sealing, sodium buildup, and soil erosion. Crops' ability to absorb nutrients from the soil is hampered by increased soil salinity and sodium buildup, which also deteriorates the soil. The long-term effects of reusing wastewater in agriculture have been examined for these causes.

Furthermore, helminths, which include nematodes and tapeworms, are intestinal parasites that are spread through the fecal–oral pathway and are primarily found in soils contaminated by wastewater. Since blood deficiency and behavioral or cognitive development are related, helminth infections are already well-known. Utilizing raw or partially treated sewage effluent and sludge for agricultural irrigation is a significant global source of helminth infections. Heavy metal pollution from foundries, mining, and other metal-based industries reaches crops that are irrigated with wastewater. Numerous health issues can arise from crop irrigation using wastewater that contains heavy metals like arsenic, cadmium, lead, and mercury. Human osteoporosis, for instance, is brought on by excessive cadmium consumption. Significant health risks have been linked to the rice crop's absorption of heavy metals from untreated paper mill effluent. Japan experiences an outbreak of Itai-itai disease as a result of irrigating rice paddies with highly contaminated water containing heavy metals.

The WHO issued recommendations for threshold contaminant levels in wastewater and released the third edition of its guidelines for the safe use of wastewater in crop irrigation in response to these pervasive health risks. The degree of chemical contamination, microorganisms, trace elements, and nutrient availability have all been taken into consideration when classifying wastewater quality for agricultural reuse. Depending on the kind of source—household sewage, chemical, pharmaceutical, paper, or textile industry effluents—the degree of contamination varies greatly. International reports provide standard measurements of irrigation water quality, with recommended limits for metals, COD, BOD, nitrogen, and phosphorus as well as trace elements and other elements. Based on its standards and guidelines for water quality, researchers examined the state of wastewater reuse for agriculture. It is clear from these suggestions and guidelines that more awareness is needed in order to treat wastewater safely.

10.3.1. Health Risks of Agricultural Wastewater Reuse

Depending on the hygienic and financial circumstances of a specific area, wastewater contains different types and concentrations of pathogens and chemicals. Compared to developed countries, the concentration of viruses, protozoan parasites, and helminths in wastewater can vary between 10 and 1000 times. The main categories of sanitary interest and enteric pathogens present in wastewater used for agricultural irrigation are listed in Table 10.6.

Table 10.6. *Chemical and Biological Risks Associated with the Use of Raw Wastewater in Agriculture*

Type of Risk	Pathogen	
Biological	Bacteria	*E. coli*, *Vibrio cholerae*, *Salmonella* spp., *Shigella* spp.
	Helminths	*Ascaris*, *Ancylostoma*, *Tenia* spp.
	Protozoans	*Intestinal Giardia*, *Crysptospridium*, *Entamoeba* spp.
	Virus	Hepatitis A and E, Adenovirus, Rotavirus, Norovirus
	Schistosoma	Blood-flukes
Chemical	Substance of sanitary interest	
	Heavy Metals	Arsenic, Cadmium, Mercury
	Hydrocarbons	Dioxins, Furans, PCBs
	Pesticides	Aldrin, DDT

Wastewater-borne diseases can also be chronic or acute. Acute risk corresponds to the possibility of becoming ill in the short-term when exposed to low infectious doses of a pollutant, whereas chronic risk refers to the presence of pollutants of a chemical nature that affect human health after long periods of exposure. Additionally, microbial diseases can be directly or indirectly transmitted by water (Table 10.7). Globally, such diseases have significantly contributed to premature mortality, especially in developing countries.

Table 10.7. *Some Water-Borne Diseases Related to Wastewater*

Disease	Cause
Typhoid fever	*Salmonella typhi*
Paratyphoid fever	*Salmonella paratyphi*
Gastroenteritis	*Salmonella typhimurium*

Cholera	*Vibrio cholerae*
Bacillary dysentery	*Shigella dysenteriae*
Amebiasis	*Entamoeba histolytica*
Giardiasis	*Giardia duodenalis*
Cryptosporidiosis	*Cryptosporidium*
Cyclosporiasis	*Cyclospora cayetanensis*
Infectious hepatitis	Hepatitis A
Gastroenteritis	Enterovirus, parvovirus, rotavirus
Infantile paralysis	Poliovirus
Leptospirosis	*Leptospira icterohaemorrhagiae*
Ear infections	*Pseudomonas aeruginosa*
Scabies	*Sarcoptes scabiei*
Trachoma	*Chlamydia trachomatis*
Schistosomiasis	*Schistosoma*
Malaria	*Plasmodium*
Yellow fever	Flavivirus
Dengue	Flavivirus

Emerging contaminants (ECs) are additional substances found in irrigated wastewater that could be harmful to human health. ECs are molecules that exhibit biological activity on a variety of organisms. Their bioaccumulation and persistence in the environment are facilitated by their unique physicochemical characteristics. Among other medications, ECs include antibiotics, analgesics, and antihypertensive medications. Moreover, endocrine disrupters (EDs) and certain ECs are related. Due to a lack of knowledge regarding their accumulation in soil, water, air, and plant and animal tissue, these complex substances were previously not regarded as contaminants. But starting in the 1990s, the amount of these substances in water sources started to be measured. According to estimates, 500 tons of analgesics have been dumped into superficial water sources in countries like Greece, France, Italy, Germany, Canada, Brazil, and Spain. Salicylic acid and diclofenac concentrations in these water sources are 0.22 μg/L and 3.02 μg/L, respectively. ECs are frequently added to aquatic media by a variety of human activities, which may leave toxic residues and have negative effects on aquatic life and, eventually, humans. Municipal wastewater treatment plant effluents are categorized as one of the main sources of EC because traditional treatment methods are unable to effectively stop these compounds from being released into the environment. Other sources of EC pollution include farming and agriculture, which are sources of diffuse pollution from pesticides and antibiotics, respectively.

Sludge digested from wastewater treatment systems used as fertilizers or fert-irrigation, or runoff from soils containing animal excrement, can also be a source of EC contamination in superficial water. ECs can contaminate groundwater through landfill leachate, manure container leaks, or agricultural land treated or untreated wastewater sprayed via spray irrigation. The impact of ECs on human health is still not entirely known. Nonetheless, it is well known that a large number of these substances modify aquatic organisms' immune and endocrine systems. All substances that have an impact on the endocrine system are generally referred to as endocrine disruptors (EDs). It has been demonstrated that these EDs cause hormonal changes in a few fish species and amphibians; some of these changes are linked to the thyroid gland's stimulating hormone (TSH), luteinizing hormone (LH), and follicle stimulating hormone (FSH) secretion. Other medications that alter endocrine activity include fluoxetine, carbamazepine, and clofibrate. Furthermore, in wastewater treatment facilities, ED medications are difficult to remove. Thus, EDs find their way into surface waters and water sources meant for human consumption, thereby exposing people to their harmful effects on a long-term basis.

The development of resistant microorganisms may be facilitated by EC contamination. Since antibiotics are used so frequently to treat pathogenic microorganisms in humans and animals as well as to preserve food, their production and consumption have increased. This has led to high volumes of antibiotic discharge into water bodies, which has resulted in microbial resistance. Salmonella, Escherichia, Aeromonas, Pseudomonas, and Staphylococcus are a few of the microorganisms that stand out among those that have shown resistance. The existence of resistant microorganisms in water bodies is therefore a significant concern in terms of legal policies and functionality as it relates to public health and wastewater treatment and reuse systems.

10.3.2. Limitations Associated with Agricultural Wastewater Reuse

Negative consequences on the environment, particularly on soil, can arise from the use of treated or untreated wastewater in agriculture. There is evidence of changes in the physicochemical parameters of soil in the scientific literature. Furthermore, changes in the composition and amount of microbial biomass in soil have been noted in recent studies, along with an increase in microbial activity brought on by the reuse of agricultural wastewater. Changes in soil microbiota and physicochemical parameters can impact fertility and productivity, which can disrupt soil sustainability due to insufficient irrigation using wastewater.

The effects of wastewater reuse in agriculture are reviewed, including how it affects microbial diversity and physicochemical parameters like pH, organic matter, nutrients, salinity, and contaminants. The various research studies on the impact of wastewater on soil are shown in Table 10.8.

Table 10.8. *The Influence of Agricultural Reuse on the Soil's Physicochemical and Microbiological Parameters*

Parameter	Associated Effects on the Soil and the Environment	
	Physicochemical Properties	**Microbiological Properties**
pH	Increases the availability of nutrients and metals	Increases the richness and diversity of the microbial community
	Mineralization of organic matter	
	Improves the cation exchange capacity	
Organic matter	Soil structure stabilization	Selection of specific populations and soil microhabitats
	Formation of aggregates	
	Water retention	
	Improves nutrient content	
	Buffer Capacity	
	Cation exchange capacity	
	Enzymatic activity	
	Increase in TOC	
	Increases the availability of contaminants	
Nutrients	Increase in organic soil matter	Perturbation of the metabolic activity of microbial soil communities
	Water retention	
	Leaching to groundwater	
	Improves nutrient content	
	Risk of eutrophication of aquatic environments	

Salinity	Soil salinization or sodification	Changes in soil microhabitats and variation in the richness and diversity of the microbial community
	Decreased stability of aggregates	
	Changes in soil structure in the long term	
	Permeability of soil and water retention	
	Increased soil compaction	
	Variation in soil pH	
	Negative impact on soil fertility	
	Dynamics in organic and inorganic compounds	
	Heavy metal leaching	
Contaminants	Soil toxicity and leaching	Increased tolerance to microbial contaminants. Antimicrobial resistance. Reduction of microbial biomass and changes in its structure
	Accumulation in soils	
	Negative impact on soil fertility	
	Potential contamination of the food chain	
	Mineralization of organic matter	
	Changes in enzyme activity	
	Decomposition of fallen leaves	
	Limiting soil fertility	

Numerous investigations have documented fluctuations in soil pH as a consequence of irrigating with municipal wastewater treatment system effluents at varying treatment levels (primary, secondary, and preliminary). Furthermore, three variables are associated with variations in soil pH: (i) soil cover type; (ii) soil texture; and (iii) irrigation period. The cation exchange capacity (CEC), the availability of nutrients and metals, and the mineralization of organic matter are all impacted by variations in soil pH. Furthermore, because an increase in free metals is unrelated to changes in soil pH and because metal concentrations and availability may have an impact on the substrate of microbial communities, various researchers believe that pH incidence plays a critical role in determining the diversity and number of species of soil microorganisms. Furthermore, soil structure and nutrient storage depend on organic matter. The organic matter content influences the soil's ability to hold onto water, which influences drainage characteristics and resistance to compaction. This is accomplished through the formation and stabilization of aggregates, such as sand, lime, and clay.

Important macro- and micronutrients (N, P, and S) for plant growth are also deposited in organic matter, which raises the cation exchange capacity (CEC) and, in turn, the fertility of the soil. Several studies have shown an increase in total organic carbon (TOC) and nitrogen (N) in those soils irrigated with domestic wastewater, depending on the amount of organic matter contributed. Additionally, this phenomenon increases the amount of organic matter that is available. As a result, certain bacterial populations may be more prevalent in the soil. The percentage of soil bacteria linked to stable aggregates (clay particles) ranges from 40% to 70%.

The concentration levels, organic matter composition, and soil texture all affect the stability of aggregates in the soil as well as the organic matter's ability to retain water when irrigated with wastewater. Therefore, the aggregate stability of sandy-clay soils that are irrigated with wastewater is increased. On the other hand, clayey soils cause their aggregates to become less stable. Additionally, because of the buildup of sodium in the exchange complex, using wastewater for irrigation for longer than 20 years may negatively alter the structure of the soil. The authors of a study on sugarcane that was irrigated with treated wastewater for a full year observed a rise in the amount of organic matter in the soil, which supported the reuse of wastewater in the studied areas (Valle del Cauca, Cali, Colombia).

Studies have shown that when wastewater is irrigated for 1 to 20 years, there is an increase in the various forms of nitrogen (N-NO_3, NH_4-N, or Total N). However, soil microbial communities can be impacted, particularly the activities connected to the cycle of these elements, even though there are already benefits in agricultural production and a decrease in chemical agents (fertilizers) due to the increase in N and P contributed by wastewater. Over 90% of the nitrogen in the soil is in organic form. Plants primarily absorb ammonium and nitrate, along with a few organic nitrogen compounds. Since relatively small amounts of nitrite may have toxic effects on plant growth, it is generally accepted that nitrite is an intermediate product in the conversion of ammonium to nitrate in the soil. Plants are able to absorb these intermediate products of complex organic nitrogen substances.

The plan's metabolism and the quality of the plant product can both be impacted by organic nitrogen feeding. Similarly, excessive nitrogen application—from fertilizer, sewage, or other sources—can cause vegetables to accumulate high nitrate levels, which can be extremely harmful to living things when ingested. The buildup of inorganic nitrogen in the soil has additional effects that may impact the breakdown of carbon compounds through biodegradation. Furthermore, there could be negative consequences from the soil's overabundance of nutrients. Certain nutrients, like nitrate and phosphorus, can leach into groundwater or be included in runoff, which can eutrophicate or contaminate other habitats.

Wastewater that has been irrigated can lead to sodification, or an excess of interchangeable sodium relative to other cations, or salinization, an increase in the concentration of soluble salts in the soil. When soluble salts are concentrated in the root zone, osmotic stress is created, which restricts the ability of plants to absorb water and nutrients, leading to salinity issues. Because of the reduction in permeability caused by a high interchangeable sodium content, sodicity consequently has a negative effect on the stability of aggregates and soil structure. Expansive and dispersive processes on clays result in sodicity as a result of aggregates being destroyed by elevated Na^+ concentrations. Various studies have observed that variations in sodicity lead to a rise in soil compaction and a decrease in water infiltration rate.

Thus, changes in soil salinity or sodicity have an impact on the soil microbiota. Reduced osmotic potential and modifications to soil structure are the main factors affecting microbial communities. A different study evaluated how salinity affected the composition, activity, and community of soil microorganisms. According to their findings, soil microbiota is metabolically stressed by increased salinity content. Furthermore, higher salinity soils tend to have a lower carbon-nitrogen biomass ratio, which indicates that bacteria predominate in the microbial biomass of saline soils. Moreover, pollutants (metals and pharmaceutical compounds) that are disposed of through various media, like wastewater, which builds up in the soil as a result of irrigation, contribute to an increase in soil degradation. In soils that have not been exposed to human activity, metal concentrations typically depend on the parent material (stone) and can exist in the soil at levels that are not harmful to living things.

But as a result of industrialization and population growth, there are now more of these polluting agents in wastewater and, by extension, irrigated soils. The list of potential polluting agents that may have accumulated in soil as a result of wastewater irrigation is led by metals like Fe, Cr, Zn, Pb, Ni, Cd, and Cu, all of which are abundant in wastewater. These elements affect a soil's phytotoxicity potential, which has an impact on plant growth and pollution. They can also limit fertility and/or alter soil microbial communities. The mineralization of organic matter, alterations in soil enzyme activity, decomposition of litter, decreases in microbial biomass, and modifications in microbial structure are additional ecosystem processes impacted by metal pollution. Furthermore, pharmaceuticals or other ECs may interact with the accumulated metals in a soil to intensify the possible effects on the soil. Strong co-occurrence patterns between soil metal concentrations and antibiotic resistance under specific environmental conditions have also been reported in a number of studies.

The chemical characteristics of the pollutant type, the type and age of the vegetation cover, the makeup of rhizosphere microorganisms, temperature, pH of the nutritional environment, soil texture, and structure are some of the variables that determine the fate and impact of these compounds (emerging metals and/or polluting agents). In contrast to high-mobility compounds, some researchers have found that low-mobility compounds accumulate in soils with irrigation periods ranging from one to 100 years. Additionally, because high-mobility compounds may leach and contaminate groundwater sources, researchers from all over the world have drawn attention to the risks that these compounds pose. For instance, it was found that high-mobility compounds in some amoxicillin-degradation products contaminated the groundwater of agricultural fields that were irrigated by wastewater. A different study came to an end.

10.3.3. Assessment of the Risk Associated with the Use of Wastewater in Agriculture

The risks associated with various exposure routes, exposed groups, and concentrations of different physicochemical and microbiological parameters limit the use of wastewater in agriculture. Therefore, throughout the process chain, there is exposure to the soil as a means of receiving wastewater, the irrigation technique, the kind of irrigated crop, the products consumed, farmers and their families, and final consumers. The WHO guidelines from 1989 acknowledged that the primary threat to human health is human parasites, and the primary approach for mitigating this risk was suggested to be the development of wastewater treatment systems. Thus, the idea of "zero risk" could only be realized through the use of primary, secondary, and disinfection treatment technologies. While this is technically possible, it is not a workable solution in the real-world and financial context of developing nations.

The WHO's 2006 guidelines recognized that risk reduction strategies needed to be adaptable and tailored to the local context, suppressing effluent quality thresholds for the first time. This led to a clear formulation of the need to know the extent of the risk associated with this kind of practice and the conceptual underpinnings for its estimation. Consequently, the notion of "multiple barriers" was presented. Rather than concentrating solely on treatment infrastructure for the purpose of improving wastewater quality for reuse, it suggested a number of obstacles throughout the reuse chain. The health-based goals, which are estimated using a standard disease measure chosen in relation to the Disability

Adjusted Life Year (DALY), were elevated by the WHO guidelines (2006). A quantitative measure of disease burden, disability-adjusted life years (DALY) is the total amount of healthy life that is lost due to premature death or because of a disability. The target that was set was $\leq 10^{-6}$ DALY/person, which represents the approximate disease burden linked to mild diarrhea.

In 1992, a review of the data accumulated in the period from 1975 to 1989 led to the reformulation of the quality criteria of wastewater of the state of California (United States). Based on the above, a comparative study was carried out of the possible risks of enteric virus infection with secondary and tertiary effluents from treatment systems, as opposed to four exposure scenarios for wastewater use (irrigation of food crops, golf courses, recreational reservoirs and the recharge of aquifers). The analyzes of this study showed that the annual risk of exposing a tertiary effluent with chlorine disinfection, with a viral unit content of 100 L, entails an associated risk in golf courses and recreational reservoirs in a range of 10^{-2} to 10^{-7}, while in crop irrigation and aquifer recharge, it may have an associated risk between 10^{-6} and 10^{-11}. These are determining results for the formulation of mitigation strategies and the prioritization of the investment.

The most often assessed methods have been quantitative microbiological risk assessments linked to viruses in lettuce crops. The effect of two risk factors: (i) the virus death rates in lettuce cultivation; and (ii) the density function linked to the presence of human enterovirus in irrigation water. Researchers found that changes in the density function had little effect on the estimated infection rates when using the Monte Carlo simulation method. The anticipated infection rates, however, were more accurate than the rates of virus decay.

Water and public health managers can perform Quantitative Microbiological Risk Assessments with the aid of this tool. Using particular dose-response models to define the pathogen of interest and the exposure scenario, RIRA was created to simulate a broad variety of scenarios. The primary benefit of RIRA lies in its adaptable and general framework, which facilitates risk assessments in compliance with the primary guidelines on recycled water and local context scenarios.

The developed a QMRA model to estimate the burden of norovirus disease associated with the consumption of irrigated lettuce with untreated gray water, a practice commonly performed in Australia and not endorsed by normative guidelines. The estimated annual disease burden fluctuated over a range of 2×10^{-8} and 5×10^{-4} depending on the source of gray water and of how thoroughly the consumer washes the product at home. The model predicted disease loads of 4×10^{-9} and 3×10^{-6} for bath and washing waters respectively. Using these results, the authors recommended the use of bath water that conforms to normative standards in Australia (threshold value 10^{-6} DALY per person). In addition, in Australia, a QMRA model was developed to know the risk of irrigation with wastewater in other types of vegetables such as lettuce, broccoli, cabbage, Asian vegetables and cucumber. Norovirus concentration was used, using faecal dumping rates in black wastewater and the annual norovirus disease burden after irrigation with treated wastewater. The annual estimates of disease burden showed that the primary treatment scenarios evaluated fluctuated within a range of 10^{-5} to 10^{-3} DALY per person, exceeding all mean values suggested by the WHO and Australian regulations (threshold $\leq 10^{-6}$ DALY per person). However, in the advanced treatment scenarios, most of the cucumber consumption scenarios obtained mean values of disease burden that met the threshold. In general, lettuce consumption posed the greatest risks, while cucumber consumption had the lowest risks. This research was relevant because it was the first QMRA to consider viral accumulation by irrigation using wastewater.

Hamilton and Mok conducted an experiment to determine the volume of water collected in Asian vegetables and lettuce after irrigation by sprinkling with wastewater. The proposed objective contributed to the decrease in the knowledge gap, associated with the estimation of rotavirus microbiological risk in high consumption products in China. Four vegetables were evaluated. The predicted annual probability of infection was 7×10^{-4} for the consumption of *Bok choy*, 4×10^{-3} for *choy* and 2×10^{-3} for him *gai Lan* and lettuce. Likewise, the annual average disease burden ranged between 5×10^{-6} DALY per person and 3×10^{-5} DALY per person for the consumption of *Bok choy* and *Suma choy*, respectively. The disease burden for *gai Lan* and lettuce was 2×10^{-5} DALY per person. This was the first presentation of water retention measurements for Asian vegetables, as well as the first viral risk assessment for the consumption of vegetables from wastewater in China. This research is significant because China is home to one-fifth of the world's population, and because of the availability of data on rotavirus concentrations, documented as the predominant cause of diarrheal disease in children.

Another study was conducted in Ghana to assess the risk associated with water used in the irrigation of vegetables. Pollution parameters evaluated were fecal coliforms and helminth eggs. The water quality was monitored during two months and their concentration levels ranged from 3 to 4 log units of fecal coliforms in 100 mL and from 6 to 15 eggs of helminth per L. Regarding the evaluated product of consumption (lettuce) the concentration of fecal coliforms ranged from 7×10^{2} and 1.8×10^{3} in 10 g and the helminth eggs ranged from 6 to 9 per 100 g. The annual risk of infection was 10^{-2} for Ascaris and 10^{-1} for *E. coli.* This study is relevant for its development in the risk assessment for possible infections caused by helminths and *E. coli.*

REFERENCES

1. Adrover, M., Farrús, E., Moyà, G., & Vadell, J. (2012). Chemical properties and biological activity in soils of Mallorca following twenty years of treated wastewater irrigation. *Environmental Management, 95*, S188–S192.

2. Angelakis, A., & Snyder, S. (2015). Wastewater treatment and reuse: Past, present, and future. *Water, 7*, 4887–4895.

3. Asano, T., Burton, F., Leverenz, H., Tsuchihashi, R., & Tchobanoglous, G. (2007). *Water Reuse: Issues, Technologies and Applications*. McGraw Hill Professional.

4. Barker, S. F., O'Toole, J., Sinclair, M. I., Leder, K., Malawaraarachchi, M., & Hamilton, A. J. (2013). A probabilistic model of norovirus disease burden associated with greywater irrigation of home-produced lettuce in Melbourne, Australia. *Water Research, 47*, 1421–1432.

5. Becerra, C., Lopes, A., Vaz, I., Silva, E., Manaia, C., & Nunes, O. (2015). Wastewater reuse in irrigation: A microbiological perspective on implications in soil fertility and human and environmental health. *Environmental International, 75*, 117–135.

6. Carr, R. (2005). WHO guidelines for safe wastewater use—more than just numbers. *Irrigation and Drainage, 54*, 103–111.

7. Cooper, P. (2001). Historical aspect of wastewater treatment. In *Decentralised Sanitation Reuse: Concepts, System and Implementation* (pp. 11–38). IWA Publishing.

8. Drechsel, P., Scott, A., Sally, R., Redwood, M., & Bachir, A. (2010). *Wastewater Irrigation and Health: Assessing and Mitigating Risk in Low-Income Countries*. International Water Management Institute, Earthscan.

9. Ellis, R. J., Neish, B., Trett, M. W., Best, J. G., Weightman, A. J., Morgan, P., & Fry, J. C. (2001). Comparison of microbial and meiofaunal community analyses for determining impact of heavy metal contamination. *Journal of Microbiological Methods, 45*, 171–185.

10. González, M. I., & Chiroles, S. (2011). Safe use and microbiological risks of wastewater for agriculture. *Revista Cubana de Salud Pública, 37*, 61–73.

11. Halliwell, D., Barlow, K., & Nash, D. (2001). A review of the effects of wastewater sodium on soil physical properties and their implications for irrigation systems. *Australian Journal of Soil Research, 39*, 1259–1267.

12. Henze, M., & Comeau, Y. (2008). Wastewater characterization. In *Biological Wastewater Treatment: Principles, Modeling and Design* (pp. 33–52). IWA Publishing.

13. Kamizoulis, G. (2008). Setting health based targets for water reuse (in agriculture). *Desalination, 218*, 154–163.
14. Kong, W. D., Zhu, Y. G., Fu, B. J., Marschner, P., & He, J. Z. (2006). The veterinary antibiotic oxytetracycline and Cu influence functional diversity of the soil microbial community. *Environmental Pollution, 143*, 129–137.
15. Lado, M., & Ben-Hur, M. (2009). Treated domestic sewage irrigation effects on soil hydraulic properties in arid and semiarid zones: A review. *Soil & Tillage Research, 106*, 152–163.
16. Mara, D., & Kramer, A. (2008). The 2006 WHO guidelines for wastewater and grey-water use in agriculture: A practical interpretation. In *Efficient Management of Wastewater* (pp. 1–17). Springer.
17. Mok, H. F., Barker, S. F., & Hamilton, A. J. (2014). A probabilistic quantitative microbial risk assessment model of norovirus disease burden from wastewater irrigation of vegetables in Shepparton, Australia. *Water Research, 54*, 347–362.
18. Oliveira, S., & Von Sperling, M. (2008). Reliability analysis of wastewater treatment plants. *Water Research, 42*, 1182–1194.
19. Pérez, F., Madera, C., Echeverri, A., & Urrutia, N. (2015). Wastewater reuse: Impact on the chemical and macronutritional attributes of an Inceptisol irrigated with treated domestic wastewater. *Ingeniería y Competitividad, 17*, 19–28.
20. Petterson, S. R., Ashbolt, N. J., & Sharma, A. (2001). Microbial risks from wastewater irrigation of salad crops: A screening-level risk assessment. *Water Environment Research, 73*, 667–672.
21. Sun, H., Terhonen, E., Koskinen, K., Paulin, L., Kasanen, R., & Asiegbu, F. O. (2014). Bacterial diversity and community structure along different peat soils in boreal forest. *Applied Soil Ecology, 74*, 37–45.
22. Travis, M. J., Wiel-Shafran, A., Weisbrod, N., Adar, E., & Gross, A. (2010). Greywater reuse for irrigation: Effect on soil properties. *Science of the Total Environment, 408*, 2501–2508.
23. Tzanakakis, V., Koo-Oshima, S., Haddad, M., Apostolidis, N., & Angelakis, A. (2014). The history of land application and hydroponic systems for wastewater treatment and reuse. In *Evolution of Sanitation and Wastewater Technologies through the Centuries* (p. 457). IWA Publishing.
24. Tzanakakis, V., Paranychianaki, N., & Angelakis, A. (2007). Soil as a wastewater treatment system: Historical development. *Water Science and Technology: Water Supply, 7*, 67–75.

INDEX

R

S

MIX
Papier | Fördert gute Waldnutzung
FSC® C083411

Kolibri 360 GmbH
Ferdinand-Jühlke-Straße 7
99095 Erfurt, Deutschland
produktsicherheit@kolibri360.de